Lattice Points

Mathematics and Its Applications (*East European Series*)

Managing Editor:

M. HAZEWINKEL

Centre for Mathematics and Computer Science, Amsterdam, The Netherlands

Editorial Board:

A. BIALYNICKI-BIRULA, *Institute of Mathematics Warsaw University, Poland*
H. KURKE, *Humboldt University Berlin, G.D.R.*
J. KURZWEIL, *Mathematics Institute, Academy of Sciences, Prague, Czechoslovakia*
L. LEINDLER, *Bolyai Institute, Szeged, Hungary*
L. LOVÁSZ, *Bolyai Institute, Szeged, Hungary*
D. S. MITRINOVIĆ, *University of Belgrade, Yugoslavia*
S. ROLEWICZ, *Polish Academy of Sciences, Warsaw, Poland*
BL. H. SENDOV, *Bulgarian Academy of Sciences, Sofia, Bulgaria*
I. T. TODOROV, *Bulgarian Academy of Sciences, Sofia, Bulgaria*
H. TRIEBEL, *Friedrich Schiller University Jena, G.D.R.*

Ekkehard Krätzel

Friedrich Schiller University Jena, G.D.R.

Lattice Points

Kluwer Academic Publishers

Dordrecht/Boston/London

Library of Congress Cataloging in Publication Data

Krätzel, Ekkehard, 1935–
Lattice points.

(Mathematics and its applications (East European Series))
Bibliography: p.
Includes index.
1. Lattice theory. I. Title. II. Series: Mathematics and its applications (D. Reidel Publishing Company). East European series.
QA171.5.K73 1988 512′.7 88-7356
ISBN 90-277-2733-3

Distributors for the Socialist Countries
VEB Deutscher Verlag der Wissenschaften, Berlin

Distributors for the U.S.A. and Canada
Kluwer Academic Publishers,
101 Philip Drive, Norwell, MA 02061, U.S.A.

Distributors for all remaining countries
Kluwer Academic Publishers Group,
P.O. Box 322, 3300 AH Dordrecht, Holland

Printed in the German Democratic Republic

Contents

Series Editor's Preface

Approach your problems from the right end and begin with the answers. Then one day, perhaps you will find the final question.

'The Hermit Clad in Crane Feathers' in R. van Gulik's *The Chinese Maze Murders*.

It isn't that they can't see the solution. It is that they can't see the problem.

G.K. Chesterton. *The Scandal of Father Brown* 'The point of a Pin'.

Growing specialization and diversification have brought a host of monographs and textbooks on increasingly specialized topics. However, the "tree" of knowledge of mathematics and related fields does not grow only by putting forth new branches. It also happens, quite often in fact, that branches which were thought to be completely disparate are suddenly seen to be related.

Further, the kind and level of sophistication of mathematics applied in various sciences has changed drastically in recent years: measure theory is used (non-trivially) in regional and theoretical economics; algebraic geometry interacts with physics; the Minkowsky lemma, coding theory and the structure of water meet one another in packing and covering theory; quantum fields, crystal defects and mathematical programming profit from homotopy theory; Lie algebras are relevant to filtering; and prediction and electrical engineering can use Stein spaces. And in addition to this there are such new emerging subdisciplines as "experimental mathematics", "CFD", "completely integrable systems", "chaos, synergetics and large-scale order", which are almost impossible to fit into the existing classification schemes. They draw upon widely different sections of mathematics. This programme, Mathematics and Its Applications, is devoted to new emerging (sub)disciplines and to such (new) interrelations as exampla gratia:

- a central concept which plays an important role in several different mathematical and/or scientific specialized areas;
- new applications of the results and ideas from one area of scientific endeavour into another;
- influences which the results, problems and concepts of one field of enquiry have and have had on the development of another.

The Mathematics and Its Applications programme tries to make available a careful selection of books which fit the philosophy outlined above. With such books, which are stimulating rather than definitive, intriguing rather than encyclopaedic, we hope to contribute something towards better communication among the practitioners in diversified fields.

Because of the wealth of scholarly research being undertaken in the Soviet Union, Eastern Europe, and Japan, it was decided to devote special attention to the work emanating from these particular regions. Thus it was decided to start three regional series under the umbrella of the main MIA programme.

As the author writes in his preface, this book is devoted to a special part of number theory, namely the old and venerable question of finding out how many lattice points there are in (large) closed domains in Euclidean space. That sounds quite specialized and not at all a topic for the MIA series. However, though this book does not deal with them, questions of estimating how many lattice points there are in given domains are likely to crop up in quite varying parts of mathematics, and there are important relations with even more parts e.g. automorphic forms and Kleinian groups, though there the relations seem mostly to point in the direction of applications of these fields to lattice point problems instead of the otherway. Traditionally, lattice points estimates — the geometry, of numbers — are important in exponential sums. They obviously have relevance to packing and covering problems, and hence to the many applications areas (including structure of matter) of these fields. Further, it is not difficult to see that they are important in computational geometry, and combinatorial optimization problems are important potential application areas of 'lattice point theory'. As said already, this book is not about these application fields. Instead it provides a thorough treatise and the conceptional background of the field, indispensable for those who used (to adapt) these and similar results in various situations. Incidentally, the field is also famous for its many unsolved problems and conjectures; and it is a field in which much is happening with some hundreds of papers a year.

The unreasonable effectiveness of mathematics in science ...

Eugene Wigner

Well, if you know of a better 'ole, go to it.

Bruce Bairnsfather

What is now proved was once only imagined.

William Blake

As long as algebra and geometry proceeded along separate paths, their advance was slow and their applications limited.

But when these sciences joined company they drew from each other fresh vitality and thenceforward marched on at a rapid pace towards perfection.

Joseph Louis Lagrange.

Bussum, February 1988

Michiel Hazewinkel

Preface

This book is devoted to a special problem of number theory, that is the estimation of the number of lattice points in large closed domains of ordinary Euclidean spaces. This problem is closely related to the problem of estimation of exponential sums. Van der Corput developed an important method of dealing with such sums at the beginning of our century. His method has initiated a new period of intensive development during the last few years. Beside this method some interesting applications of this theory to the circle and sphere problems, to Dirichlet's divisor problem and some generalizations of these problems are considered in this book. The distribution of powerful numbers and finite Abelian groups are also investigated. It is, however, impossible to describe all the interesting problems. Furthermore, the estimations proved are not the very best in all cases. Further improvements can be obtained, however, by refinements of the method; whereupon many technical difficulties arise. The object of this book is to acquaint the reader with the fundamental results and methods so that he is able to follow up the original papers.

I wish to express my gratitude to G. Horn, H. Menzer and L. Schnabel for reading the manuscript and for correcting a large number of mistakes. I should also like to this opportunity to thank M. Fritsch for valuable assistance in checking the English.

Jena, 1987 **Ekkehard Krätzel**

Notation

The following notations and conventions will be used throughout the book.

If f and g are two functions and $g \geqq 0$, we write

$$f = O(g) \quad \text{or} \quad f \ll g$$

if there is an absolute constant c such that $|f| \leqq cg$, and the inequality will be valid over the entire range of definition of the functions. An equation of the form

$$f = h + O(g)$$

means that $f - h = O(g)$. The expression $g \gg f$ means the same as $f \ll g$, but it will be used only when both f and g are non-negative. If we have both $f \ll g$ and $f \gg g$, then we write

$$f \asymp g .$$

In the next notations functions of a single variable are considered. If

$$\lim_{x \to \infty} \frac{f(x)}{g(x)} = 1 ,$$

we write

$$f(x) \sim g(x) \qquad \text{as } x \to \infty .$$

If

$$\lim_{x \to \infty} \frac{f(x)}{g(x)} = 0 ,$$

we write

$$f(x) = o(g(x)) \qquad \text{as } x \to \infty ,$$

where $g(x)$ may be supposed to be positive. Again, an equation of the form

$$f(x) = h(x) + o(g(x)) \qquad \text{as } x \to \infty$$

means that $f(x) - h(x) = o(g(x))$ as $x \to \infty$. Whenever it can be done without causing confusion, we omit the additional notation "as $x \to \infty$".

We use the notation

$$f(x) = \Omega(g(x))$$

if the relation $f(x) = o(g(x))$ does not hold. If this is the case, then a positive constant K exists such that

$$|f(x)| > Kg(x),$$

$g(x) > 0$, for values of x surpassing all limit. We write

$$f(x) = \Omega_+(g(x))$$

if

$$f(x) > Kg(x)$$

is satisfied by values of x surpassing all limit. Similarly, we write

$$f(x) = \Omega_-(g(x))$$

in the case of

$$f(x) < -Kg(x).$$

If we have both $f(x) = \Omega_+(g(x))$ and $f(x) = \Omega_-(g(x))$, we write

$$f(x) = \Omega_\pm(g(x)).$$

Also, an equation of the form

$$f(x) = h(x) + \Omega(g(x))$$

means that $f(x) - h(x) = \Omega(g(x))$. Analogous notations are used for the other Ω-symbols.

If x is a real number, $[x]$ denotes the greatest integer not exceeding x, and $||x||$ denotes the distance from x to the nearest integer. This means that

$$||x|| = \min(x - [x], 1 - (x - [x])).$$

The abbreviation

$$\psi(x) = x - [x] - \frac{1}{2}$$

is used throughout the book. Because of the periodicity we obtain at once the so-called trivial estimation

$$\psi(x) = O(1).$$

Sometimes we use the notation $\mathrm{e}(x)$ for the function

$$\mathrm{e}(x) = \mathrm{e}^{2\pi i x}.$$

Let A be a bounded set. The number of elements of A will be denoted by $\# A$ or

$$\# A = \sum_{x \in A} 1 .$$

It may happen that the summation conditions of a sum are rather complicated. Then we write these conditions separately and use the notation $SC(\Sigma)$. For example, instead of

$$F(x) = \sum_{a \leqq n \leqq x} f(n)$$

we write

$$F(x) = \Sigma f(n) , \qquad SC(\Sigma): \ a \leqq n \leqq x .$$

Similarly, for

$$F(x) = \int_a^x f(t) \, dt$$

we write the integration conditions separately such that we get the representation

$$F(x) = \int f(t) \, dt , \qquad IC(\int): \ a \leqq t \leqq x .$$

It is also possible that we have both summation and integration. Then we use for the expression

$$F(x) = \sum_{a \leqq g(n,t) \leqq x} \int f(n, t) \, dt$$

the notation

$$F(x) = \Sigma \int f(n, t) \, dt , \qquad SC(\Sigma \int): \ a \leqq g(n, t) \leqq x .$$

A sum

$$\sum_{a < n \leqq b}{}' f(n)$$

means that the possible term $f(b)$, b being an integer, gets the factor 1/2.

Similarly, in the sum

$$\sum_{a \leqq n \leqq b}{}'' f(n)$$

both terms, $f(a)$ and $f(b)$, get factors 1/2.

Chapter 1

Introduction

Let a Euclidean space with respect to a Cartesian coordinate system be given. Then we consider the set of points whose coordinates are integers. These points form a lattice and therefore they are called lattice points. Lattice point theory is then concerned with the estimation of the number of lattice points in large closed domains. Historically, the first problems which were investigated in this connection were the circle problem introduced by C. F. Gauss and the problem of estimation of the number of lattice points under a hyperbola, firstly considered by P. G. L. Dirichlet. At the present time there is a widespread interest not only for two-dimensional but also for many-dimensional problems. In this introduction we shall show that the number of lattice points in a closed domain can be approximated by its volume in general. We shall develop some important summation formulas, and at last it will be seen that the problem of estimation of lattice points is connected with the problem of estimation of some exponential sums.

1.1. LATTICE POINTS IN PLANE DOMAINS

We begin with the classical circle problem. Given the Cartesian coordinate system (ξ, η). Consider the circle $\xi^2 + \eta^2 = x$ $(x \geqq 1)$. In order to estimate the number

$$G = \# \{(n, m): n, m \in \mathbf{Z}, n^2 + m^2 \leqq x\} \quad (\mathbf{Z} \text{ denotes the set of integers})$$

of lattice points within it, let us use a simple geometrical argument.

We associate each lattice point $P_i = (\xi_i, \eta_i)$ of the circle with the square

$$Q_i = \left\{(\xi, \eta): |\xi - \xi_i| < \frac{1}{2}, |\eta - \eta_i| < \frac{1}{2}\right\}$$

having the area 1. Therefore, the set of lattice points of the circle is associated with the union of the squares Q_i (see Fig. 1). These squares are included in the circle $\xi^2 + \eta^2 = \left(\sqrt{x} + \frac{1}{2}\sqrt{2}\right)^2$. Hence, the number of lattice points is smaller than the area of this circle, and we obtain

$$G \leqq \pi\left(\sqrt{x} + \frac{1}{2}\sqrt{2}\right)^2 < \pi x + 2\pi\sqrt{x}.$$

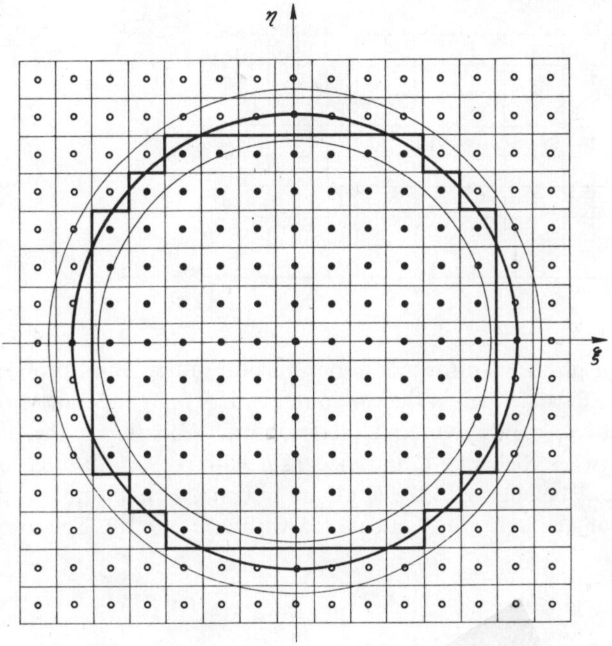

Fig. 1

Otherwise, the squares include the circle $\xi^2 + \eta^2 = \left(\sqrt{x} - \frac{1}{2}\sqrt{2}\right)^2$ such that

$$G \geqq \pi\left(\sqrt{x} - \frac{1}{2}\sqrt{2}\right)^2 > \pi x - 2\pi\sqrt{x}.$$

Altogether we have

$$|G - \pi x| < 2\pi\sqrt{x}\,. \tag{1.1}$$

Inequality (1.1) shows that the number of lattice points inside and on a circle is given by the area in the first approximation, and the error is of order of the length. It is possible to extend this idea to more general curves. Let J be a closed continuous curve with the area F and the length $l \geqq 1$, and let G be the number of lattice points inside or on the curve. Then, by applying the same conception, H. Steinhaus [1] proved a theorem of V. Jarník, namely

$$|G - F| < l\,.$$

However, we do not pursue this conception, we shall proceed on another line. It is well known that every continuous curve with a length possesses a representation by a continuous function of bounded variation. Each function of this type can be expressed by a sum of two continuous, monotonical functions. For this reason we shall take into consideration only functions which are both continuous and monotonical.

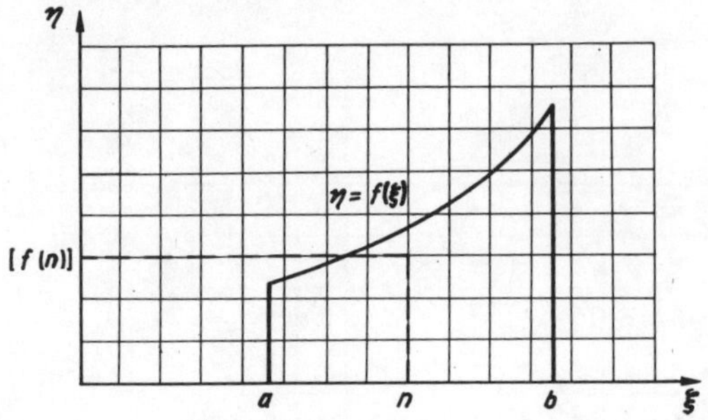

Fig. 2

We are now in a position to formulate our general lattice point problem in the plane. Let the curve C be defined in the closed interval $[a, b]$ by the function $\eta = f(\xi)$ which is non-negative, continuous and monotonical. We ask for the number G of lattice points between the curve C and the axis $\eta = 0$ (see Fig. 2). It is convenient that the lattice points belonging to the left-hand bound $\xi = a$ are not counted, and each lattice point on the axis $\eta = 0$ is counted with a factor $1/2$. Then we get

$$G = \sum_{a<n\leqq b} \left\{ [f(n)] + \frac{1}{2} \right\}.$$

Using the function $\psi(x) = x - [x] - 1/2$, we obtain

$$G = \sum_{a<n\leqq b} f(n) - \sum_{a<n\leqq b} \psi(f(n)).$$

Now there are two problems:

1. *Find an asymptotic representation of the sum* $\sum_{a<n\leqq b} f(n)$.
2. *Find a non-trivial estimation of the sum* $\sum_{a<n\leqq b} \psi(f(n))$.

This introductory chapter is devoted to the first problem which can be easily solved in general. Concerning the second problem only an important transformation formula is considered here. Estimates of such sums will be given in the second chapter.

Theorem 1.1. *Let the function $f(t)$ for $a \leqq t \leqq b$ be non-negative, continuous and monotonic. Then*

$$\left| \sum_{a<n\leqq b} f(n) - \int_a^b f(t)\,\mathrm{d}t + \psi(b)\,f(b) - \psi(a)\,f(a) \right| \leqq \frac{1}{2}\,|f(b) - f(a)| \tag{1.2}$$

and

$$\left| \sum_{a<n\leqq b} f(n) - \int_a^b f(t)\,\mathrm{d}t \right| \leqq \max_{a\leqq t\leqq b} f(t)\,. \tag{1.3}$$

Proof. Suppose that $f(t)$ is monotonically increasing. Then

$$\sum_{a<n\leqq b} f(n) - \int_a^b f(t)\,\mathrm{d}t$$

$$= \sum_{n=[a]+1}^{[b]-1} \int_n^{n+1} \{f(n) - f(t)\}\,\mathrm{d}t + f([b]) - \int_a^{[a]+1} f(t)\,\mathrm{d}t - \int_{[b]}^{b} f(t)\,\mathrm{d}t .$$

Now we have the following inequalities

$$\sum_{a<n\leqq b} f(n) - \int_a^b f(t)\,\mathrm{d}t \leqq -(b-[b]-1)\,f([b]) + (a-[a]-1)\,f(a)$$

$$\leqq -\psi(b)\,f(b) + \psi(a)\,f(a) + \frac{1}{2}\{f(b) - f(a)\}$$

and similarly

$$\sum_{a<n\leqq b} f(n) - \int_a^b f(t)\,\mathrm{d}t \geqq (a-[a])\,f([a]+1) - (b-[b])\,f(b)$$

$$\geqq -\psi(b)\,f(b) + \psi(a)\,f(a) - \frac{1}{2}\{f(b) - f(a)\} .$$

Thus inequalities (1.2) and (1.3) follow at once.

Theorem 1.2 (Partial summation). *Let $f(n)$ and $g(n)$ be number-theoretical functions, $G(x) = \sum_{n\leqq x} g(n)$ for $x \geqq 1$ and $G(x) = 0$ for $x < 1$. Then we have for $0 \leqq a < b$*

$$\sum_{a<n\leqq b} f(n)\,g(n) = f([b])\,G(b) - f([a]+1)\,G(a) + \sum_{a<n\leqq b-1} \{f(n) - f(n+1)\}\,G(n) . \tag{1.4}$$

Moreover, if $f(t)$ is a continuous function in $[a, b]$ and has a continuous derivative in (a, b), then

$$\sum_{a<n\leqq b} f(n)\,g(n) = f(b)\,G(b) - f(a)\,G(a) - \int_a^b f'(t)\,G(t)\,\mathrm{d}t . \tag{1.5}$$

Proof.

$$\sum_{a<n\leqq b} f(n)\,g(n) = \sum_{n=[a]+1}^{[b]} f(n)\,\{G(n) - G(n-1)\}$$

$$= \sum_{n=[a]+1}^{[b]} f(n)\,G(n) - \sum_{n=[a]}^{[b]-1} f(n+1)\,G(n)$$

$$= f([b])G(b) - f([a]+1)\,G(a) + \sum_{a<n\leqq b-1} \{f(n) - f(n+1)\}\,G(n).$$

This is formula (1.4). If $f(t)$ has a continuous derivative, we deduce (1.5) from (1.4):

$$\sum_{a<n\leqq b} f(n)\,g(n) = f([b])\,G(b) - f([a]+1)\,G(a) - \sum_{a<n\leqq b-1} G(n) \int_n^{n+1} f'(t)\,dt$$

$$= f([b])\,G(b) - f([a]+1)\,G(a) - \int_{[a]+1}^{[b]} f'(t)\,G(t)\,dt$$

$$= f(b)\,G(b) - f(a)\,G(a) - \int_a^b f'(t)\,G(t)\,dt\,.$$

Theorem 1.3 (Euler-Maclaurin sum formula). *Let $f(t)$ be a continuous function in $[a, b]$ $(0 \leqq a < b)$ with a continuous derivative in (a, b). Then*

$$\sum_{a<n\leqq b} f(n) = \int_a^b f(t)\,dt - \psi(b)\,f(b) + \psi(a)\,f(a) + \int_a^b f'(t)\,\psi(t)\,dt\,.$$

Proof. Putting $g(n) = 1$ in equation (1.5), the sum becomes

$$\sum_{a<n\leqq b} f(n) = [b]\,f(b) - [a]\,f(a) - \int_a^b f'(t)\,[t]\,dt\,.$$

By using the identity

$$\int_a^b f'(t)\left(t - \frac{1}{2}\right)dt = f(b)\left(b - \frac{1}{2}\right) - f(a)\left(a - \frac{1}{2}\right) - \int_a^b f(t)\,dt$$

in this equation the Euler-Maclaurin sum formula follows immediately.

Theorem 1.4. *Let $f(t)$ for $t \geqq a$ be positive, continuous, monotonically decreasing and let $\lim_{t\to\infty} f(t) = 0$. Then there exists*

$$\lim_{b\to\infty}\left\{\sum_{a<n\leqq b} f(n) - \int_a^b f(t)\,dt\right\} = \gamma\,, \tag{1.6}$$

and

$$\left|\sum_{a<n\leqq b} f(n) - \int_a^b f(t)\,dt - \gamma + \psi(b)\,f(b)\right| \leqq \frac{1}{2}\,f(b)\,. \tag{1.7}$$

Furthermore, if $f(t)$ has a continuous derivative for $t \geqq a$, then

$$\gamma = \psi(a)\,f(a) + \int_a^\infty f'(t)\,\psi(t)\,dt\,. \tag{1.8}$$

Proof. We consider the sequence (A_n) with $n > a$ and

$$A_n = \sum_{a<\nu\leqq n} \int_\nu^{\nu+1} \{f(\nu) - f(t)\}\,dt\,.$$

Clearly,

$$0 \leqq A_n \leqq \sum_{a<\nu\leqq n} \int_\nu^{\nu+1} \{f(\nu) - f(\nu+1)\}\,\mathrm{d}t = f([a]+1) - f(n+1) \leqq f([a]+1).$$

Thus it can be seen that the sequence (A_n) is monotonically increasing and bounded. Therefore the limit (1.6) exists. Moreover, we have

$$\sum_{a<n\leqq b} f(n) - \int_a^b f(t)\,\mathrm{d}t = \sum_{n=[a]+1}^{[b]} \int_n^{n+1} \{f(n) - f(t)\}\,\mathrm{d}t - \int_a^{[a]+1} f(t)\,\mathrm{d}t + \int_b^{[b]+1} f(t)\,\mathrm{d}t$$

$$= \gamma - \sum_{n=[b]+1}^{\infty} \int_n^{n+1} \{f(n) - f(t)\}\,\mathrm{d}t + \int_b^{[b]+1} f(t)\,\mathrm{d}t$$

$$\leqq \gamma - \psi(b)\, f(b) + \frac{1}{2} f(b)$$

and on the other hand

$$\sum_{a<n\leqq b} f(n) - \int_a^b f(t)\,\mathrm{d}t \geqq \gamma - \sum_{n=[b]+1}^{\infty} \int_n^{n+1} \{f(n) - f(n+1)\}\,\mathrm{d}t + \int_b^{[b]+1} f(t)\,\mathrm{d}t$$

$$\geqq \gamma - (b-[b])\, f([b]+1) \geqq \gamma - \psi(b)\, f(b) - \frac{1}{2} f(b)\,.$$

Thus assertion (1.7) follows. Taking $b \to \infty$ in the Euler-Maclaurin sum formula, we deduce the representation (1.8) of γ. This proves the theorem.

We now turn to the consideration of the sum over the ψ-function. The transformation formula (1.9) will be often applied in Chapters 3 and 5 implicitly.

Theorem 1.5. *Let $f(t)$ in $[a, b]$ $(0 \leqq a < b)$ be a non-negative, strictly decreasing function with a continuous derivative in (a, b). If $f^{-1}(t)$ denotes the inverse function of $f(t)$, then*

$$\sum_{a<n\leqq b} \psi(f(n)) - \int_a^b f'(t)\,\psi(t)\,\mathrm{d}t - \psi(a)\,\psi(f(a))$$

$$= \sum_{f(b)<m\leqq f(a)} \psi(f^{-1}(m)) - \int_{f(b)}^{f(a)} \frac{\psi(t)}{f'(f^{-1}(t))}\,\mathrm{d}t - \psi(b)\,\psi(f(b))\,. \tag{1.9}$$

Proof. We count the number of lattice points in the domain D (see Fig. 3) in two ways, where again the lattice points on the axis get the factor 1/2 and so the point (0, 0) gets the factor 1/4. Then

$$\sum_{\substack{a<n\leqq b\\ m\leqq f(n)}} 1 + \left([a]+\frac{1}{2}\right)\left([f(a)]+\frac{1}{2}\right) = \sum_{\substack{f(b)<m\leqq f(a)\\ n\leqq f^{-1}(m)}} 1 + \left([b]+\frac{1}{2}\right)\left([f(b)]+\frac{1}{2}\right),$$

$$\sum_{a<n\leqq b}\{f(n)-\psi(f(n))\}+\{a-\psi(a)\}\{f(a)-\psi(f(a))\}$$
$$=\sum_{f(b)<m\leqq f(a)}\{f^{-1}(m)-\psi(f^{-1}(m))\}+\{b-\psi(b)\}\{f(b)-\psi(f(b))\},$$

$$\sum_{a<n\leqq b}\psi(f(n))-\sum_{a<n\leqq b}f(n)-\psi(b)\,f(b)+\psi(a)\,f(a)$$
$$=\sum_{f(b)<m\leqq f(a)}\psi(f^{-1}(m))-\sum_{f(b)<m\leqq f(a)}f^{-1}(m)$$
$$-a\psi(f(a))+b\psi(f(b))+af(a)-bf(b)+\psi(a)\,\psi(f(a))-\psi(b)\,\psi(f(b)).$$

By applying the Euler-Maclaurin sum formula, (1.9) follows.

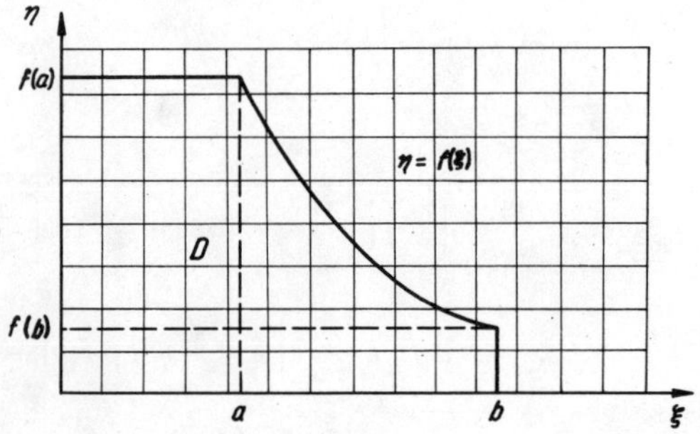

Fig. 3

Finally we make some remarks on the *Euler-Maclaurin sum formula*. If t is no integer, then the function $\psi(t)$ has the Fourier representation

$$\psi(t)=-\frac{1}{\pi}\sum_{n=1}^{\infty}\frac{\sin 2\pi nt}{n}. \tag{1.10}$$

The series converges uniformly in any closed interval not containing any integer. If t is an integer, the series also converges, and the value of the series is 0. The partial sums

$$\sum_{n=1}^{N}\frac{\sin 2\pi nt}{n}$$

are uniformly bounded. Consequently, under the conditions of Theorem 1.3 we can use representation (1.10) in the Euler-Maclaurin sum formula, and we may interchange summation and integration:

$$\sum_{a<n\leqq b}f(n)=\int_a^b f(t)\,\mathrm{d}t-\psi(b)\,f(b)+\psi(a)\,f(a)-\frac{1}{\pi}\int_a^b f'(t)\sum_{n=1}^{\infty}\frac{\sin 2\pi nt}{n}\,\mathrm{d}t$$
$$=\int_a^b f(t)\,\mathrm{d}t-\psi(b)\,f(b)+\psi(a)\,f(a)-\frac{1}{\pi}\sum_{n=1}^{\infty}\frac{1}{n}\int_a^b f'(t)\sin 2\pi nt\,\mathrm{d}t.$$

Integration by parts gives

$$\sum_{a \leqq n \leqq b}'' f(n) = \lim_{N\to\infty} \sum_{n=-N}^{+N} \int_a^b f(t)\, \mathrm{e}^{2\pi i n t}\, \mathrm{d}t\,. \tag{1.11}$$

It is noteworthy that under suitable conditions the equation (1.11) is also correct for $a = -\infty$ and $b = +\infty$. The resulting formula is called *Poisson sum formula*

$$\sum_{n=-\infty}^{+\infty} f(n) = \lim_{N\to\infty} \sum_{n=-N}^{+N} \int_{-\infty}^{+\infty} f(t)\, \mathrm{e}^{2\pi i n t}\, \mathrm{d}t\,. \tag{1.12}$$

It can be easily seen that the following conditions are sufficient:

(a) The series

$$F(x) = \sum_{n=-\infty}^{+\infty} f(n+x)$$

converges uniformly for $0 \leqq x \leqq 1$.

(b) The function $F(x)$ satisfies the Fourier conditions, that is, $F(x)$ is the sum of its Fourier series.

For, if $F(x)$ is represented by its Fourier series

$$F(x) = \lim_{N\to\infty} \sum_{n=-N}^{+N} c_n\, \mathrm{e}^{2\pi i n x}\,,$$

condition (a) enables us to calculate the Fourier coefficients in the following way:

$$c_n = \int_0^1 F(t)\, \mathrm{e}^{-2\pi i n t}\, \mathrm{d}t = \sum_{k=-\infty}^{+\infty} \int_0^1 f(k+t)\, \mathrm{e}^{-2\pi i n t}\, \mathrm{d}t$$

$$= \sum_{k=-\infty}^{+\infty} \int_k^{k+1} f(t)\, \mathrm{e}^{-2\pi i n t}\, \mathrm{d}t = \int_{-\infty}^{+\infty} f(t)\, \mathrm{e}^{-2\pi i n t}\, \mathrm{d}t\,.$$

Taking this representation of c_n in (1.13) and putting $x = 0$, formula (1.12) follows.

1.2. LATTICE POINTS IN MANY-DIMENSIONAL DOMAINS

Lattice point theory in many-dimensional domains is in principle analogous to the theory in two dimensions. We consider a bounded p-dimensional domain D ($p \geqq 1$) in a Euclidean space of the same dimension with respect to a Cartesian coordinate system. The points may be denoted by $\boldsymbol{t} = (t_1, t_2, \ldots, t_p)$ and the lattice points by $\boldsymbol{n} = (n_1, n_2, \ldots, n_p)$. We assume that any line parallel to any line of the coordinate axes meets D in a bounded number of straight-line segments. Let D be completely contained in a hyper-rectangle

$$D' = \{\boldsymbol{t}\colon a_\nu \leqq t_\nu \leqq b_\nu, \quad b_\nu - a_\nu > 1 \ (\nu = 1, 2, \ldots, p)\}$$

having the volume

$$|D'| = \prod_{\nu=1}^{p} (b_\nu - a_\nu)\,.$$

In D' the function $y = f(\boldsymbol{t})$ is considered, which is non-negative, continuous and monotonical in each variable. Then, corresponding to the two-dimensional case, the sum

$$G = \sum_{\boldsymbol{n} \in D} \left\{ [f(\boldsymbol{n})] + \frac{1}{2} \right\}$$

defines the number of lattice points in a $(p + 1)$-dimensional domain. Again, we ask for an asymptotic representation of the sum $\sum_{\boldsymbol{n} \in D} f(\boldsymbol{n})$ and a non-trivial estimation of the sum $\sum_{\boldsymbol{n} \in D} \psi(f(\boldsymbol{n}))$. Here we make some remarks on the estimate of the first sum.

Theorem 1.6 (Partial summation). *Let* $f(\boldsymbol{n})$ *and* $g(\boldsymbol{n})$ *be number-theoretical functions and*

$$G(\boldsymbol{t}) = \Sigma_1 \, g(\boldsymbol{n}) \,,$$

$$SC(\Sigma_1): \quad \boldsymbol{n} = (n_1, n_2, \dots , n_p) \,, \quad 1 \leqq n_i \leqq t_i \qquad (i = 1, 2, \dots , p) \,.$$

If one of the t_i *is less than* 1, *we set* $G(\boldsymbol{t}) = 0$. *Let*

$$\Delta_{n_i} f(\boldsymbol{n}) = f(n_1, \dots , n_p) - f(n_1, \dots , n_{i-1}, n_i + 1, n_{i+1}, \dots , n_p) \,.$$

If N_i *are integers, then*

$$\sum\nolimits_2 f(\boldsymbol{n}) \, g(\boldsymbol{n}) = \sum\nolimits_3 \sum\nolimits_4 G(\boldsymbol{n}) \left(\prod_{i=1}^{p} \Delta_{n_i}^{\alpha_i} \right) f(\boldsymbol{n}) \,, \tag{1.14}$$

$$SC(\Sigma_2): \quad \boldsymbol{n} = (n_1, n_2, \dots , n_p) \,, \quad 1 \leqq n_i \leqq N_i \qquad (i = 1, 2, \dots , p) \,,$$

$$SC(\Sigma_3): \quad (\alpha_1, \alpha_2, \dots , \alpha_p) \in \{0, 1\}^p \,,$$

$$SC(\Sigma_4): \quad \boldsymbol{n} = (n_1, n_2, \dots , n_p) \,, \quad 1 \leqq n_i \leqq N_i - 1 \textit{ for } \alpha_i = 1$$
$$\textit{and} \quad n_i = N_i \textit{ for } \alpha_i = 0 \qquad (i = 1, 2, \dots , p) \,.$$

Moreover, if $f(\boldsymbol{t})$ *possesses continuous derivatives in each variable, we have*

$$\sum\nolimits_2 f(\boldsymbol{n}) \, g(\boldsymbol{n}) = \sum\nolimits_3 (-1)^{\alpha_1 + \dots + \alpha_p} \int G(\boldsymbol{t}) \left(\prod_{i=1}^{p} \left(\frac{\partial}{\partial t_i} \right)^{\alpha_i} \right) f(\boldsymbol{t}) \, \mathrm{d}\boldsymbol{t} \,, \tag{1.15}$$

$IC(\int)$: $\boldsymbol{t} = (t_1, \dots , t_p)$, $1 \leqq t_i \leqq N_i$, *if* $\alpha_i = 1$ *and no integration if* $\alpha_i = 0$, *there put* $t_i = N_i$.

Proof. We write

$$\begin{aligned} g(\boldsymbol{n}) = G(\boldsymbol{n}) &- \sum_{1 \leqq i \leqq p} G(n_1, \dots, n_i - 1, \dots, n_p) \\ &+ \sum_{1 \leqq i < k \leqq p} G(n_1, \dots, n_i - 1, \dots, n_k - 1, \dots, n_p) \pm \dots \\ &+ (-1)^p \, G(n_1 - 1, n_2 - 1, \dots, n_p - 1) \,. \end{aligned}$$

In order to prove the correctness of this equation, we count the number of times a term $g(\nu_1, \dots, \nu_p)$ occurs on the right-hand side. The term $g(n_1, \dots, n_p)$ stands exactly once. If r $(1 \leqq r \leqq p)$ numbers ν_i are less than the corresponding numbers n_i, while the others are equal, the number of times $g(\nu_1, \dots, \nu_p)$ is counted is given by

$$1 - \binom{r}{1} + \binom{r}{2} - \dots + (-1)^r = 0 .$$

Hence

$$\begin{aligned}\textstyle\sum_2 f(\boldsymbol{n})\, g(\boldsymbol{n}) &= \textstyle\sum_2 f(\boldsymbol{n}) \{G(\boldsymbol{n}) - \sum_i G(n_1, \dots, n_i - 1, \dots, n_p) \pm \dots \\ &\quad + (-1)^p G(n_1 - 1, \dots, n_p - 1)\} \\ &= \textstyle\sum G(\boldsymbol{n}) \left(\prod_{i=1}^{p} \Delta_{n_i} \right) f(\boldsymbol{n}) .\end{aligned}$$

This gives the right-hand side of equation (1.14). Similarly to the proof of (1.5) equation (1.15) is deduced from (1.14).

Corollary. *If* $f(\boldsymbol{t})$ *denotes a real function with* $0 \leqq f(\boldsymbol{t}) \leqq F$ *such that the terms* $\left(\prod_{i=1}^{p} \Delta_{n_i}^{\alpha_i} \right) f(\boldsymbol{n})$ *in* (1.14) *or the terms* $\left(\prod_{i=1}^{p} \left(\frac{\partial}{\partial t_i} \right)^{\alpha_i} \right) f(\boldsymbol{t})$ *in* (1.15) *keep a fixed sign for each of all combinations of* α_i *with* $\alpha_1^2 + \alpha_2^2 + \dots + \alpha_p^2 > 0$ *and if* $|G(\boldsymbol{n})| \leqq G$, *then*

$$|\Sigma_2 f(\boldsymbol{n})\, g(\boldsymbol{n})| \leqq \frac{1}{2}(1 + 3^p)\, FG . \tag{1.16}$$

Proof. Equation (1.14) shows that

$$|\textstyle\sum_2 f(\boldsymbol{n})\, g(\boldsymbol{n})| \leqq G \sum_3 \left| \sum_4 \left(\prod_{i=1}^{p} \Delta_{n_i}^{\alpha_i} \right) f(\boldsymbol{n}) \right| .$$

If exactly r numbers α_i $(1 \leqq r \leqq p)$ are equal to 1, we have

$$\left| \textstyle\sum_4 \left(\prod_{i=1}^{p} \Delta_{n_i}^{\alpha_i} \right) f(\boldsymbol{n}) \right| \leqq F\, 2^{r-1}$$

and therefore

$$|\textstyle\sum_2 f(\boldsymbol{n})\, g(\boldsymbol{n})| \leqq FG \left\{ 1 + \sum_{r=1}^{p} \binom{p}{r} 2^{r-1} \right\} = \frac{1}{2}(1 + 3^p)\, FG .$$

A similar proof holds if we use (1.15) and the corresponding condition.

The next theorem shows that in general the number of lattice points of a domain D is given by its volume in the first approximation.

Theorem 1.7. *Let $f(t)$ in D' be non-negative, continuous and monotonic in each variable. If $|f(\mathbf{t})| \leqq F$ in D', then*

$$\left|\sum_{\mathbf{n}\in D} f(\mathbf{n}) - \int_D f(\mathbf{t})\, \mathrm{d}t_1 \cdot \ldots \cdot \mathrm{d}t_p\right| \leqq F\,|D'|\sum_{i=1}^{p} \frac{1}{b_i - a_i}\,.$$

Proof. Observe that the statement for $p = 1$ is proved in Theorem 1.1. Now it follows by induction and (1.3)

$$\sum_{\mathbf{n}\in D} f(\mathbf{n}) \leqq \int \sum_{(t_1,\ldots,t_{p-1},n_p)\in D} f(t_1,\ldots,t_{p-1},n_p)\,\mathrm{d}t_1 \cdot \ldots \cdot \mathrm{d}t_{p-1} + F\,|D'|\sum_{i=1}^{p-1}\frac{1}{b_i - a_i}$$

$$\leqq \int_D f(\mathbf{t})\,\mathrm{d}t_1 \cdot \ldots \cdot \mathrm{d}t_p + F\,|D'|\sum_{i=1}^{p}\frac{1}{b_i - a_i}\,.$$

A similar proof holds for the opposite sign.

1.3. LATTICE POINTS AND EXPONENTIAL SUMS

From the preceding sections we can see that for estimating the number of lattice points it remains to estimate sums of the type

$$\sum_{\mathbf{n}\in D} \psi(f(\mathbf{n}))\,.$$

By means of the Fourier representation of $\psi(x)$ these sums can be connected with exponential sums. This will be shown in the next theorem.

Theorem 1.8. *Let H be any set of lattice points $\mathbf{n}$ in a bounded p-dimensional domain. Let $f(\mathbf{n})$ be a real function defined on H. Let $C(f(\mathbf{n}))$ be defined by*

$$C(f(\mathbf{n})) = \sum_{\nu=-\infty}^{+\infty} c_\nu\, \mathrm{e}^{2\pi i \nu f(\mathbf{n})}$$

where $c_0 = 0$ and

$$c_\nu = \frac{1}{z}\left(\frac{-z}{2\pi\, i\nu}\right)^{s+1} (\mathrm{e}^{-2\pi i\nu/z} - 1)^s$$

for $\nu \neq 0$, $z > 0$, $s = 1, 2, \ldots$ Then

$$-\sum_{\mathbf{n}\in H}\frac{s}{2z} - \sum_{\mathbf{n}\in H} C(-f(\mathbf{n})) \leqq \sum_{\mathbf{n}\in H}\psi(f(\mathbf{n})) \leqq \sum_{\mathbf{n}\in H}\frac{s}{2z} + \sum_{\mathbf{n}\in H} C(f(\mathbf{n}))\,. \tag{1.17}$$

In particular, the estimation

$$\sum_{\boldsymbol{n}\in H} \psi(f(\boldsymbol{n})) \ll \sum_{\boldsymbol{n}\in H} \frac{1}{z} + \sum_{\nu=1}^{\infty} \min\left(\frac{z^s}{\nu^{s+1}}, \frac{1}{\nu}\right) \left|\sum_{\boldsymbol{n}\in H} \mathrm{e}^{2\pi \mathrm{i}\nu f(\boldsymbol{n})}\right| \tag{1.18}$$

holds.

Proof. It is easily seen that

$$\psi(x) \leqq \psi(x-y) + y$$

if $y \geqq 0$. We set $x = f(\boldsymbol{n})$, $y = t_1 + t_2 + \ldots + t_s$ and obtain by multiplying with z^s and by integrating

$$\psi(f(\boldsymbol{n})) \leqq z^s \int_0^{1/z} \ldots \int_0^{1/z} \psi(f(\boldsymbol{n}) - t_1 - \ldots - t_s)\, \mathrm{d}t_1 \cdot \ldots \cdot \mathrm{d}t_s + \frac{s}{2z}.$$

We get by (1.10)

$$\psi(f(\boldsymbol{n})) \leqq -\frac{z^s}{2\pi \mathrm{i}} \sum_{\nu=1}^{\infty} \frac{1}{\nu} \int_0^{1/z} \ldots \int_0^{1/z} \{\mathrm{e}^{2\pi \mathrm{i}\nu(f(\boldsymbol{n})-t_1-\ldots-t_s)}$$

$$- \mathrm{e}^{-2\pi \mathrm{i}\nu(f(\boldsymbol{n})-t_1-\ldots-t_s)}\}\, \mathrm{d}t_1 \cdot \ldots \cdot \mathrm{d}t_s + \frac{s}{2z} = C(f(\boldsymbol{n})) + \frac{s}{2z}.$$

This proves the right-hand side of (1.17). The left-hand side of (1.17) follows in a similar manner by applying the inequality $\psi(x+y) - y \leqq \psi(x)$ provided that $y \geqq 0$.

In order to prove (1.18), we deduce from (1.17)

$$\sum_{\boldsymbol{n}\in H} \psi(f(\boldsymbol{n})) \ll \sum_{\boldsymbol{n}\in H} \frac{1}{z} + \sum_{\nu=1}^{\infty} |c_\nu| \left|\sum_{\boldsymbol{n}\in H} \mathrm{e}^{2\pi \mathrm{i} f(\boldsymbol{n})}\right|.$$

Because of

$$|c_\nu| \ll \min\left(\frac{z^s}{\nu^{s+1}}, \frac{1}{\nu}\right)$$

we obtain (1.18) at once.

Chapter 2

Estimates of exponential sums

Many problems in lattice point theory and, generally speaking, in number theory lead to the estimation of exponential sums

$$\sum_{\boldsymbol{n}\in D} e^{2\pi i f(\boldsymbol{n})},$$

where f is a real function of $\boldsymbol{n} = (n_1, n_2, \dots, n_p)$ and D is a suitable domain in a p-dimensional Euclidean space. H. Weyl [1] was the first who introduced a method of estimating such sums for $p = 1$ in a fundamental paper. His method based on the inequality

$$\left|\sum_{a<n\leq b} e^{2\pi i n\vartheta}\right| \leq \min\left(b-a, \frac{1}{|\sin \pi\vartheta|}\right) \qquad (\vartheta \text{ real}, \neq 0)$$

and on a repeated application of Schwarz's inequality. However, at that time G. H. Hardy and J. E. Littlewood [1] proved the estimate

$$\sum_{a<n\leq b} e^{2\pi i n^2\vartheta} \ll (b-a)\sqrt{|\vartheta|} + \frac{1}{\sqrt{|\vartheta|}} \qquad (\vartheta \text{ real}, \neq 0),$$

where this exponential sum is associated with the elliptic theta-function. Owing to this result J. G. van der Corput developed a new method of approximating exponential sums for $p = 1$. His method makes use of two processes A and B. Each process transforms a given exponential sum into another sum. By applying the Euler-Maclaurin sum formula the process B yields an approximate functional equation. By a trivial estimate of the new sum one gets a first non-trivial estimate of the given sum. Process B, when applied twice, transforms the sum into itself.

Process A corresponds to a suitably chosen application of Schwarz's inequality, it can be applied as many times as desired. A suitably chosen sequence of applications of the two processes may lead to better and better estimates of the exponential sum. On the other hand R. A. Rankin [1] has shown that there does exist a limiting case with a best possible estimate for the method which cannot be improved upon.

E. C. Titchmarsh generalized van der Corput's method to two-dimensional exponential sums. In most problems this method provides sharper estimates than those given by van der Corput's method. However, it is very hard to extend Titchmarsh's method to multiple exponential sums. For this purpose we make use of a result of I. M. Vinogradov. In order to handle the sphere problem he sharpened a transformation formula of J. G. van der Corput, which was advantageously used for multiple exponential sums by G. Kolesnik and E. Krätzel.

2.1. ONE-DIMENSIONAL EXPONENTIAL SUMS

2.1.1. The method of van der Corput

In this section we shall investigate *the simplest case of van der Corput's method.* By means of the Euler-Maclaurin sum formula an exponential sum

$$\sum_{a<n\leqq b} e^{2\pi i f(n)}$$

will be replaced by an integral

$$\int_a^b e^{2\pi i f(t)}\,dt\,.$$

The estimation of the integral leads to a non-trivial estimate of the above sum.

Lemma 2.1. *Let $f(t)$ be a real function in $[a, b]$ with a continuous derivative $f'(t)$. Let $f'(t)$ be monotonic, and let $|f'(t)| \leqq \vartheta < 1$, where ϑ is a fixed constant. Then*

$$\sum_{a<n\leqq b} e^{2\pi i f(n)} = \int_a^b e^{2\pi i f(t)}\,dt + O(1)\,. \tag{2.1}$$

Proof. If $b - a \leqq 1$, the result is trivial. So we may suppose that $b - a > 1$. Applying the Euler-Maclaurin sum formula, we obtain

$$\sum_{a<n\leqq b} e^{2\pi i f(n)} = \int_a^b e^{2\pi i f(t)}\,dt + 2\pi i \int_a^b \psi(t)\, f'(t)\, e^{2\pi i f(t)}\,dt + O(1)\,. \tag{2.2}$$

In the second term on the right-hand side of (2.2) we use the Fourier expansion (1.10) of $\psi(t)$ and integrate term-by-term. Then

$$\begin{aligned} 2\pi i \int_a^b \psi(t)\, f'(t)\, e^{2\pi i f(t)}\,dt &= -2i \sum_{\nu=1}^{\infty} \frac{1}{\nu} \int_a^b f'(t) \sin 2\pi\nu t\, e^{2\pi i f(t)}\,dt \\ &= \sum_{\substack{\nu=-\infty \\ \nu\neq 0}}^{+\infty} \frac{1}{\nu} \int_a^b f'(t)\, e^{2\pi i (f(t)-\nu t)}\,dt\,. \end{aligned}$$

Since the function $f'(t)/(f'(t) - \nu)$ is monotonic, we may apply the second mean-value theorem to the real and imaginary parts of the integral. Hence we obtain that

$$\int_a^b \frac{f'(t)}{f'(t)-\nu}\,(f'(t)-\nu)\,e^{2\pi i(f(t)-\nu t)}\,dt \ll \left|\frac{f'(a)}{f'(a)-\nu}\right| + \left|\frac{f'(b)}{f'(b)-\nu}\right| \ll \frac{1}{\nu}$$

and

$$2\pi i \int_a^b \psi(t)\, f'(t)\, e^{2\pi i f(t)}\,dt \ll \sum_{\nu=1}^{\infty} \frac{1}{\nu^2} \ll 1\,.$$

Therefore result (2.1) follows from (2.2).

Lemma 2.2. *Let $f(t)$ be a real function in $[a, b]$, twice continuously differentiable, and let $|f''(t)| \geqq \lambda_2 > 0$. Then*

$$\int_a^b e^{if(t)}\,dt \ll \frac{1}{\sqrt{\lambda_2}}\,. \tag{2.3}$$

Proof. Assume that $f''(t) \geqq \lambda_2 > 0$. *Then* $f'(t)$ is strictly increasing. Consequently, the interval $[a, b]$ is divided into at most two parts, where $f'(t)$ is either positive or negative throughout the interior of each part. Hence, we may suppose without loss of generality that either $f'(t) > 0$ throughout the interval (a, b) or $f'(t) < 0$. Let $f'(t) > 0$. Moreover, we assume that $b - a > 1/\sqrt{\lambda_2}$, since otherwise (2.3) is trivial. Then

$$\int_a^{a+1/\sqrt{\lambda_2}} e^{if(t)}\,dt \ll \frac{1}{\sqrt{\lambda_2}} \tag{2.4}$$

and, by the second mean-value theorem,

$$\int_{a+1/\sqrt{\lambda_2}}^{b} e^{if(t)}\,dt \ll \frac{1}{f'(a+1/\sqrt{\lambda_2})}\,.$$

Now there exists a number c with $a < c < a + 1/\sqrt{\lambda_2}$ such that

$$\frac{f'(a+1/\sqrt{\lambda_2}) - f'(a)}{a + 1/\sqrt{\lambda_2} - a} = f''(c) \geqq \lambda_2\,.$$

Hence $f'(a + 1/\sqrt{\lambda_2}) \geqq f'(a) + \sqrt{\lambda_2} \geqq \sqrt{\lambda_2}$ and

$$\int_{a+1/\sqrt{\lambda_2}}^{b} e^{if(t)}\,dt \ll \frac{1}{\sqrt{\lambda_2}}\,. \tag{2.5}$$

From (2.4) and (2.5) we deduce (2.3). The proof in case of $f'(t) < 0$ is similar.

Theorem 2.1. *Let $f(t)$ be a real function in $[a, b]$, twice continuously differentiable, and let $|f''(t)| \geqq \lambda_2 > 0$. Then*

$$\sum_{a<n\leqq b} e^{2\pi i f(n)} \ll \frac{|f'(b) - f'(a)| + 1}{\sqrt{\lambda_2}}. \tag{2.6}$$

Proof. If $\lambda_2 \geqq 1$, the result is trivial since we have

$$\sum_{a<n\leqq b} e^{2\pi i f(n)} \ll (b-a)\sqrt{\lambda_2} \ll \frac{1}{\sqrt{\lambda_2}} \int_a^b |f''(t)|\, dt = \frac{|f'(b) - f'(a)|}{\sqrt{\lambda_2}}.$$

Therefore we suppose that $\lambda_2 < 1$. Let $f''(t) \geqq \lambda_2 > 0$. Let $[\alpha, \beta]$ be a subinterval of $[a, b]$ with $h - 1/2 \leqq f'(t) \leqq h + 1/2$, where h is an integer. Thus we obtain from (2.1) and (2.3)

$$\sum_{\alpha<n\leqq\beta} e^{2\pi i f(n)} = \sum_{\alpha<n\leqq\beta} e^{2\pi i (f(n)-hn)} = \int_\alpha^\beta e^{2\pi i (f(t)-ht)}\, dt + O(1) \ll \frac{1}{\sqrt{\lambda_2}}.$$

The interval $[a, b]$ can be divided into at most $O(|f'(b) - f'(a)| + 1)$ subintervals of the above type. This proves estimate (2.6).

In lattice point theory one is interested in the sum $\Sigma\, \psi(f(n))$. Now it is very easy to obtain an estimate for this sum by means of Theorem 1.8 and (2.6).

Theorem 2.2. *Let $f(t)$ be a real function in $[a, b]$, twice continuously differentiable, and let $|f''(t)| \geqq \lambda_2 > 0$. Then*

$$\sum_{a<n\leqq b} \psi(f(n)) \ll \frac{|f'(b) - f'(a)|}{\lambda_2^{2/3}} + \frac{1}{\sqrt{\lambda_2}}. \tag{2.7}$$

Proof. We deduce from (2.6)

$$\sum_{a<n\leqq b} e^{2\pi i \nu f(n)} \ll |f'(b) - f'(a)| \sqrt{\frac{\nu}{\lambda_2}} + \frac{1}{\sqrt{\nu\lambda_2}}.$$

Taking (1.18) with $s = 1$, we obtain

$$\sum_{a<n\leqq b} \psi(f(n)) \ll \frac{b-a}{z} + \sum_{\nu=1}^{\infty} \min\left(\frac{z}{\nu^2}, \frac{1}{\nu}\right)\left(|f'(b) - f'(a)| \sqrt{\frac{\nu}{\lambda_2}} + \frac{1}{\sqrt{\nu\lambda_2}}\right)$$

$$\ll \frac{b-a}{z} + |f'(b) - f'(a)| \sqrt{\frac{z}{\lambda_2}} + \frac{1}{\sqrt{\lambda_2}}.$$

Because of $|f'(b) - f'(a)| = \int_a^b |f''(t)|\, dt \geqq (b-a)\lambda_2$ we get

$$\sum_{a<n\leqq b} \psi(f(n)) \ll |f'(b) - f'(a)| \left(\frac{1}{z\lambda_2} + \sqrt{\frac{z}{\lambda_2}}\right) + \frac{1}{\sqrt{\lambda_2}}.$$

The first two terms are of the same order if we put $z = \lambda_2^{-1/3}$, and estimate (2.7) follows.

The following version of the theorem is more useful for the applications.

Theorem 2.3. *Let $f(t)$ be a real function in $[a, b]$, twice continuously differentiable. Let $f''(t)$ be monotonic and be either positive or negative throughout. Then*

$$\sum_{a<n\leqq b} \psi(f(n)) \ll \int_a^b |f''(t)|^{1/3}\,dt + \frac{1}{\sqrt{|f''(a)|}} + \frac{1}{\sqrt{|f''(b)|}}. \tag{2.8}$$

Proof. We may suppose without loss of generality that $|f''(t)|$ is monotonically increasing. We divide the interval $(a, b]$ into subintervals $(t_\nu, t_{\nu+1}]$ $(\nu = 0, 1, \dots, N-1)$ and $(t_N, b]$ such that $t_0 = a$,

$$2^\nu |f''(a)| \leqq |f''(t)| \leqq 2^{\nu+1} |f''(a)| \leqq f''(b)$$

if $t_\nu < t \leqq t_{\nu+1}$, and

$$N = \left[\frac{\log|f''(b)| - \log|f''(a)|}{\log 2}\right].$$

Applying (2.7), we obtain

$$\begin{aligned}
\sum_{a<n\leqq b} \psi(f(n)) &= \sum_{\nu=0}^{N-1} \sum_{t_\nu<n\leqq t_{\nu+1}} \psi(f(n)) + \sum_{t_N<n\leqq b} \psi(f(n)) \\
&\ll \sum_{\nu=0}^{N-1} \frac{|f'(t_{\nu+1}) - f'(t_\nu)|}{(2^\nu |f''(a)|)^{2/3}} + \frac{|f'(b) - f'(t_N)|}{(2^N |f''(a)|)^{2/3}} + \sum_{\nu=0}^{N} \frac{1}{\sqrt{2^\nu |f''(a)|}} \\
&\ll \sum_{\nu=0}^{N-1} (2^\nu |f''(a)|)^{-2/3} \int_{t_\nu}^{t_{\nu+1}} |f''(t)|\,dt \\
&\quad + (2^N |f''(a)|)^{-2/3} \int_{t_N}^{b} |f''(t)|\,dt + \frac{1}{\sqrt{|f''(a)|}} \\
&\ll \int_a^b |f''(t)|^{1/3}\,dt + \frac{1}{\sqrt{|f''(a)|}}.
\end{aligned}$$

This proves the theorem.

Van der Corput's Theorem 2.3 should be applied only if the integral of (2.8) is of higher order than both the other terms. And this integral allows a geometrical interpretation.

Theorem 2.4. *Let $f(t)$ be a real function in $[a, b]$, twice continuously differentiable. Let $f''(t)$ be monotonic and either positive or negative throughout. Let l denote the length of the curve $y = f(t)$, and let r_{max} and r_{min} be the maximum and minimum of the absolute value of the radius of curvature, respectively. Then*

$$\int_a^b |f''(t)|^{1/3}\, dt \ll \min(r_{max}^{2/3}, l r_{min}^{-1/3}) . \tag{2.9}$$

Proof. The length of the curve is given by

$$l = \int_a^b \sqrt{1 + f'^2(t)}\, dt$$

and the absolute value $r(t)$ of the radius of curvature by

$$r(t) = \frac{(1 + f'^2(t))^{3/2}}{|f''(t)|} .$$

If $r_{min} > 0$, we find easily

$$\int_a^b |f''(t)|^{1/3}\, dt = \int_a^b (r(t))^{-1/3} \sqrt{1 + f'^2(t)}\, dt \leqq l r_{min}^{-1/3} . \tag{2.10}$$

If $r_{max} < \infty$, we obtain

$$\int_a^b |f''(t)|^{1/3}\, dt = \int_a^b |f''(t)| \frac{(r(t))^{2/3}}{1 + f'^2(t)}\, dt$$

$$\leqq r_{max}^{2/3} \int_a^b |f''(t)| \min\left(1, \frac{1}{f'^2(t)}\right) dt \ll r_{max}^{2/3} . \tag{2.11}$$

Result (2.9) now follows from (2.10) and (2.11).

Corollary. *Let $1 \ll r_{max} < \infty$. Then*

$$\sum_{a < n \leqq b} \psi(f(n)) \ll r_{max}^{2/3} .$$

This estimate follows immediately from (2.8) and (2.9) since

$$\frac{1}{\sqrt{|f''(a)|}} + \frac{1}{\sqrt{|f''(b)|}} \ll \sqrt{r_{max}} \ll r_{max}^{2/3} .$$

2.1.2. *The application of Weyl's steps*

The process under discussion corresponds to a suitably chosen application of Schwarz's inequality. It may be called *Weyl's step.*

Theorem 2.5. *Let $f(t)$ be a real function in $[a, b]$ and H an integer with $1 \leqq H \leqq b - a$. Then*

$$\sum_{a<n\leqq b} e^{2\pi i f(n)} \ll \frac{b-a}{\sqrt{H}} + \left\{\frac{b-a}{H}\sum_{h=1}^{H-1}\left|\sum_{a<n\leqq b-h} e^{2\pi i(f(n+h)-f(n))}\right|\right\}^{1/2}. \tag{2.12}$$

Proof. By Schwarz's inequality we have

$$\begin{aligned}
H^2\left|\sum_{a<n\leqq b} e^{2\pi i f(n)}\right|^2 &= \left|\sum_{m=1}^{H}\sum_{a<n+m\leqq b} e^{2\pi i f(n+m)}\right|^2 \\
&= \left|\sum_{a-H<n\leqq b-1}\sum_{\substack{a<n+m\leqq b\\ 1\leqq m\leqq H}} e^{2\pi i f(n+m)}\right|^2 \\
&\leqq \sum_{a-H<n'\leqq b-1} 1 \sum_{a-H<n\leqq b-1}\left|\sum_{\substack{a<n+m\leqq b\\ 1\leqq m\leqq H}} e^{2\pi i f(n+m)}\right|^2 \\
&< 2(b-a)\sum_{\substack{a<n+m\leqq b\\ a<n+m'\leqq b\\ 1\leqq m,\, m'\leqq H}} e^{2\pi i(f(n+m)-f(n+m'))} \\
&\leqq 2H(b-a)^2 + 4(b-a)\sum_{\substack{a<n+m\leqq b\\ a<n+m'\leqq b\\ 1\leqq m'<m\leqq H}} e^{2\pi i(f(n+m)-f(n+m'))} \\
&= 2H(b-a)^2 + 4(b-a)\sum_{h=1}^{H-1}\sum_{\substack{m-m'=h\\ 1\leqq m'<m\leqq H}}\sum_{a<n\leqq b-h} e^{2\pi i(f(n+h)-f(n))} \\
&\leqq 2H(b-a)^2 + 4H(b-a) + \sum_{h=1}^{H-1}\left|\sum_{a<n\leqq b-h} e^{2\pi i(f(n+h)-f(n))}\right|.
\end{aligned}$$

This proves (2.12).

Applying this result repeatedly and using Theorem 2.1, it is easy to prove the following estimate.

Theorem 2.6. *Let $f(t)$ be a real function with continuous derivatives up to the k-th order in $[a, b]$. Let $k \geqq 2$, $K = 2^k$ and $0 < \lambda_k \leqq |f^{(k)}(t)| \ll \lambda_k$. Then*

$$\sum_{a<n\leqq b} e^{2\pi i f(n)} \ll (b-a)\,\lambda_k^{\frac{1}{K-2}} + (b-a)^{1-\frac{4}{K}}\,\lambda_k^{-\frac{1}{K-2}}. \tag{2.13}$$

Proof. If $\lambda_k \geqq 1$, estimate (2.13) is trivial. So we assume that $\lambda_k < 1$. If $k = 2$, the result stated is contained in Theorem 2.1. Now suppose the theorem to be true for all integers up to $k - 1$. Because of

$$|f^{(k-1)}(t+h) - f^{(k-1)}(t)| = \int_t^{t+h} |f^{(k)}(\tau)|\, d\tau \tag{2.14}$$

we have

$$h\lambda_k \leqq |f^{(k-1)}(t+h) - f^{(k-1)}(t)| \ll h\lambda_k\,.$$

Hence, by (2.12) and (2.13) with $k - 1$ instead of k,

$$\sum_{a<n\leqq b} e^{2\pi i f(n)}$$

$$\ll \frac{b-a}{\sqrt{H}} + \left\{\frac{b-a}{H}\sum_{h=1}^{H-1}\left((b-a)(h\lambda_k)^{\frac{2}{K-4}} + (b-a)^{1-\frac{8}{K}}(h\lambda_k)^{-\frac{2}{K-4}}\right)\right\}^{\frac{1}{2}}$$

$$\ll \frac{b-a}{\sqrt{H}} + (b-a)(H\lambda_k)^{\frac{1}{K-4}} + (b-a)^{1-\frac{4}{K}}(H\lambda_k)^{-\frac{1}{K-4}}.$$

If we put $H = \left[\lambda_k^{-\frac{2}{K-2}}\right] + 1$, provided that $H \leqq b - a$, the first two terms are of the same order. Therefore result (2.13) follows on condition that

$$H \leqq 2\lambda_k^{-\frac{2}{K-2}} \leqq b - a\,.$$

Otherwise, we obtain by trivial estimation

$$\sum_{a<n\leqq b} e^{2\pi i f(n)} \leqq b - a < \sqrt{2(b-a)}\,\lambda_k^{-\frac{1}{K-2}} \ll (b-a)^{1-\frac{4}{K}}\lambda_k^{-\frac{1}{K-2}}.$$

The last inequality holds if $k \geqq 3$. Thus (2.13) also holds in this case.

Let us now consider a stronger version of Theorem 2.6, which may be applied for multiplicative problems.

Theorem 2.7. *Let u be a fixed positive constant and $1 \leqq a < b \leqq au$. Let $f(t)$ be a real function in $[a, b]$ with continuous derivatives up to the k-th order. Let $k \geqq 2$, $K = 2^k$ and*

$$|f^{(\nu)}(t)| \asymp \frac{\lambda}{a^\nu} \qquad (\nu = 2, 3, \ldots, k)\,.$$

If $\lambda \gg a$, then

$$\sum_{a<n\leqq b} e^{2\pi i f(n)} \ll a^{1-\frac{k}{K-2}}\lambda^{\frac{1}{K-2}}. \tag{2.15}$$

Proof. If $k = 2$, we apply Theorem 2.1 with $\lambda_2 = \lambda/a^2$ and

$$|f'(b) - f'(a)| = \int_a^b |f''(t)|\,dt \ll \frac{\lambda}{a}\,.$$

Then, by (2.6),

$$\sum_{a<n\leqq b} e^{2\pi i f(n)} \ll \sqrt{\lambda} + \frac{a}{\sqrt{\lambda}} \tag{2.16}$$

for any positive λ. But if $\lambda \gg a$ is satisfied, (2.15) follows for $k = 2$ at once. Now suppose the theorem to be true for all integers up to $k - 1$. We apply Theorem 2.5.

Let us first consider the sum in (2.12) only for $\lambda h \leqq (b-a)^2$. Then, by (2.14),

$$|f''(t+h) - f''(t)| \asymp \frac{h\lambda}{a^3}.$$

Hence, estimate (2.16) with $h\lambda/a$ for λ gives

$$\frac{b-a}{H} \sum_{\lambda h \leqq (b-a)^2} \left| \sum_{a<n\leqq b-h} e^{2\pi i(f(n+h)-f(n))} \right| \ll \frac{a}{H} \sum_{\lambda h \ll a^2} a \sqrt{\frac{a}{\lambda h}} \ll \frac{a^2}{H} \left(\frac{a^3}{\lambda^2}\right)^{1/2}. \tag{2.17}$$

For the second part of the sum in (2.12) we use the induction argument. We have, by (2.14),

$$|f^{(k-1)}(t+h) - f^{(k-1)}(t)| \asymp \frac{h\lambda}{a^k}.$$

Then we obtain by (2.15) with $k - 1$ for k and $h\lambda/a$ for λ

$$\begin{aligned} &\frac{b-a}{H} \sum_{(b-a)^2/\lambda < h < H} \left| \sum_{a<n\leqq b-h} e^{2\pi i(f(n+h)-f(n))} \right| \\ &\ll \frac{a}{H} \sum_{h=1}^{H} a^{1-\frac{2(k-1)}{K-4}} \left(\frac{h\lambda}{a}\right)^{\frac{2}{K-4}} \ll a^{2-\frac{2K}{K-4}} (H\lambda)^{\frac{2}{K-4}}. \end{aligned} \tag{2.18}$$

Hence, we get from (2.12) and (2.17), (2.18)

$$\sum_{a<n\leqq b} e^{2\pi i f(n)} \ll \frac{a}{\sqrt{H}} \left\{1 + \left(\frac{a^3}{\lambda^2}\right)^{\frac{1}{4}}\right\} + a^{1-\frac{k}{K-4}} (H\lambda)^{\frac{1}{K-4}}.$$

If $\lambda^2 \gg a^3$, the terms are of the same order by putting $H = \left[\left(\frac{a^k}{\lambda}\right)^{\frac{2}{K-2}}\right]$. Therefore estimate (2.15) holds provided that $1 \leqq H \leqq b - a$. Clearly, if $\lambda > a^k$, the theorem is trivial. If $H > b - a$, we find by trivial estimation for $k \geqq 3$

$$\left| \sum_{a<n\leqq b} e^{2\pi i f(n)} \right| \leqq b - a < H \leqq \left(\frac{a^k}{\lambda}\right)^{\frac{2}{K-2}} \ll a^{1-\frac{k}{K-2}} \lambda^{\frac{1}{K-2}}.$$

It remains the case $\lambda^2 \ll a^3$. Here we obtain by (2.16)

$$\sum_{a<n\leqq b} e^{2\pi i f(n)} \ll \sqrt{\lambda} \ll a^{1-\frac{k}{K-2}} \lambda^{\frac{2}{3}\frac{k}{K-2}-\frac{1}{6}} \ll a^{1-\frac{k}{K-2}} \lambda^{\frac{1}{K-2}},$$

and the result again holds.

Theorem 2.8. *Let $u > 0$, $1 \leqq a < b \leqq au$. Let $f(t)$ be a real function in $[a, b]$ with continuous derivatives up to the k-th order. Let $k \geqq 2$, $K = 2^k$ and*

$$|f^{(\nu)}(t)| \asymp \frac{\lambda}{a^\nu} \qquad (\nu = 2, 3, \dots, k).$$

If $\lambda \gg a$, then

$$\sum_{a<n\leqq b} \psi(f(n)) \ll (a^{K-k-1}\lambda)^{\frac{1}{K-1}}. \tag{2.19}$$

Proof. Applying (1.18), we obtain from (2.15)

$$\sum_{a<n\leqq b} \psi(f(n)) \ll \frac{a}{z} + \sum_{\nu=1}^{\infty} \min\left(\frac{z}{\nu^2}, \frac{1}{\nu}\right) a^{1-\frac{k}{K-2}} (\lambda\nu)^{\frac{1}{K-2}}$$

$$\ll \frac{a}{z} + a^{1-\frac{k}{K-2}} (\lambda z)^{\frac{1}{K-2}}.$$

If $z = \left(\frac{a^k}{\lambda}\right)^{\frac{1}{K-1}}$, (2.19) follows at once, provided that $\lambda \leqq a^K$. Otherwise the theorem is trivial.

2.1.3. *Transformation of exponential sums*

We now turn to the transformation of exponential sums. The resulting formula is based on an application of the Euler-Maclaurin sum formula and an asymptotic expansion of some integrals. The transformation of exponential sums is also referred to as *van der Corput transform*.

Lemma 2.3. *Let $f(t)$ and $g(t)$ be real functions with continuous and strictly monotonical derivatives in $[a, b]$. Let*

$$\alpha = \min f'(t), \qquad \beta = \max f'(t), \qquad |g(t)| \leqq G, \qquad |g'(t)| \leqq G_1,$$

and let η be any positive constant less than 1. Then

$$\sum_{a<n\leqq b} g(n)\, e^{2\pi i f(n)} = \sum_{\alpha-\eta<\nu\leqq\beta+\eta} \int_a^b g(t)\, e^{2\pi i (f(t)-\nu t)}\, dt$$
$$+ O(G \log(\beta - \alpha + 2)) + O(G_1). \tag{2.20}$$

Proof. Suppose that $f'(t)$ is strictly decreasing. If k is an integer such that $\eta - 1 \leqq \alpha - k < \eta$, we can replace the function $f(t)$ by $f(t) - kt$ in (2.20). Hence we may assume without loss of generality that $\eta - 1 \leqq \alpha < \eta$. Applying the Euler-Maclaurin sum formula, we obtain

$$\sum_{a<n\leqq b} g(n)\, e^{2\pi i f(n)}$$

$$= \int_a^b g(t)\, e^{2\pi i f(t)}\, dt + \int_a^b \psi(t)\,\{2\pi\, i f'(t)\, g(t) + g'(t)\}\, e^{2\pi i f(t)}\, dt + O(G)\,. \quad (2.21)$$

In the second integral on the right-hand side of (2.21) we use the Fourier expansion of $\psi(t)$ in the form

$$\psi(t) = \frac{1}{2\pi i} \sum_{\nu=1}^{\infty} \frac{1}{\nu} (e^{-2\pi i\nu t} - e^{2\pi i\nu t})\,.$$

Integrating term-by-term and splitting up the sum into five parts, we get

$$\int_a^b \psi(t)\,\{2\pi\, i f'(t)\, g(t) + g'(t)\}\, e^{2\pi i f(t)}\, dt = \sum_{k=1}^{5} I_k\,,$$

$$I_1 = \sum_{1\leqq\nu\leqq\beta+\eta} \frac{1}{2\pi i\nu} \int_a^b \{2\pi\, i f'(t)\, g(t) + g'(t)\}\, e^{2\pi i(f(t)-\nu t)}\, dt\,,$$

$$I_2 = \sum_{\nu>\beta+\eta} \frac{1}{\nu} \int_a^b \frac{f'(t)\, g(t)}{f'(t)-\nu} (f'(t)-\nu)\, e^{2\pi i(f(t)-\nu t)}\, dt\,,$$

$$I_3 = \sum_{\nu>\beta+\eta} \frac{1}{2\pi i\nu} \int_a^b \frac{g'(t)}{f'(t)-\nu} (f'(t)-\nu)\, e^{2\pi i(f(t)-\nu t)}\, dt\,,$$

$$I_4 = -\sum_{\nu=1}^{\infty} \frac{1}{\nu} \int_a^b \frac{f'(t)\, g(t)}{f'(t)+\nu} (f'(t)+\nu)\, e^{2\pi i(f(t)+\nu t)}\, dt\,,$$

$$I_5 = -\sum_{\nu=1}^{\infty} \frac{1}{2\pi i\nu} \int_a^b \frac{g'(t)}{f'(t)+\nu} (f'(t)+\nu)\, e^{2\pi i(f(t)+\nu t)}\, dt\,.$$

For the term I_1 we easily obtain

$$I_1 = \sum_{1\leqq\nu\leqq\beta+\eta} \left\{ \int_a^b g(t)\, e^{2\pi i(f(t)-\nu t)}\, dt + \frac{1}{2\pi i\nu} \left[g(t)\, e^{2\pi i(f(t)-\nu t)} \right]_a^b \right\}$$

$$= \sum_{1\leqq\nu\leqq\beta+\eta} \int_a^b g(t)\, e^{2\pi i(f(t)-\nu t)}\, dt + O(G \log(\beta+2))\,.$$

In I_2 the function $f'(t)/(\nu - f'(t))$ is strictly decreasing. Since $g'(t)$ is strictly monotonic, we can divide the interval $[a, b]$ into at most two subintervals such that $g(t)$ is monotonic in each of these subintervals. Twice applying the second mean-value theorem to the real and imaginary parts, we get

$$I_2 \ll G \sum_{\nu > \beta + \eta} \frac{\beta}{\nu(\nu - \beta)} < G \sum_{\beta + \eta < \nu \leqq 2\beta} \frac{1}{\nu - \beta} + G \sum_{\nu > 2\beta} \frac{2\beta}{\nu^2} \ll G \log(\beta + 2).$$

Similarly, the third term is

$$I_3 \ll G_1 \sum_{\nu > \beta + \eta} \frac{1}{\nu(\nu - \beta)} \ll G_1 .$$

In the same way we obtain

$$I_4 \ll G \log(\beta + 2), \qquad I_5 \ll G_1 .$$

Substituting these estimates into (2.21), we obtain (2.20) in case of $\eta - 1 \leqq \alpha < \eta$, and the lemma is proved.

Lemma 2.4. *Let $f(t)$ be a real function with continuous derivatives upt to the third order in $[a, b]$. Let $|f''(t)| \asymp \lambda_2$, $|f'''(t)| \ll \lambda_3$ throughout the interval. Let $f'(c) = 0$, where $a \leqq c \leqq b$. If*

$$\varepsilon = \begin{cases} e^{\pi i/4} & for \quad f''(t) > 0, \\ e^{-\pi i/4} & for \quad f''(t) < 0, \end{cases}$$

then

$$\int_a^b e^{if(t)}\, dt = \varepsilon \sqrt{\frac{2\pi}{f''(c)}}\, e^{if(c)} + O\left(\min\left(\frac{1}{|f'(a)|}, \frac{1}{\sqrt{\lambda_2}}\right)\right)$$

$$+ O\left(\min\left(\frac{1}{|f'(b)|}, \frac{1}{\sqrt{\lambda_2}}\right)\right) + O(r), \tag{2.22}$$

$$r = \lambda_2^{-4/5} \lambda_3^{1/5} . \tag{2.23}$$

Moreover, if $f(t)$ possesses continuous derivatives up to the fourth order with $|f^{(4)}(t)| \ll \lambda_4$ and $\lambda_3^2 \asymp \lambda_2 \lambda_4$, (2.23) *can be replaced by*

$$r = \lambda_2^{-1} \lambda_3^{1/3} . \tag{2.24}$$

Remark. In most cases of the applications we have $\lambda_2^3 \gg \lambda_3^2$ such that (2.24) is a better estimate than (2.23).

Proof. Since $f'(t)$ is a strictly monotonical function, there exists only one value c with $f'(c) = 0$. Suppose that $f''(t) > 0$. Let δ denote a positive fixed value. If $c < b$, we write

$$\int_c^b e^{if(t)}\, dt = \int_c^{c+\delta} e^{if(t)}\, dt + \int_{c+\delta}^b e^{if(t)}\, dt ,$$

provided that $\delta \leqq b - c$. Clearly, by the second mean-value theorem we have

$$\int_{c+\delta}^{b} e^{if(t)}\,dt \ll \frac{1}{f'(c+\delta)} = \frac{1}{\int_c^{c+\delta} f''(t)\,dt} \ll \frac{1}{\delta\lambda_2}.$$

By means of the Taylor expansion

$$f(t) = f(c) + \frac{1}{2} f''(c)\,(t-c)^2 + \frac{1}{6} f'''(c + \vartheta(t-c))\,(t-c)^3 \quad (0 < \vartheta < 1)$$

we obtain

$$\int_c^{c+\delta} e^{if(t)}\,dt = e^{if(c)} \int_c^{c+\delta} e^{\frac{i}{2} f''(c)(t-c)^2} \{1 + O(\lambda_3 (t-c)^3)\}\,dt$$

$$= \sqrt{\frac{2}{f''(c)}}\, e^{if(c)} \int_0^{\infty} e^{i\tau^2}\,d\tau$$

$$+ O\left(\frac{1}{\sqrt{\lambda_2}} \int_{\frac{1}{2}\delta^2 f''(c)}^{\infty} \frac{1}{\sqrt{\tau}}\, e^{i\tau}\,d\tau\right) + O(\delta^4 \lambda_3)$$

$$= \varepsilon \sqrt{\frac{\pi}{2 f''(c)}}\, e^{if(c)} + O\left(\frac{1}{\delta\lambda_2}\right) + O(\delta^4 \lambda_3).$$

Hence

$$\int_c^b e^{if(t)}\,dt = \varepsilon \sqrt{\frac{\pi}{2 f''(c)}}\, e^{if(c)} + O\left(\frac{1}{\delta\lambda_2}\right) + O(\delta^4 \lambda_3).$$

If $\delta > b - c$, we have

$$\int_c^b e^{if(t)}\,dt = e^{if(c)} \int_c^b e^{\frac{i}{2} f''(c)(t-c)^2}\,dt + O(\delta^4 \lambda_3).$$

On the right-hand side we now integrate from c up to $c + \delta$ and subtract the integral from b up to $c + \delta$. The new integral can be estimated by

$$\int_b^{c+\delta} e^{\frac{i}{2} f''(c)(t-c)^2}\,dt \ll \frac{1}{\sqrt{f''(c)}} \ll \frac{1}{\sqrt{\lambda_2}}$$

and

$$\int_b^{c+\delta} e^{\frac{i}{2}f''(c)(t-c)^2}\,dt = \int_b^{c+\delta} \frac{if''(c)(t-c)}{if''(c)(t-c)}\, e^{\frac{i}{2}f''(c)(t-c)^2}\,dt$$

$$\ll \frac{1}{f''(c)(b-c)} \ll \frac{1}{\lambda_2(b-c)} \ll \frac{1}{\int_c^b f''(t)\,dt} = \frac{1}{f'(b)}.$$

Therefore, without any condition on δ, we have

$$\int_c^b e^{if(t)}\,dt = \varepsilon\sqrt{\frac{\pi}{2f''(c)}}\, e^{if(c)} + O\left(\min\left(\frac{1}{f'(b)}, \frac{1}{\sqrt{\lambda_2}}\right)\right)$$
$$+ O\left(\frac{1}{\delta\lambda_2}\right) + O(\delta^4\lambda_3).$$

If we choose $\delta = (\lambda_2\lambda_3)^{-1/5}$, we obtain

$$\int_c^b e^{if(t)}\,dt = \varepsilon\sqrt{\frac{\pi}{2f''(c)}}\, e^{if(c)} + O\left(\min\left(\frac{1}{f'(b)}, \frac{1}{\sqrt{\lambda_2}}\right)\right) + O(r), \tag{2.25}$$

where r is given by (2.23). Similarly, we get for $a < c$

$$\int_a^c e^{if(t)}\,dt = \varepsilon\sqrt{\frac{\pi}{2f''(c)}}\, e^{if(c)} + O\left(\min\left(\frac{1}{|f'(a)|}, \frac{1}{\sqrt{\lambda_2}}\right)\right) + O(r). \tag{2.26}$$

This proves (2.22) for $a \leqq c \leqq b$ with the error term (2.23).

In order to prove (2.22) with the error term (2.24), we use the Taylor expansion

$$f(t) = f(c) + \frac{1}{2}f''(c)(t-c)^2 + \frac{1}{6}f'''(c)(t-c)^3 + \frac{1}{24}f^{(4)}(c+\vartheta(t-c))(t-c)^4.$$

Then

$$\int_c^{c+\delta} e^{if(t)}\,dt = e^{if(c)}\int_0^{\delta} e^{\frac{i}{2}f''(c)t^2+\frac{i}{6}f'''(c)t^3}\{1+O(\lambda_4 t^4)\}\,dt = e^{if(c)}\int_0^{\delta} e^{\frac{i}{2}f''(c)t^2}\,dt$$

$$+ e^{if(c)}\sum_{\nu=1}^{\infty}\frac{1}{\nu!}\int_0^{\delta} e^{\frac{i}{2}f''(c)t^2}\left(\frac{i}{6}f'''(c)t^3\right)^{\nu} dt + O(\delta^5\lambda_4)$$

$$= \varepsilon\sqrt{\frac{\pi}{2f''(c)}}\, e^{if(c)} + O\left(\frac{1}{\delta\lambda_2}\right) + O\left(\frac{1}{\delta\lambda_2}\sum_{\nu=1}^{\infty}\frac{1}{\nu!}(\delta^3\lambda_3)^{\nu}\right) + O(\delta^5\lambda_4)$$

$$= \varepsilon\sqrt{\frac{\pi}{2f''(c)}}\, e^{if(c)} + O\left(\frac{1}{\delta\lambda_2}\right) + O\left(\frac{1}{\delta\lambda_2}e^{\delta^3\lambda_3}\right) + O(\delta^5\lambda_4).$$

Taking $\delta = (\lambda_2\lambda_4)^{-1/6} \asymp \lambda_3^{-1/3}$, we obtain (2.25) from this estimate, where r is now given by (2.24). Clearly, (2.26) follows in the same way. Hence, (2.22) is proved for $a \leqq c \leqq b$ with the error term (2.24).

Theorem 2.9. *Let $f(t)$ be a real function with continuous derivatives up to the third order in $[a, b]$. Let $|f''(t)| \asymp \lambda_2$, $|f'''(t)| \ll \lambda_3$ throughout the interval. Let $\varphi(t)$ be defined by $f'(\varphi) = t$. If $\alpha = \min f'(t)$, $\beta = \max f'(t)$ and*

$$\varepsilon = \begin{cases} e^{\pi i/4} & \text{for} \quad f''(t) > 0\,, \\ e^{-\pi i/4} & \text{for} \quad f''(t) < 0\,, \end{cases}$$

then

$$\sum_{a<n\leqq b} e^{2\pi i f(n)} = \varepsilon \sum_{\alpha<\nu\leqq\beta} \frac{1}{\sqrt{|f''(\varphi(\nu))|}}\, e^{2\pi i(f(\varphi(\nu)) - \nu\varphi(\nu))} + O\left(\frac{1}{\sqrt{\lambda_2}}\right) + O(\log((b-a)\lambda_2 + 2)) + O(R)\,, \tag{2.27}$$

$$R = (b-a)(\lambda_2\lambda_3)^{1/5}\,. \tag{2.28}$$

Moreover, if $f(t)$ possesses continuous derivatives up to the fourth order with $|f^{(4)}(t)| \ll \lambda_4$ and $\lambda_3^2 \asymp \lambda_2\lambda_4$, (2.28) can be replaced by

$$R = (b-a)\lambda_3^{1/3}\,. \tag{2.29}$$

Proof. We use Lemma 2.3 with $g(t) = 1$ and $f''(t) > 0$. We have

$$\beta - \alpha = f'(b) - f'(a) = \int_a^b f''(t)\,dt \ll (b-a)\lambda_2\,.$$

Hence

$$\sum_{a<n\leqq b} e^{2\pi i f(n)} = \sum_{\alpha-\eta<\nu\leqq\beta+\eta} \int_a^b e^{2\pi i(f(t)-\nu t)}\,dt + O(\log((b-a)\lambda_2 + 2))\,.$$

Applying Lemma 2.4 with $2\pi(f(t) - \nu t)$ instead of $f(t)$, the number c is given by $f'(c) - \nu = 0$. Thus $c = \varphi(\nu)$. (2.22) then gives

$$\begin{aligned} \sum_{a<n\leqq b} e^{2\pi i f(n)} &= \sum_{\alpha<\nu\leqq\beta} \frac{1}{\sqrt{f''(\varphi(\nu))}}\, e^{2\pi i(f(\varphi(\nu))-\nu\varphi(\nu))} + O\left(\frac{1}{\sqrt{\lambda_2}}\right) \\ &\quad + O\left(\sum_{\alpha+1<\nu<\beta-1} \frac{1}{\nu-\alpha}\right) + O\left(\sum_{\alpha+1<\nu<\beta-1} \frac{1}{\beta-\nu}\right) \\ &\quad + O((\beta-\alpha)r) + O(\log((b-a)\lambda_2 + 2)) \\ &= \varepsilon \sum_{\alpha<\nu\leqq\beta} \frac{1}{\sqrt{f''(\varphi(\nu))}}\, e^{2\pi i(f(\varphi(\nu))-\nu\varphi(\nu))} + O\left(\frac{1}{\sqrt{\lambda_2}}\right) \\ &\quad + O(\log((b-a)\lambda_2 + 2))) + O((b-a)\lambda_2 r)\,. \end{aligned}$$

Since in both cases $R = (b - a)\,\lambda_2 r$, formula (2.27) follows with (2.28) or (2.29).

We now come to a theorem in which the van der Corput transform will be connected with Weyl's steps. In many applications this theorem gives better estimates than Theorem 2.1. For the sake of simplicity let us consider a simple, but sufficiently general case of the previous theorems.

Theorem 2.10. *Let $u > 0$, $1 \leqq a < b < au$. Let $f(t)$ be a real function with continuous derivatives up to the k-th order in $[a, b]$. Suppose that $|f^{(\nu)}(t)| \asymp \lambda/a^\nu$ ($\nu = 1, 2, 3, 4$). Let the function $\varphi(\tau)$ be defined by $f'(\varphi(\tau)) = \tau$ and assume that*

$$|\varphi^{(\nu)}(\tau)| \asymp \lambda \left(\frac{a}{\lambda}\right)^{\nu+1} \qquad (\nu = 2, 3, \ldots, k)\,.$$

If $\lambda \gg a$, $k \geqq 4$, $K = 2^k$, then

$$\sum_{a<n\leqq b} e^{2\pi i f(n)} \ll \left(a^k \lambda^{\frac{K}{2}-k}\right)^{\frac{1}{K-2}}, \tag{2.30}$$

$$\sum_{a<n\leqq b} \psi(f(n)) \ll \left(a^K \lambda^{K-2k}\right)^{\frac{1}{3K-2k-4}}. \tag{2.31}$$

Proof. Suppose that $f''(t) > 0$. Then $f'(t)$ is strictly increasing. By means of (2.27) with the remainder (2.29) we obtain

$$\sum_{a<n\leqq b} e^{2\pi i f(n)} = \varepsilon \sum_{f'(a)<\nu\leqq f'(b)} \frac{1}{\sqrt{f''(\varphi(\nu))}}\, e^{2\pi i F(\nu)} + O\left(\frac{a}{\sqrt{\lambda}}\right) + O(\lambda^{1/3})\,.$$

$F(t)$ is defined by

$$F(t) = f(\varphi(t)) - t\varphi(t)$$

such that

$$F^{(m)} = -\varphi^{(m-1)}(t)\,.$$

Thus we have

$$|F^{(m)}(t)| \asymp \lambda \left(\frac{a}{\lambda}\right)^m,$$

and we can use Theorem 2.7. Hence, by (2.15),

$$\sum_{f'(a)<\nu\leqq t} e^{2\pi i F(\nu)} \ll \left(\frac{\lambda}{a}\right)^{1-\frac{k}{K-2}} \lambda^{\frac{1}{K-2}},$$

where $t \leqq f'(b)$. The function

$$\frac{d}{dt} \frac{1}{\sqrt{f''(\varphi(t))}} = -\frac{1}{2} f'''(\varphi(t))\,(f''(\varphi(t)))^{-5/2}$$

is either positive or negative throughout. Partial summation now leads to

$$\sum_{a<n\leqq b} e^{2\pi i f(n)} = \frac{\varepsilon}{\sqrt{f''(b)}} \sum_{f'(a)<v\leqq f'(b)} e^{2\pi i F(v)}$$

$$- \varepsilon \int_{f'(a)}^{f'(b)} \left(\frac{d}{dt}\frac{1}{f''(\varphi(t))}\right) \sum_{f'(a)<v\leqq t} e^{2\pi i F(v)}\, dt + O\left(\frac{a}{\sqrt{\lambda}}\right) + O(\lambda^{1/3})$$

$$\ll \left(a^k \lambda^{\frac{K}{2}-k}\right)^{\frac{1}{K-2}} + \frac{a}{\sqrt{\lambda}} + \lambda^{1/3}.$$

The second term is always smaller than the first one, whereas the third one is only smaller for $k \geqq 5$. Therefore estimate (2.30) holds for $k \geqq 5$. For $k = 4$ we obtain

$$\sum_{a<n\leqq b} e^{2\pi i f(n)} \ll (a\lambda)^{2/7} + \lambda^{1/3} \ll (a\lambda)^{2/7},$$

provided that $\lambda \ll a^6$. Otherwise the theorem is trivial. Thus (2.30) also holds in this case.

The estimate (2.31) is a simple corollary of (2.30). We have by (1.18)

$$\sum_{a<n\leqq b} \psi(f(n)) \ll \frac{a}{z} + \left(a^k (\lambda z)^{\frac{K}{2}-k}\right)^{\frac{1}{K-2}}.$$

Both the terms on the right-hand side are of the same order if

$$z = \left(a^{K-k-2} \lambda^{-\frac{K}{2}+k}\right)^{\frac{2}{3K-2k-4}}.$$

This leads to result (2.31).

The preceding developments lead to sufficiently good approximations of one-dimensional exponential sums, and we can proceed to the theory of exponent pairs by repeated applications of van der Corput's transform and Weyl's steps in the next section. But for extending the method to many-dimensional sums we need a *stronger form of van der Corput's transform*.

Lemma 2.5. *Let $a < c < b$ and let $f(t)$ be a real function in $[a, c]$ and $[c, b]$, respectively, with continuous derivatives up to the third order. Let $f'(c) = 0$ and $|f''(t)| \asymp \lambda_2$, $0 < |f'''(t)| \ll \lambda_3$. Suppose that the function*

$$f'^6(t) - 8f''(c)\, f''^2(t)\, (f(t) - f(c))^3$$

only has a bounded number of points of zero. Let $g(t)$ *be a real function in* $[a, c]$ *and* $[c, b]$, *respectively, with a continuous and monotonical derivative and* $|g(t)| \leqq G$, $|g'(t)| \leqq G_1$. *Let*

$$\varepsilon = \begin{cases} e^{\pi i/4} & \text{for} \quad f''(t) > 0\,, \\ e^{-\pi i/4} & \text{for} \quad f''(t) < 0\,. \end{cases}$$

Then

$$\int_c^b g(t)\, e^{if(t)}\, dt = \varepsilon g(c) \sqrt{\frac{\pi}{2\,|f''(c)|}}\, e^{if(c)} + O\left(\frac{G_1}{\lambda_2}\right) + G\left\{O\left((b-c)\frac{\lambda_3^2}{\lambda_2^3}\right) + O\left(\frac{\lambda_3}{\lambda_2^2}\right) + O\left(\min\left(\frac{1}{|f'(b)|}, \frac{1}{\sqrt{\lambda_2}}\right)\right)\right\}, \tag{2.32}$$

$$\int_a^c g(t)\, e^{if(t)}\, dt = \varepsilon g(c) \sqrt{\frac{\pi}{2\,|f''(c)|}}\, e^{if(c)} + O\left(\frac{G_1}{\lambda_2}\right) + G\left\{O\left((c-a)\frac{\lambda_3^2}{\lambda_2^3}\right) + O\left(\frac{\lambda_3}{\lambda_2^2}\right) + O\left(\min\left(\frac{1}{|f'(a)|}, \frac{1}{\sqrt{\lambda_2}}\right)\right)\right\}. \tag{2.33}$$

Proof. In order to prove (2.32), suppose that $f''(t) > 0$. The $f'(t)$ increases monotonically. Because of $f'(c) = 0$ the function $f(t)$ is also monotonically increasing. Substituting $z = f(t)$, the integral becomes

$$\int_c^b g(t)\, e^{if(t)}\, dt = \int_{f(c)}^{f(b)} \frac{g(t)}{f'(t)}\, e^{iz}\, dz = I_1 + I_2\,, \tag{2.34}$$

$$I_1 = \int_{f(c)}^{f(b)} \frac{g(c)\, e^{iz}}{\sqrt{2f''(c)\,(z - f(c))}}\, dz\,,$$

$$I_2 = \int_{f(c)}^{f(b)} e^{iz} \left\{\frac{g(t)}{f'(t)} - \frac{g(c)}{\sqrt{2f''(c)\,(z - f(c))}}\right\} dz\,.$$

The integral I_1 may be written in the form

$$I_1 = \frac{g(c)\, e^{if(c)}}{\sqrt{2f''(c)}} \int_0^{f(b)-f(c)} \frac{1}{\sqrt{z}}\, e^{iz}\, dz = \varepsilon g(c) \sqrt{\frac{\pi}{2\,|f''(c)|}}\, e^{if(c)} + I_3\,,$$

where I_3 is given by

$$I_3 = -\frac{g(c)\,e^{if(c)}}{\sqrt{2f''(c)}} \int\limits_{f(b)-f(c)}^{\infty} \frac{1}{\sqrt{z}}\, e^{iz}\, dz\,.$$

By partial integration, we obtain the estimate

$$I_3 \ll \frac{G}{\sqrt{f''(c)\,(f(b)-f(c))}}\,.$$

Because of $f(b) - f(c) = \frac{1}{2}(b-c)^2 f''(c + \vartheta(b-c)) \gg (b-c)^2 \lambda_2$ $(0 < \vartheta < 1)$ we have

$$I_3 \ll \frac{G}{(b-c)\,\lambda_2} \ll \frac{G}{\int_c^b f''(t)\,dt} = \frac{G}{f'(b)}\,.$$

Otherwise we get

$$I_3 = -g(c)\,e^{if(c)} \sqrt{\frac{2}{f''(c)}} \int\limits_0^{\infty} e^{i\left(z-\sqrt{f(b)-f(c)}\right)^2}\, dz \ll \frac{G}{\sqrt{\lambda_2}}\,.$$

Hence

$$I_1 = \varepsilon g(c) \sqrt{\frac{\pi}{2f''(c)}}\, e^{if(c)} + O\left(G \min\left(\frac{1}{f'(b)}, \frac{1}{\sqrt{\lambda_2}}\right)\right). \tag{2.35}$$

In order to estimate I_2, we write

$$I_2 = I_4 + I_5\,,$$

$$I_4 = \int\limits_{f(c)}^{f(b)} \frac{g(t)-g(c)}{f'(t)}\, e^{iz}\, dz = \int\limits_c^b (g(t)-g(c))\, e^{if(t)}\, dt\,,$$

$$I_5 = \int_{f(c)}^{f(b)} F(z)\, e^{iz}\, dz\,,$$

where

$$F(z) = g(c)\left\{\frac{1}{f'(t)} - \frac{1}{\sqrt{2f''(c)\,(z-f(c))}}\right\}.$$

Since

$$\frac{d}{dt}\,\frac{g(t)-g(c)}{t-c} = \frac{g'(t)(t-c)-g(t)+g(c)}{(t-c)^2} = \frac{g'(t)-g'(c+\vartheta(t-c))}{t-c}$$

$$(0 < \vartheta < 1)$$

and $g'(t)$ is monotonic, the right-hand side is either positive or negative throughout. Therefore, the function $\frac{g(t)-g(c)}{t-c}$ is monotonic. We may suppose that it is monotonically increasing. Then, by the second mean-value theorem, we have

$$I_4 \ll G_1 \left| \int_c^\xi (t-c)\, e^{if(t)}\, dt \right| \qquad (c < \xi < b).$$

The function $f''(t)$ is monotonic, since $f'''(t)$ retains its sign throughout the interval. Therefore, the function

$$\frac{d}{dt} \frac{t-c}{f'(t)} = \frac{f'(t)-(t-c)\,f''(t)}{f'^2(t)} = \frac{t-c}{f'^2(t)} \{f''(t+\vartheta(t-c)) - f''(t)\}$$
$$(0 < \vartheta < 1)$$

is either positive or negative throughout. Hence, by a new application of the second mean-value theorem,

$$I_4 \ll \frac{G_1}{\lambda_2}.$$

Note that the function

$$F'(z) = g(c) \left\{ -\frac{f''(t)}{f'^3(t)} + \frac{1}{\sqrt{8f''(c)}\,(z-f(c))^{3/2}} \right\}$$

$$= \frac{g(c)}{\sqrt{8f''(c)}} \frac{f'^3(t) - \sqrt{8f''(c)}\, f''(t)\,(f(t)-f(c))^{3/2}}{f'^3(t)\,(f(t)-f(c))^{3/2}}$$

has only a bounded number of points of zero. The function $F(z)$ is therefore partly monotonic. Hence, by the second mean-value theorem,

$$I_5 \ll \max |F(z)|.$$

By means of the estimates

$$f(t) - f(c) = \frac{1}{2} f''(c)\,(t-c)^2 + O(\lambda_3(t-c)^3) \gg \lambda_2(t-c)^2,$$

$$f'(t) = f''(c)\,(t-c) + O(\lambda_3(t-c)^2) \gg \lambda_2(t-c)$$

we obtain

$$|F(z)| = \frac{|g(c)|}{f'(t)} \left| 1 - \frac{f'(t)}{\sqrt{2f''(c)\,(f(t)-f(c))}} \right| \leqq \frac{|g(c)|}{f'(t)} \left| 1 - \frac{f'^2(t)}{2f''(c)\,(f(t)-f(c))} \right|$$
$$\ll G \frac{\lambda_2\lambda_3(t-c)^3 + \lambda_3^2(t-c)^4}{f''(c)\,\lambda_2^2(t-c)^3} \ll G \left\{ \frac{\lambda_3}{\lambda_2^2} + (b-c)\frac{\lambda_3^2}{\lambda_2^3} \right\}.$$

Hence

$$I_2 \ll G\left\{\frac{\lambda_3}{\lambda_2^2} + (b-c)\frac{\lambda_3^2}{\lambda_2^3}\right\} + \frac{G_1}{\lambda_2}. \tag{2.36}$$

Substituting (2.35) and (2.36) into (2.34), we obtain the asymptotic representation (2.32). Similarly (2.33) can be proved.

Theorem 2.11. *Let $f(t)$ be a real function with continuous derivatives up to the third order in $[a, b]$. Let $|f''(t)| \asymp \lambda_2$, $0 < |f'''(t)| \ll \lambda_3$ throughout the interval. Suppose that for each $c \in [a, b]$ the function*

$$F(t, c; f) = (f'(t) - f'(c))^6 - 8f''(c)f''^2(t)(f(t) - f(c) - f'(c)(t-c))^3 \tag{2.37}$$

only has a bounded number of points of zero. Let $\varphi(t)$ be defined by $f'(\varphi) = t$. Let $g(t)$ be a real function with a continuous and monotonical derivative in $[a, b]$ and $|g(t)| \leqq G$, $|g'(t)| \leqq G_1$. If $\alpha = \min f'(t)$, $\beta = \max f'(t)$,

$$T(z) = \begin{cases} 0 & \text{for } f'(z) \in \mathbf{Z}, \\ \min\left(\dfrac{1}{\|f'(z)\|}, \dfrac{1}{\sqrt{\lambda_2}}\right) & \text{for } f'(z) \notin \mathbf{Z} \end{cases}$$

($\mathbf{Z}$ denotes the set of integers),

$$\varepsilon = \begin{cases} e^{\pi i/4} & \text{for } f''(t) > 0, \\ e^{-\pi i/4} & \text{for } f''(t) < 0 \end{cases}$$

and

$$\Delta = (b-a)^2\frac{\lambda_3^2}{\lambda_2^2} + (b-a)\frac{\lambda_3^2}{\lambda_2^3} + (b-a)\frac{\lambda_3}{\lambda_2} + \frac{\lambda_3}{\lambda_2^2} + \log((b-a)\lambda_2 + 2),$$

then

$$\begin{aligned}\sum_{a<n\leqq b} g(n)\, e^{2\pi i f(n)} = \varepsilon \sum_{\alpha\leqq\nu\leqq\beta}'' \frac{g(\varphi(\nu))}{\sqrt{|f''(\varphi(\nu))|}}\, e^{2\pi i (f(\varphi(\nu)) - \nu\varphi(\nu))} \\ + G\{O(T(a)) + O(T(b)) + O(\Delta)\} + \\ + G_1\left\{O(b-a) + O\left(\frac{1}{\lambda_2}\right)\right\}.\end{aligned} \tag{2.38}$$

Proof. Suppose that $f''(t) > 0$. Then $f'(t)$ is strictly increasing. We use (2.20) with $\eta = 1/2$ and obtain that

$$\begin{aligned}\sum_{a<n\leqq b} g(n)\, e^{2\pi i f(n)} = \sum_{f'(a)\leqq\nu\leqq f'(b)} \int_a^b g(t)\, e^{2\pi i (f(t) - \nu t)}\, dt \\ + G\{O(T(a)) + O(T(b)) + O(\log((b-a)\lambda_2 + 2))\} \\ + O(G_1).\end{aligned} \tag{2.39}$$

We get the new term $O(GT(b))$ if an integer k exists such that $k + 1/2 \leqq f'(b) < k + 1$, because then in the sum on the right-hand side of (2.39) the term

$$A = \int_a^b g(t)\, e^{2\pi i(f(t)-(k+1)t)}\, dt$$

is omitted. This term can be estimated in two ways. The function $g'(t)$ is monotonic. Therefore, the interval of integration is divided into at most three subintervals corresponding to the possibilities $g'(t) < 0$, $g'(t) = 0$, $g'(t) > 0$. Suppose that $g'(t) > 0$ in (a_1, b_1) with $a \leqq a_1 < b_1 \leqq b$. Then, by the second mean-value theorem,

$$A_1 = \int_{a_1}^{b_1} g(t)\, e^{2\pi i(f(t)-(k+1)t)}\, dt \ll G \left| \int_{a_1}^{\xi} e^{2\pi i(f(t)-(k+1)t)}\, dt \right| \qquad (a_1 < \xi < b_1)\,.$$

Because of the monotonicity of the function $k + 1 - f'(t)$ a new application of the second mean-value theorem gives

$$A_1 \ll \frac{G}{k+1-f'(b)} = \frac{G}{\|f'(b)\|}\,.$$

Otherwise Lemma 2.2 shows that

$$A_1 \ll \frac{G}{\sqrt{\lambda_2}}\,.$$

Analogous results hold in cases of $g'(t) < 0$ and $g'(t) = 0$. Hence

$$A \ll G \min\left(\frac{1}{\|f'(b)\|}, \frac{1}{\sqrt{\lambda_2}}\right).$$

This leads to the term $O(GT(b))$.

Similarly, we get the new term $O(GT(a))$ if an integer k exists such that $k < f'(a) < k + 1/2$, because then in the sum on the right-hand side of (2.39) the term $\int_a^b g(t)\, e^{2\pi i(f(t)-kt)}\, dt$ is omitted.

We now apply Lemma 2.5 to the integrals on the right-hand side of (2.39). The conditions of this lemma are satisfied by the functions $2\pi(f(t) - vt)$ instead of $f(t)$ and $g(t)$. The number c in Lemma 2.5 is defined by $f'(c) - v = 0$ here. Hence

$$\sum_{f'(a)<v<f'(b)} \int_a^b g(t)\, e^{2\pi i(f(t)-vt)}\, dt$$

$$= \varepsilon \sum_{f'(a)<v<f'(b)} \left\{ \frac{g(\varphi(v))}{\sqrt{f''(\varphi(v))}}\, e^{2\pi i(f(\varphi(v))-v\varphi(v))} + O\left(G(b-a)\frac{\lambda_3^2}{\lambda_2^3}\right) + O\left(G\frac{\lambda_3}{\lambda_2^2}\right)\right.$$

$$\left. + O\left(\frac{G_1}{\lambda_2}\right) + O\left(G \min\left(\frac{1}{f'(b)-v}, \frac{1}{\sqrt{\lambda_2}}\right)\right) + O\left(G \min\left(\frac{1}{v-f'(a)}, \frac{1}{\sqrt{\lambda_2}}\right)\right)\right\}.$$

If $f'(b)$ is an integer, we obtain

$$\sum_{f'(a)<\nu<f'(b)} \min\left(\frac{1}{f'(b)-\nu}, \frac{1}{\sqrt{\lambda_2}}\right) \leqq \sum_{f'(a)<\nu\leqq f'(b)-1} \frac{1}{f'(b)-\nu}$$
$$\ll \log(f'(b)-f'(a)+2)$$
$$\ll \log((b-a)\lambda_2+2).$$

Otherwise we get

$$\sum_{f'(a)<\nu<f'(b)} \min\left(\frac{1}{f'(b)-\nu}, \frac{1}{\sqrt{\lambda_2}}\right)$$
$$= \sum_{f'(a)<\nu\leqq[f'(b)]-1} \frac{1}{f'(b)-\nu} + \min\left(\frac{1}{f'(b)-[f'(b)]}, \frac{1}{\sqrt{\lambda_2}}\right)$$
$$\ll \log((b-a)\lambda_2+2) + \min\left(\frac{1}{\|f'(b)\|}, \frac{1}{\sqrt{\lambda_2}}\right).$$

This gives the estimate

$$\sum_{f'(a)<\nu<f'(b)} \min\left(\frac{1}{f'(b)-\nu}, \frac{1}{\sqrt{\lambda_2}}\right) \ll \log((b-a)\lambda_2+2) + T(b)$$

and similarly

$$\sum_{f'(a)<\nu<f'(b)} \min\left(\frac{1}{\nu-f'(a)}, \frac{1}{\sqrt{\lambda_2}}\right) \ll \log((b-a)\lambda_2+2) + T(a).$$

Hence

$$\sum_{f'(a)<\nu<f'(b)} \int_a^b g(t)\, e^{2\pi i(f(t)-\nu t)}\, dt$$
$$= \varepsilon \sum_{f'(a)<\nu<f'(b)} \frac{g(\varphi(\nu))}{\sqrt{f''(\varphi(\nu))}}\, e^{2\pi i(f(\varphi(\nu))-\nu\varphi(\nu))}$$
$$+ G\{O(T(a)) + O(T(b)) + O(\Delta)\} + G_1\left\{O(b-a) + O\left(\frac{1}{\lambda_2}\right)\right\}. \tag{2.40}$$

If $f'(b)$ is an integer, in addition to that we must take into consideration the term $\nu = f'(b)$. We obtain by (2.33)

$$\int_a^b g(t)\, e^{2\pi i(f(t)-f'(b)t)}\, dt$$
$$= \frac{\varepsilon}{2} \frac{g(b)}{\sqrt{f''(b)}}\, e^{2\pi i(f(b)-f'(b)b)} + G\left\{O\left((b-a)\frac{\lambda_3^2}{\lambda_2^3}\right) + O\left(\frac{\lambda_3}{\lambda_2^2}\right)\right.$$
$$\left. + O\left(\min\left(\frac{1}{f'(b)-f'(a)}, \frac{1}{\sqrt{\lambda_2}}\right)\right)\right\} + O\left(\frac{G_1}{\lambda_2}\right).$$

Since $\frac{1}{f'(b)-f'(a)} \leqq \frac{1}{\|f'(a)\|}$ this result can be written as

$$\int_a^b g(t)\,\mathrm{e}^{2\pi\mathrm{i}(f(t)-f'(b)t)}\,\mathrm{d}t = \frac{\varepsilon}{2}\frac{g(b)}{\sqrt{f''(b)}}\,\mathrm{e}^{2\pi\mathrm{i}f(b)} + O\left(\frac{G_1}{\lambda_2}\right)$$
$$+ G\{O(T(a)) + O(\Delta)\}\,. \tag{2.41}$$

Similarly, if $f'(a)$ is an integer, we obtain by (2.32)

$$\int_a^b g(t)\,\mathrm{e}^{2\pi\mathrm{i}(f(t)-f'(a)t)}\,\mathrm{d}t = \frac{\varepsilon}{2}\frac{g(a)}{\sqrt{f''(a)}}\,\mathrm{e}^{2\pi\mathrm{i}f(a)} + O\left(\frac{G_1}{\lambda_2}\right)$$
$$+ G\{O(T(b)) + O(\Delta)\}\,. \tag{2.42}$$

If we substitute (2.40)—(2.42) into (2.39), result (2.38) follows.

In the applications there are usually some simplifications which relax the error in (2.38). So we mostly have $|f'''(t)| \asymp \lambda_3$ and $|g'(t)| \asymp G_1$ such that

$$(b-a)\,\lambda_3 \ll \left|\int_a^b f'''(t)\,\mathrm{d}t\right| = |f''(b) - f''(a)| \ll \lambda_2\,,$$

$$(b-a)\,G_1 \ll \left|\int_a^b g'(t)\,\mathrm{d}t\right| = |g(b) - g(a)| \ll G\,.$$

Moreover, if $|f'(b) - f'(a)| \gg 1$, then

$$\frac{\lambda_3}{\lambda_2^2} \ll \frac{1}{(b-a)\,\lambda_2} \ll \frac{1}{|f'(b)-f'(a)|} \ll 1\,.$$

Under these additional conditions the transformation formula takes the simple form

$$\sum_{a<n\leqq b} g(n)\,\mathrm{e}^{2\pi\mathrm{i}f(n)} = \varepsilon \sum_{\alpha\leqq\nu\leqq\beta} \frac{g(\varphi(\nu))}{\sqrt{|f''(\varphi(\nu))|}}\,\mathrm{e}^{2\pi\mathrm{i}(f(\varphi(\nu))-\nu\varphi(\nu))}$$
$$+ G\{O(T(a)) + O(T(b)) + O(\log(|f'(b) - f'(a)| + 1))\}\,. \tag{2.43}$$

2.1.4. *The method of exponent pairs*

Van der Corput's theory of exponent pairs, or exponent systems, as originally developed, consists of a repeated application of Weyl's steps and van der Corput transforms. In his explanation of the method E. C. Titchmarsh uses the symbols A and B to denote the two processes. Later E. Phillips introduced a great simplification into the theory by showing that only one exponent pair need be retained at each stage.

Therefore we use his form of the theory. We take into consideration such functions whose derivatives can be approximated by power functions.

Definition 2.1. *A pair (k, l) of real numbers is called an exponent pair if $0 \leqq k \leqq 1/2 \leqq l \leqq 1$, and if, corresponding to every positive number s, there exist two numbers r and c depending only on s (r an integer greater than 4 and $0 < c < 1/2$) such that the inequality*

$$\sum_{a<n\leqq b} e^{2\pi i f(n)} \ll z^k a^l \tag{2.44}$$

holds with respect to s and u when the following conditions are satisfied:

$$u > 0\,, \qquad 1 \leqq a < b < au\,, \qquad y > 0\,, \qquad z = ya^{-s} > 1\,;$$

$f(t)$ being any real function with differential coefficients of the first r orders in $[a, b]$ and

$$\left| f^{(\nu+1)}(t) - y\,\frac{d^\nu}{dt^\nu}\,t^{-s} \right| < (-1)^\nu\, cy\,\frac{d^\nu}{dt^\nu}\,t^{-s} \tag{2.45}$$

for $a \leqq t \leqq b$ and $\nu = 0, 1, \ldots, r-1$.

Similarly to the proceeding in the proof of Theorem 2.7 we can construct an exponent pair from another one by applying a Weyl's step. This leads to the so-called A-process.

Theorem 2.12 (A-process). *If $(\varkappa, \lambda)$ is an exponent pair, then so is*

$$(k, l) = A(\varkappa, \lambda) = \left(\frac{\varkappa}{2(\varkappa+1)}\,, \frac{1}{2} + \frac{\lambda}{2(\varkappa+1)}\right).$$

Proof. Suppose (2.45) to be true for a suitable value of r. We apply Theorem 2.5, and we get from (2.12)

$$\sum_{a<n\leqq b} e^{2\pi i f(n)} \ll \frac{a}{\sqrt{H}} + \left\{\frac{a}{H}\sum_{h=1}^{H-1}\left|\sum_{a<n\leqq b-h} e^{2\pi i F(n)}\right|\right\}^{1/2} \tag{2.46}$$

with

$$F(t) = f(t) - f(t+h)\,.$$

Using (2.45), we have

$$F^{(\nu+1)}(t) = -\int_t^{t+h} f^{(\nu+2)}(\tau)\,d\tau = -\int_t^{t+h} (1+\varepsilon_1 c)\,\frac{d^{\nu+1}}{d\tau^{\nu+1}}\,\tau^{-s}\,d\tau\,,$$

where $|\varepsilon_1| < 1$. We suppose that $H < ac$. Then $t \leqq \tau < t + H < t + ac \leqq (1+c)\,t$. If $0 < \gamma < 1/2$, we now choose c $(0 < c < 1/2)$ such that

$$F^{(\nu+1)}(t) = -y\int_t^{t+h} (1+\varepsilon_2\gamma)\,\frac{d^{\nu+1}}{dt^{\nu+1}}\,t^{-s}\,d\tau = -y(1+\varepsilon_3\gamma)\,h\,\frac{d^{\nu+1}}{dt^{\nu+1}}\,t^{-s}\,,$$

where $|\varepsilon_2|, |\varepsilon_3| < 1$. Hence

$$\left| F^{(\nu+1)}(t) - ysh \frac{\mathrm{d}^\nu}{\mathrm{d}t^\nu} t^{-s-1} \right| < (-1)^\nu \gamma ysh \frac{\mathrm{d}^\nu}{\mathrm{d}t^\nu} t^{-s-1} .$$

Therefore, $F(t)$ satisfies the condition (2.45) and z is to replace by zh/a in (2.44).

We now split up the sum on the right-hand side of (2.46) into two parts according to $h \leqq a/z$ and $h > a/z$.

In the first case we have $|F'(t)| \ll 1$ and $|F''(t)| \gg zh/a^2$. Then Theorem 2.1 shows that

$$\sum_{1 \leqq h \leqq a/z} \left| \sum_{a<n \leqq b-h} \mathrm{e}^{2\pi \mathrm{i} F(n)} \right| \ll \sum_{1 \leqq h \leqq a/z} \frac{a}{\sqrt{zh}} \ll \frac{a^{3/2}}{z} . \tag{2.47}$$

In the second case there is, because of $zh/a > 1$, an exponent pair $(\varkappa, \lambda)$ such that

$$\sum_{a/z<h<H} \left| \sum_{a<n \leqq b-h} \mathrm{e}^{2\pi \mathrm{i} F(n)} \right| \ll \sum_{a/z<h<H} \left(\frac{zh}{a}\right)^{\varkappa} a^{\lambda} \ll z^{\varkappa} a^{\lambda-\varkappa} H^{\varkappa+1} . \tag{2.48}$$

Substituting (2.47) and (2.48) in (2.46), we get

$$\sum_{a<n \leqq b} \mathrm{e}^{2\pi \mathrm{i} f(n)} \ll \frac{a}{\sqrt{H}} \left\{ 1 + \left(\frac{a}{z^2}\right)^{1/4} \right\} + (z^{\varkappa} a^{1+\lambda-\varkappa} H^{\varkappa})^{1/2} .$$

Suppose that $z^2 \geqq a$. Then both the terms are of the same order if

$$H = \left[(a^{\varkappa+1-\lambda} z^{-\varkappa})^{\frac{1}{\varkappa+1}} \right],$$

provided that $H < ac$. Then

$$\sum_{a<n \leqq b} \mathrm{e}^{2\pi \mathrm{i} f(n)} \ll z^{\frac{\varkappa}{2(\varkappa+1)}} a^{\frac{1}{2}+\frac{\lambda}{2(\varkappa+1)}} = z^k a^l .$$

If $H \geqq ac$, we have by trivial estimating that

$$\sum_{a<n \leqq b} \mathrm{e}^{2\pi \mathrm{i} f(n)} \ll b - a \ll H \ll a^{1-\frac{\lambda}{\varkappa+1}} \ll a^{\frac{1}{2}+\frac{\lambda}{2(\varkappa+1)}} \ll z^k a^l .$$

In case $z^2 < a$ we use Theorem 2.1. We have from (2.45) $|f'(t)| \ll z$ and $|f''(t)| \gg z/a$. Hence, by (2.6),

$$\sum_{a<n \leqq b} \mathrm{e}^{2\pi \mathrm{i} f(n)} \ll \sqrt{za} \ll z^{\frac{1}{2}-\frac{\lambda}{\varkappa+1}} a^{\frac{1}{2}+\frac{\lambda}{2(\varkappa+1)}} \ll z^{\frac{\varkappa}{2(\varkappa+1)}} a^{\frac{1}{2}+\frac{\lambda}{2(\varkappa+1)}} = z^k a^l .$$

This completes the proof of the theorem.

In a similar manner we can construct an exponent pair from another one by applying the van der Corput transform. This leads to the so-called B-process.

Theorem 2.13 (*B*-process). *If* $(\varkappa, \lambda)$ *is an exponent pair with* $\varkappa + 2\lambda \geqq 3/2$, *then so is*

$$(k, l) = B(\varkappa, \lambda) = \left(\lambda - \frac{1}{2}, \varkappa + \frac{1}{2}\right).$$

Proof. The proof is closely related to the corresponding one of Theorem 2.10. We assume $f''(t) > 0$ such that $f'(t)$ is strictly increasing. Suppose (2.45) to be true for a suitable value of r. We can use Theorem 2.9 with $\lambda_2 = z/a$, $\lambda_3 = z/a^2$, $\lambda_4 = z/a^3$. Because of $\lambda_3^2 = \lambda_2\lambda_4$ we have from (2.27) and (2.29) that

$$\sum_{a<n\leqq b} e^{2\pi i f(n)} = \varepsilon \sum_{f'(a)<v\leqq f'(b)} \frac{1}{\sqrt{f''(\varphi(v))}} e^{2\pi i F(v)} + O\left(\sqrt{\frac{a}{z}}\right) + O((za)^{1/3}).$$

$F(t)$ is defined by

$$F(t) = f(\varphi(t)) - t\varphi(t), \qquad f'(\varphi(t)) = t.$$

In what follows ε_i are bounded parameters with $\varepsilon_i \to 0$ for $c \to 0$. We have

$$f^{(\nu+1)}(t) = (1 + \varepsilon_1 c)\, y \frac{d^\nu}{dt^\nu} t^{-s} \qquad (|\varepsilon_1| < 1).$$

For $\nu = 0,1$ we obtain

$$t = f'(\varphi(t)) = (1 + \varepsilon_1 c)\, y(\varphi(t))^{-s},$$

$$F'(t) = -\varphi(t) = (1 + \varepsilon_2)\, y^{1/s} t^{-1/s},$$

$$F''(t) = -\varphi'(t) = \frac{-1}{f''(\varphi(t))} = -(1 + \varepsilon_3)\, y^{1/s} \frac{d}{dt} t^{-s}.$$

Now let $\nu \geqq 2$. Then

$$F^{(\nu+1)}(t) = \frac{-1}{(f''(\varphi))^{2\nu-1}} \Sigma\, \omega(\boldsymbol{n}) f^{(n_1+1)}(\varphi) \ldots f^{(n_{\nu-1}+1)}(\varphi)$$

with some constants $\omega(\boldsymbol{n}) = \omega(n_1, \ldots, n_{\nu-1})$ and

$$SC(\Sigma):\ 1 \leqq n_1 \leqq \ldots \leqq n_{\nu-1} \leqq \nu, \qquad n_1 + \ldots + n_{\nu-1} = 2\nu - 2.$$

We now introduce the functions

$$g(t) = \begin{cases} \dfrac{y}{1-s} t^{1-s} & \text{for } s \neq 1 \\ y \log t & \text{for } s = 1, \end{cases}$$

$$G(t) = -g(\chi(t)) + t\chi(t), \qquad g'(\chi(t)) = t.$$

Then

$$\chi(t) = y^{1/s} t^{-1/s},$$

$$G^{(\nu+1)}(t) = y^{1/s} \frac{\mathrm{d}^\nu}{\mathrm{d}t^\nu} t^{-1/s}$$

$$= \frac{1}{(g''(\chi))^{2\nu-1}} \Sigma\, \omega(\boldsymbol{n})\, g^{(n_1+1)}(\chi) \ldots g^{(n_{\nu-1}+1)}(\chi) .$$

Because of $\varphi(t) = (1 + \varepsilon_4)\, \chi(t)$ we have

$$f^{(\nu+1)}(\varphi) = (1 + \varepsilon_5)\, g^{(\nu+1)}(\chi)$$

and therefore

$$-F^{(\nu+1)}(t) = \frac{1 + \varepsilon_6}{(g''(\chi))^{2\nu-1}} \Sigma\, \omega(\boldsymbol{n})\, g^{(n_1+1)}(\chi) \ldots g^{(n_{\nu-1}+1)}(\chi) ,$$

$$-F^{(\nu+1)}(t) - G^{(\nu+1)}(t) = \varepsilon_6 G^{(\nu+1)}(t) .$$

If $0 < \gamma < 1/2$ we can choose c $(0 < c < 1/2)$ such that $|\varepsilon_6| < \gamma$. Hence

$$\left| -F^{(\nu+1)}(t) - y^{1/s} \frac{\mathrm{d}^\nu}{\mathrm{d}t^\nu} t^{-1/s} \right| < (-1)^\nu \gamma y^{1/s} \frac{\mathrm{d}^\nu}{\mathrm{d}t^\nu} t^{-1/s} .$$

Therefore, $-F(t)$ satisfies the condition (2.45) and z is to replace by a and a by z in (2.44). Thus there is an exponent pair $(\varkappa, \lambda)$ such that

$$\sum_{f'(a) < \nu \leqq t} \mathrm{e}^{2\pi i F(\nu)} \ll a^\varkappa z^\lambda ,$$

where $t \leqq f'(b)$. Thus we find by partial summation that

$$\sum_{a<n\leqq b} \mathrm{e}^{2\pi i f(n)} \ll z^{\lambda-1/2} a^{\varkappa+1/2} + \sqrt{\frac{a}{z}} + (za)^{1/3} \ll z^k a^l + (za)^{1/3} \ll z^k a^l ,$$

which proves the theorem, provided that $k \geqq 1/3$.

If $k < 1/3$ and $a \geqq z^{\frac{1-3k}{3l-1}}$, then

$$(za)^{\frac{1}{3}} \leqq z^k a^{\frac{1}{3}} \left(a^{\frac{3l-1}{1-3k}} \right)^{\frac{1-3k}{3}} = z^k a^l .$$

If $k < 1/3$ and $a < z^{\frac{1-3k}{3l-1}}$, then

$$\sum_{a<n\leqq b} \mathrm{e}^{2\pi i f(n)} \ll a \ll a^l \left(z^{\frac{1-3k}{3l-1}} \right)^{1-l} \ll z^k a^l .$$

This completes the proof of the theorem.

The inequality (2.44) can be used immediately to estimate a sum $\Sigma\, \psi(f(n))$ by means of Theorem 1.8. The formula (1.18)

$$\sum_{a<n\leqq b} \psi(f(n)) \ll \frac{a}{w} + \sum_{m=1}^{\infty} \min\left(\frac{w}{m^2}, \frac{1}{m}\right) \left| \sum_{a<n\leqq b} e^{2\pi i m f(n)} \right|$$

shows that the function $f(t)$ must be replaced by $mf(t)$. Hence we must replace z by mz. We obtain

$$\sum_{a<n\leqq b} \psi(f(n)) \ll \frac{a}{w} + (wz)^k a^l \ll (a^{k+l} z^k)^{\frac{1}{k+1}}$$

if we put $w = (z^{-k} a^{1-l})^{\frac{1}{k+1}}$.

We get a somewhat simpler expression by assuming that the exponent pair comes from an A-process. Then the following theorem is clear.

Theorem 2.14. *Let* $u > 0$, $1 \leqq a < b \leqq au$. *Let* $f(t)$ *satisfy the conditions of the method of exponent pairs. If* $(k, l) = A(\varkappa, \lambda)$, *then*

$$\sum_{a<n\leqq b} \psi(f(n)) \ll a^{2(k+l-1/2)} z^{2k}.$$

Special exponent pairs

Obviously, (0, 1) is an exponent pair since

$$\sum_{a<n\leqq b} e^{2\pi i f(n)} \ll b - a \ll a = z^0 a^1.$$

From (0, 1) we find the next exponent pair

$$B(0, 1) = \left(\frac{1}{2}, \frac{1}{2}\right).$$

The B-process when applied twice brings us back to the first pair. Therefore, we must apply to (1/2, 1/2) at least one A-process. Then a B-process can only follow an A-process. However, if

$$(k, l) = A(\varkappa, \lambda) = \left(\frac{\varkappa}{2(\varkappa+1)}, \frac{1}{2} + \frac{\varkappa}{2(\varkappa+1)}\right),$$

we obtain

$$k + 2l = 1 + \frac{\varkappa + 2\lambda}{2(\varkappa+1)} \geqq \frac{3}{2}$$

because of $\lambda \geqq 1/2$. This shows that the restriction $k + 2l \geqq 3/2$ by applying the B-process to (k, l) is unimportant. Now we notice some special exponent pairs which will be useful later.

$$A\left(\frac{1}{2}, \frac{1}{2}\right) = \left(\frac{1}{6}, \frac{4}{6}\right),$$

$$A\left(\frac{1}{6}, \frac{4}{6}\right) = \left(\frac{1}{14}, \frac{11}{14}\right), \qquad BA\left(\frac{1}{6}, \frac{4}{6}\right) = \left(\frac{2}{7}, \frac{4}{7}\right),$$

$$A^2\left(\frac{1}{6}, \frac{4}{6}\right) = \left(\frac{1}{30}, \frac{26}{30}\right), \qquad BA^2\left(\frac{1}{6}, \frac{4}{6}\right) = \left(\frac{11}{30}, \frac{16}{30}\right),$$

$$A^3\left(\frac{1}{6}, \frac{4}{6}\right) = \left(\frac{1}{62}, \frac{57}{62}\right), \qquad BA^3\left(\frac{1}{6}, \frac{4}{6}\right) = \left(\frac{13}{31}, \frac{16}{31}\right),$$

$$ABA\left(\frac{1}{6}, \frac{4}{6}\right) = \left(\frac{2}{18}, \frac{13}{18}\right), \qquad BA^2BA\left(\frac{1}{6}, \frac{4}{6}\right) = \left(\frac{13}{40}, \frac{22}{40}\right).$$

In several problems to which the method of exponent pairs can be applied the best estimates of error terms arise when an exponent pair (k, l) is chosen for which $k + l$ is as small as possible. The exponents of the error terms in many lattice point problems are given by $k + l - 1/2$ which is plausible from Theorem 2.14. Therefore we give some suitable values here. In his investigations on exponent pairs R. A. Rankin [1] showed that it is useful to separate the operations into groups of the form ABA^ν ($\nu = 0, 1, 2$). So one needs the formulas

$$AB(k, l) = \left(\frac{2l - 1}{2(2l + 1)}, \frac{1}{2} + \frac{2k + 1}{2(2l + 1)}\right),$$

$$ABA(k, l) = \left(\frac{l}{2(2k + l + 2)}, \frac{1}{2} + \frac{2k + 1}{2(2k + l + 2)}\right),$$

$$ABA^2(k, l) = \left(\frac{k + l + 1}{2(7k + l + 5)}, \frac{1}{2} + \frac{4k + 2}{2(7k + l + 5)}\right).$$

We notice some exponent pairs.

1. $AB(0, 1) = \left(\frac{1}{6}, \frac{4}{6}\right), \qquad k + l - \frac{1}{2} = \frac{1}{3}.$

2. Nieland's [1] and Titchmarsh's [1, 11] exponent pair

$$ABA^2\left(\frac{1}{6}, \frac{4}{6}\right) = \left(\frac{11}{82}, \frac{57}{82}\right), \qquad k + l - \frac{1}{2} = \frac{27}{28} = 0{,}329268\ldots$$

3. $(ABA^2)(ABA)\left(\frac{1}{6}, \frac{4}{6}\right) = ABA^2\left(\frac{2}{18}, \frac{13}{18}\right) = \left(\frac{33}{234}, \frac{161}{234}\right),$

$$k + l - \frac{1}{2} = \frac{77}{234} = 0{,}329059\ldots$$

4. $(ABA^2)^2\left(\frac{1}{6}, \frac{4}{6}\right) = (ABA^2)\left(\frac{11}{82}, \frac{57}{82}\right) = \left(\frac{75}{544}, \frac{376}{544}\right)$,

$$k + l - \frac{1}{2} = \frac{179}{544} = 0{,}329\,044\ldots$$

5. Phillips's [1] exponent pair

$$(ABA)^2 (ABA)^2 \left(\frac{1}{6}, \frac{4}{6}\right) = (ABA^2)(ABA)\left(\frac{2}{18}, \frac{13}{18}\right)$$

$$= (ABA^2)\left(\frac{13}{106}, \frac{75}{106}\right) = \left(\frac{97}{696}, \frac{480}{696}\right),$$

$$k + l - \frac{1}{2} = \frac{229}{696} = 0{,}329\,022\ldots$$

6. Rankin's [1] exponent pair

$$(ABA^2)(ABA)(AB)^2(ABA)(AB)^2(ABA)(ABA^2)\left(\frac{1}{6}, \frac{4}{6}\right)$$

$$= (ABA^2)(ABA)(AB)^2(ABA)(AB)^2(ABA)\left(\frac{11}{82}, \frac{57}{82}\right)$$

$$= (ABA^2)(ABA)(AB)^2(ABA)(AB)^2\left(\frac{57}{486}, \frac{347}{486}\right)$$

$$= (ABA^2)(ABA)(AB)^2(ABA)(AB)\left(\frac{52}{590}, \frac{445}{590}\right)$$

$$= (ABA^2)(ABA)(AB)^2(ABA)\left(\frac{150}{1480}, \frac{1087}{1480}\right)$$

$$= (ABA^2)(ABA)(AB)^2\left(\frac{1087}{8694}, \frac{6127}{8694}\right) = (ABA^2)(ABA)(AB)\left(\frac{445}{5\,237}, \frac{3977}{5\,237}\right)$$

$$= (ABA^2)(ABA)\left(\frac{2717}{26\,382}, \frac{19\,318}{26\,382}\right) = (ABA^2)\left(\frac{19\,318}{155\,032}, \frac{109\,332}{155\,032}\right)$$

$$= \left(\frac{141\,841}{1\,019\,718}, \frac{703\,527}{1\,019\,718}\right)$$

$$k + l - \frac{1}{2} = \frac{335\,509}{1\,019\,718} = 0{,}329\,021\ldots$$

Sometimes we make use of exponent pairs of the form $A^n(k, l)$. It is easily seen that

$$A^n(k, l) = \left(\frac{k}{2k(2^n - 1) + 2^n}, 1 - \frac{1 - l + kn}{2k(2^n - 1) + 2^n}\right)$$

for $n = 1, 2, \ldots$ For $n = 1$ Theorem 2.12 applies. Suppose the formula to be true for $n - 1$. Then

$$A^n(k, l) = AA^{n-1}(k, l) = A\left(\frac{k}{2k(2^{n-1} - 1) + 2^{n-1}}, 1 - \frac{1 - l + k(n-1)}{2k(2^{n-1} - 1) + 2^{n-1}}\right)$$

$$= \left(\frac{k}{2k(2^n - 1) + 2^n}, 1 - \frac{1 - l + kn}{2k(2^n - 1) + 2^n}\right).$$

We notice the special case

$$A^n\left(\frac{1}{2}, \frac{1}{2}\right) = \left(\frac{1}{2^{n+2} - 2}, 1 - \frac{n+1}{2^{n+2} - 2}\right).$$

We can give a clear idea of the exponent pairs if we regard each pair (k, l) as a point in a two-dimensional Euclidean space. We shall not only investigate the set of all points obtainable by applying the processes A and B to $(0, 1)$, but we also make use of a further convexity process C.

Theorem 2.15 (C-process). *If (k_1, l_1) and (k_2, l_2) are exponent pairs, then*

$$(k, l) = (tk_1 + (1 - t)\, k_2, tl_1 + (1 - t)\, l_2)$$

is an exponent pair as well, where t is any number satisfying $0 \leqq t \leqq 1$.

Proof.

$$\left|\sum_{a<n\leqq b} e^{2\pi i f(n)}\right| = \left|\sum_{a<n\leqq b} e^{2\pi i f(n)}\right|^t \left|\sum_{a<n\leqq b} e^{2\pi i f(n)}\right|^{1-t}$$

$$\ll (z^{k_1} a^{l_1})^t \, (z^{k_2} a^{l_2})^{1-t} = z^k a^l .$$

We know that the points (k, l) as exponent pairs satisfy the condition $0 \leqq k \leqq 1/2 \leqq l \leqq 1$. From the exponent pairs $P_1 = (0, 1)$ and $P_2 = (1/2, 1/2)$ and from the C-process we see that all points on the straight-line segment $l = 1 - k$, $0 \leqq k \leqq 1/2$ are exponent pairs. We consider the image of this straight-line segment under the mapping ABA^2. For $0 \leqq t \leqq 1/2$ we find

$$ABA^2(t, 1 - t) = \left(\frac{1}{6(t + 1)}, \frac{1}{2} + \frac{2t + 1}{6(t + 1)}\right).$$

Putting $k = 1/6(t + 1)$, we have

$$ABA^2(t, 1 - t) = \left(k, \frac{5}{6} - k\right), \qquad \frac{1}{9} \leqq k \leqq \frac{1}{6},$$

and by an additional B-process

$$BABA^2(t, 1 - t) = \left(\frac{1}{3} - k, \frac{1}{2} + k\right), \qquad \frac{1}{9} \leqq k \leqq \frac{1}{6}.$$

This shows that all points on the straight-line segment

$$l = \frac{5}{6} - k\,, \qquad \frac{1}{9} \leqq k \leqq \frac{2}{9}$$

are exponent pairs. Moreover, by applying the C-process we see that all points within and on the polygon formed by the points $P_1 = (0, 1)$, $P_2 = (1/2, 1/2)$, $P_3 = (2/18, 13/18)$, $P_4 = (4/18, 11/18)$ are exponent pairs. In a similar manner we can expand the domain of exponent pairs. Let S denote the set of all exponent pairs which can be obtained by applying a finite number of the processes A, B and C and which do not lie on the straight-line segment $l = 1 - k$, $0 \leqq k \leqq 1/2$. Then S is a convex, open set of points. Because of the B-process S is symmetrical to the straight line $l = k + 1/2$. The boundary of S is formed by the straight-line segment $l = 1 - k$, $0 \leqq k \leqq 1/2$ and a curve which is partly composed of segments of straight lines and joins P_1 to P_2 (see Fig. 4).

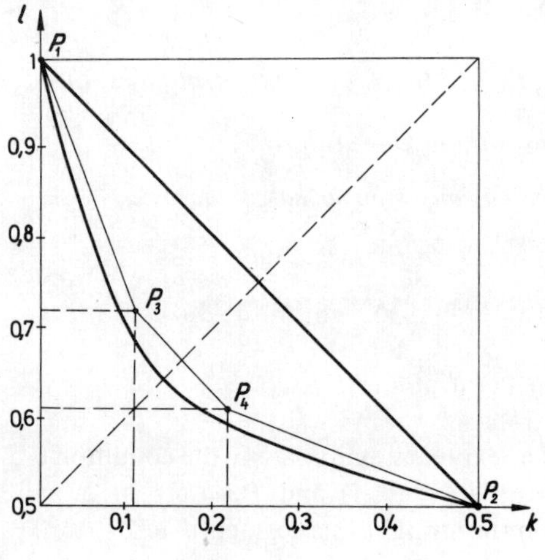

Fig. 4

2.2. DOUBLE EXPONENTIAL SUMS

E. C. Titchmarsh was the first who developed a method of estimation of double exponential sums

$$\sum_{(n_1, n_2) \in D} e^{2\pi i f(n_1, n_2)}$$

by generalizing the method of J. G. van der Corput. As usual he approximated such a sum by a sum of integrals in his investigations. Then following the same general lines of van der Corput's theory he applied a method for estimating the integrals of the type

$$\iint_{(t_1, t_2) \in D} e^{2\pi i f(t_1, t_2)}\, dt_1\, dt_2\,.$$

Though considering only the case in which the domain D is a rectangle or a parallelogram E. C. Titchmarsh [2], [3] did the decisive step for handling double exponential sums. Later some improvements, refinements and extensions to more general domains were given by S.-H. Min [1], H.-E. Richert [3], W. Haneke [1], W.-G. Nowak [6] and E. Krätzel [9], [17], [18]. In Section 2.2.1 we shall consider an approach to these basic estimates.

Another way of dealing with double exponential sums is to twice applying the van der Corput transform of one-dimensional exponential sums to double sums repeatedly. This was first done by I. M. Vinogradov [1], [2] by his treatment of the sphere problem, and is based on the very rigorous form of the van der Corput transform given by Theorem 2.11. G. Kolesnik [4] extended the special investigations of I. M. Vinogradov to general domains and E. Krätzel [17] added some refinements. In Section 2.2.2 we shall give an explanation of this method.

As seen in the preceding section, on the one hand the one-dimensional van der Corput's method depends on the transform of the exponential sums and on the other hand on the repeated use of a lemma due to H. Weyl. As will be seen later we can extend this possibility to double sums as well.

Throughout Section 2.2 the following conditions are always assumed to be true.

(A) Let D be a bounded plane domain with an area $|D|$, where the number of lattice points are of order $|D|$.

(B) Suppose that D is a subset of the rectangle

$$D' = \{(t_1, t_2): a_1 \leqq t_1 \leqq b_1, a_2 \leqq t_2 \leqq b_2\}$$

with $c_1 = b_1 - a_1 \geqq 1$, $c_2 = b_2 - a_2 \geqq 1$, $|D'| = c_1 c_2$.

(C) Any straight line parallel to any of the coordinate axes intersects D in a bounded number of line segments. For the sake of simplicity we only consider such domains D where these straight lines intersect the boundary of D in at most two points or in one line segment. We can do this without loss of generality, because each such general domain can be divided into a finite number of these special domains.

(D) Let $f(t_1, t_2)$ be a real function in D' with continuous partial derivatives of as many orders as may be required. Suppose that the functions $f_{t_1}(t_1, t_2)$, $f_{t_2}(t_1, t_2)$ are monotonic in t_1 and t_2, respectively.

(E) Intersections of D with domains of the type $f_{t_j}(t_1, t_2) \leqq c$ or $f_{t_j}(t_1, t_2) \geqq c$ $(j = 1, 2)$ are to satisfy condition (C) as well.

(F) The boundary of D can be divided into a bounded number of parts. In each part the curve of boundary is given by $t_2 =$ const or a function $t_1 = \varrho(t_2)$, which is continuous in the closed intervals described above.

Always denote the Hessian of the function $f(t_1, t_2)$ by

$$H(f) = \frac{\partial(f_{t_1}, f_{t_2})}{\partial(t_1, t_2)} = f_{t_1 t_1} f_{t_2 t_2} - f_{t_1 t_2}^2 .$$

2.2.1. *The method of Titchmarsh*

In this section we shall prove some results of the estimation of double exponential sums by means of Titchmarsh's method. We obtain the simplest results if we can assume that the Hessian of $f(t_1, t_2)$ is essentially of the same order as the single second derivatives $f_{t_1t_1}f_{t_2t_2}$ and $f^2_{t_1t_2}$. However, in many applications this property is not satisfied. Therefore, it is necessary to have a similar result when the Hessian may be small, compared with any of the mentioned single second derivatives. For this purpose we prove two basic estimates of double integrals in Section 2.2.1.1. In Section 2.2.1.2 we apply the results to double exponential sums and three-dimensional lattice point problems.

2.2.1.1. Basic Lemmas

Lemma 2.6. *Suppose that*

$$\lambda_1 \leqq |f_{t_1t_1}(t_1, t_2)| \ll \lambda_1\,, \qquad \lambda_2 \leqq |f_{t_2t_2}(t_1, t_2)| \ll \lambda_2$$
$$|f_{t_1t_2}(t_1, t_2)| \ll \sqrt{\lambda_1\lambda_2}\,, \qquad |H(f)| \gg \lambda_1\lambda_2$$

throughout the rectangle D'. For all parts of the curve of boundary let $t_2 = \text{const}$ or $t_1 = \varrho(t_2)$, where $\varrho(t)$ is partly twice differentiable and $|\varrho''(t)| \ll r$. Then

$$\iint_D e^{if(t_1,t_2)}\,dt_1\,dt_2 \ll \frac{1 + \log|D'| + |\log\lambda_1| + |\log\lambda_2|}{\sqrt{\lambda_1\lambda_2}} + \frac{c_2 r}{\lambda_2}\,. \tag{2.49}$$

Proof. The consitions, of course, imply that $f_{t_1t_1}, f_{t_2t_2}, H(f)$ are of constant sign in the rectangle. The equation $f_{t_1} = \text{const}$ defines t_1 as a one-valued function of t_2, and $f_{t_2} = \text{const}$ defines t_2 as a one-valued function of t_1. If $f_{t_1} = \text{const}$, we have

$$\frac{d}{dt_2} f_{t_2} = f_{t_2t_2} + f_{t_1t_2}\frac{dt_1}{dt_2} = \frac{H(f)}{f_{t_1t_1}}$$

so that f_{t_2} is monotonic on $f_{t_1} = \text{const}$. The curves $f_{t_1} = n_1\sqrt{\lambda_1}$, $f_{t_2} = n_2\sqrt{\lambda_2}$ $(n_1, n_2 \in \mathbf{Z})$ therefore form a network, dividing the domain D into regions

$$D_{n_1n_2} = \{(t_1, t_2)\colon (t_1, t_2) \in D,\ n_j\sqrt{\lambda_j} \leqq f_{t_j} < (n_j + 1)\sqrt{\lambda_j}\ (j = 1, 2)\}\,.$$

Assume that on the side of $f_{t_1} = n_1\sqrt{\lambda_1}$ the variable t_2 changes from t_2' to t_2'', then, by substituting $y = f_{t_2}$,

$$\sqrt{\lambda_2} = \int_{n_2\sqrt{\lambda_2}}^{(n_2+1)\sqrt{\lambda_2}} dy = \int_{t_2'}^{t_2''} \frac{H(f)}{f_{t_1t_1}}\,dt_2 \gg \lambda_2(t_2'' - t_2')$$

and

$$t_2'' - t_2' \ll \frac{1}{\sqrt{\lambda_2}}.$$

Since $\dfrac{dt_1}{dt_2} = -\dfrac{f_{t_1t_2}}{f_{t_1t_1}} \ll \sqrt{\dfrac{\lambda_2}{\lambda_1}}$ we have for the t_1-variation

$$t_1'' - t_1' \ll \frac{1}{\sqrt{\lambda_1}}.$$

Analogous results can be obtained for the other sides. Hence, in D_{n_1, n_2} the total variation of t_1 is of order $1/\sqrt{\lambda_1}$ and that of t_2 of order $1/\sqrt{\lambda_2}$. We now estimate the integrals over the single regions.

By the above remarks we can see that the area of $D_{0,0}$, by supposing it to belong to D, is of order $1/\sqrt{\lambda_1\lambda_2}$. Hence

$$\iint\limits_{D_{0,0}} e^{if(t_1t_2)}\, dt_1\, dt_2 \ll \frac{1}{\sqrt{\lambda_1\lambda_2}}.$$

Applying the second mean-value theorem with respect to t_2, we find for the regions D_{0,n_2}

$$\sum_{1 \leqq |n_2| \ll c_2\sqrt{\lambda_2}} \iint\limits_{D_{0,n_2}} e^{if(t_1,t_2)}\, dt_1\, dt_2 \ll \sum_{1 \leqq |n_2| \ll c_2\sqrt{\lambda_2}} \frac{1}{\sqrt{\lambda_1}} \frac{1}{|n_2|\sqrt{\lambda_2}}$$
$$\ll \frac{\log c_2 + |\log \lambda_2|}{\sqrt{\lambda_1\lambda_2}}$$

because of $n_2\sqrt{\lambda_2} \leqq f_{t_2} \ll c_2\lambda_2$.

Consider the region $f_{t_1} \geqq \sqrt{\lambda_1}$. Let its left-hand and right-hand boundaries be denoted by $t_1 = \alpha(t_2)$ and $t_1 = \beta(t_2)$, respectively. Then the function $\alpha(t_2)$ and $\beta(t_2)$ are either solutions of $f_{t_1} = \sqrt{\lambda_1}$ or curves of boundary. Now, integrating by parts,

$$\iint\limits_{\substack{D \\ f_{t_1} \geqq \sqrt{\lambda_1}}} e^{if(t_1,t_2)}\, dt_1\, dt_2 = I_1(\beta) - I_1(\alpha) + I_2,$$

where

$$I_1(\alpha) = -i\int \frac{e^{if(\alpha(t),\, t)}}{f_{t_1}(\alpha(t), t)}\, dt, \qquad I_2 = -i\iint f_{t_1t_1} f_{t_1}^{-2}\, e^{if(t_1,t_2)}\, dt_1\, dt_2,$$

and $I_1(\beta)$ is defined analogously to $I_1(\alpha)$.

We first estimate the integral I_2. We have

$$I_2 = -\mathrm{i} \sum\nolimits_1 \iint\limits_{D_{n_1,0}} f_{t_1 t_1} f_{t_1}^{-2} \, \mathrm{e}^{\mathrm{i} f(t_1, t_2)} \, dt_1 \, dt_2$$

$$-\mathrm{i} \sum\nolimits_2 \iint\limits_{D_{n_1,n_2}} f_{t_1 t_1} f_{t_1}^{-2} \, \mathrm{e}^{\mathrm{i} f(t_1, t_2)} \, dt_1 \, dt_2 \,,$$

$$SC(\Sigma_1)\colon 1 \leqq n_1 \ll c_1 \sqrt{\lambda_1} \,,$$
$$SC(\Sigma_2)\colon 1 \leqq n_1 \ll c_1 \sqrt{\lambda_1} \,, \qquad 1 \leqq |n_2| \ll c_2 \sqrt{\lambda_2} \,.$$

We trivially estimate the integrals over the regions $D_{n_1,0}$. Applying the second mean-value theorem with respect to the variable t_2 in the integrals over the regions D_{n_1,n_2}, we obtain

$$I_2 \ll \sum\nolimits_1 \frac{1}{n_1^2} \frac{1}{\sqrt{\lambda_1 \lambda_2}} + \sum\nolimits_2 \frac{1}{n_1^2 \, |n_2|} \frac{1}{\sqrt{\lambda_1 \lambda_2}} \ll \frac{1 + \log c_2 + |\log \lambda_2|}{\sqrt{\lambda_1 \lambda_2}} \,.$$

We now consider the integral $I_1(\alpha)$ under the assumption that $t_1 = \alpha(t_2)$ is a solution of $f_{t_1} = \sqrt{\lambda_1}$. Then

$$I_1(\alpha) = \frac{-\mathrm{i}}{\sqrt{\lambda_1}} \int \mathrm{e}^{\mathrm{i} f(\alpha(t), t)} \, dt \,.$$

We have with a certain constant $A > 0$

$$|f'(\alpha(t), t)| = |f_{t_2} - f_{t_1} \alpha'(t)| \geqq \left| f_{t_2} - f_{t_1} \frac{f_{t_1 t_2}}{f_{t_1 t_1}} \right| = |f_{t_2}| - A \sqrt{\lambda_2} \,.$$

If $|f_{t_2}| \geqq 2A \sqrt{\lambda_2}$, we obtain

$$|f'(\alpha(t), t)| \gg \sqrt{\lambda_2} \,.$$

Hence, by the second mean-value theorem,

$$I_1(\alpha) \ll \frac{1}{\sqrt{\lambda_1 \lambda_2}} \,.$$

This result also holds in the opposite case, since then the length of the interval of integration is of order $1/\sqrt{\lambda_2}$.

At last we consider the integral $I_1(\alpha)$ under the assumption that $t_1 = \alpha(t_2)$ is a part of the curve of boundary. In this case we use the notation $\alpha(t) = \varrho(t)$. We divide $I_1(\alpha)$ into three parts:

$$I_1(\alpha) = I_1(\varrho) = I_{1,1}(\varrho) + I_{1,2}(\varrho) + I_{1,3}(\varrho) \,.$$

In $I_{1,1}(\varrho)$ let

$$|f_{t_1} \varrho' + f_{t_2}| \geqq \sqrt{\lambda_2} \,,$$

in $I_{1,2}(\varrho)$ let

$$|f_{t_1}\varrho' + f_{t_2}| < \sqrt{\lambda_2}\,, \qquad |f_{t_1t_1}\varrho' + f_{t_1t_2}| \geqq \frac{1}{2}\sqrt{\lambda_1\lambda_2}\,,$$

in $I_{1,3}(\varrho)$ let

$$|f_{t_1}\varrho' + f_{t_2}| < \sqrt{\lambda_2}\,, \qquad |f_{t_1t_1}\varrho' + f_{t_1t_2}| < \frac{1}{2}\sqrt{\lambda_1\lambda_2}\,.$$

Applying the second mean-value theorem, it can be easily seen that

$$I_{1,1}(\varrho) \ll \frac{1}{\sqrt{\lambda_1\lambda_2}}\,.$$

By estimating trivially we obtain

$$I_{1,2}(\varrho) \ll \frac{1}{\sqrt{\lambda_1\lambda_2}} \int \left| \frac{f_{t_1t_1}\varrho' + f_{t_1t_2}}{f_{t_1}} \right| dt \ll \frac{|\log \eta_2 - \log \eta_1|}{\sqrt{\lambda_1\lambda_2}} = \frac{\log\left(\dfrac{\eta_2 - \eta_1}{\eta_1} + 1\right)}{\sqrt{\lambda_1\lambda_2}}\,,$$

where η_1, η_2 are suitable values with $\eta_2 > \eta_1 \geqq \sqrt{\lambda_1}$. Further, we have

$$\begin{aligned} \eta_2 - \eta_1 &\ll \max_{t_2} \int_{a_1}^{b_1} |f_{t_1t_1}(t_1, t_2)|\, dt_1 + \max_{t_1} \int_{a_2}^{b_2} |f_{t_1t_2}(t_1, t_2)|\, dt_2 \\ &\ll c_1 \sqrt{\lambda_1} + c_2 \sqrt{\lambda_1\lambda_2} \end{aligned}$$

and therefore

$$I_{1,2}(\varrho) \ll \frac{1 + \log|D'| + |\log \lambda_1| + |\log \lambda_2|}{\sqrt{\lambda_1\lambda_2}}\,.$$

In $I_{1,3}(\varrho)$ we apply the substitution

$$y = f_{t_1}(\varrho(t), t)\,\varrho'(t) + f_{t_2}(\varrho(t), t)\,.$$

Then

$$\begin{aligned} y' &= f_{t_1t_1}\varrho'^2 + 2f_{t_1t_2}\varrho' + f_{t_1}\varrho'' + f_{t_2t_2} \\ &= \frac{1}{f_{t_1t_1}} \{(f_{t_1t_1}\varrho' + f_{t_1t_2})^2 + H(f) + f_{t_1t_1}f_{t_1}\varrho''\}\,. \end{aligned}$$

In the case $|f_{t_1t_1}f_{t_1}\varrho''| \leqq \frac{1}{2}\lambda_1\lambda_2$ we have

$$\begin{aligned} |y'| &\geqq \frac{1}{|f_{t_1t_1}|} \{|H(f)| - |f_{t_1t_1}\varrho' + f_{t_1t_2}|^2 - |f_{t_1t_1}f_{t_1}\varrho''|\} \\ &\gg \frac{1}{\lambda_1}\left(\lambda_1\lambda_2 - \frac{1}{4}\lambda_1\lambda_2 - \frac{1}{2}\lambda_1\lambda_2\right) \gg \lambda_2 \end{aligned}$$

and therefore

$$I_{1,3}(\varrho) \ll \frac{1}{\sqrt{\lambda_1}} \int_0^{\sqrt{\lambda_2}} \frac{dy}{|y'|} \ll \frac{1}{\sqrt{\lambda_1 \lambda_2}}.$$

But if $|f_{t_1 t_1} f_{t_1} \varrho''| > \frac{1}{2} \lambda_1 \lambda_2$, we have

$$|f_{t_1}| \gg \frac{\lambda_1 \lambda_2}{|f_{t_1 t_1} \varrho''|} \gg \frac{\lambda_2}{r}$$

and therefore

$$I_{1,3}(\varrho) \ll \frac{c_2 r}{\lambda_2}.$$

Similar results hold if partly $t_1 = \alpha(t_2)$ is a solution of $f_{t_1} = \sqrt{\lambda_1}$ or curve of boundary. Similarly we proceed in the case $f_{t_1} < -\sqrt{\lambda_1}$. Altogether we have proved estimate (2.49).

We now turn to the case where the Hessian is possibly small. Here an additional condition is required:
(E′) Intersections of D with domains of the type

$$f_{t_2} - f_{t_1} \frac{f_{t_1 t_2}}{f_{t_1 t_1}} \geqq c \quad \text{or} \quad \leqq c$$

are to satisfy condition (C) as well.
Clearly, this condition is very harmless and is fulfilled by algebraic curves.

Lemma 2.7. *Suppose that*

$$\alpha_1 \leqq f_{t_1}(t_1, t_2) \leqq \beta_1, \qquad \gamma_1 = \beta_1 - \alpha_1,$$
$$\lambda_1 \leqq |f_{t_1 t_1}(t_1, t_2)| \ll \lambda_1, \qquad |H(f)| \geqq \Lambda.$$

Moreover, with the notation

$$u = f_{t_1}, \qquad v = f_{t_2} - f_{t_1} \frac{f_{t_1 t_2}}{f_{t_1 t_1}}$$

let

$$\left| \frac{\partial(u, v)}{\partial(t_1, t_2)} \right| \gg \Lambda.$$

For all parts of the curve of boundary let $t_2 = \text{const}$ *or* $t_1 = \varrho(t_2)$, *where* $\varrho(t)$ *is partly twice differentiable and* $|\varrho''(t)| \ll r_0$. *If on all parts* $t_1 = \varrho(t_2)$ *the condition*

$$|f_{t_1t_1} f_{t_1} \varrho''| \leqq \vartheta\, |H(f)|$$

is satisfied with some ϑ, $0 < \vartheta < 1$, *put* $r = 0$. *Otherwise put* $r = r_0$. *Then*

$$\iint\limits_D e^{if(t_1, t_2)}\, dt_1\, dt_2 \ll \frac{1 + |\log \gamma_1| + |\log \lambda_1|}{\sqrt{\Lambda}} + \frac{c_2 r \lambda_1}{\Lambda}. \tag{2.50}$$

Proof. We divide the domain D into three regions

$$D_1:\ f_{t_1}(t_1, t_2) \geqq \sqrt{\lambda_1}\,,$$
$$D_2:\ 0 \leqq f_{t_1}(t_1, t_2) < \sqrt{\lambda_1}\,,$$
$$D_3:\ f_{t_1}(t_1, t_2) < 0\,.$$

Consider the region D_1. Let us proceed as in the proof of Lemma 2.6. Denote the left-hand and right-hand boundaries of D_1 by $t_1 = \alpha(t_2)$ and $t_1 = \beta(t_2)$, respectively. Integrating by parts, we obtain

$$\iint\limits_{D_1} e^{if(t_1, t_2)}\, dt_1\, dt_2 = I_1(\beta) - I_1(\alpha) + I_2, \tag{2.51}$$

where

$$I_1(\alpha) = -i \int \frac{e^{if(\alpha(t), t)}}{f_{t_1}(\alpha(t), t)}\, dt, \tag{2.52}$$

$$I_2 = -i \iint f_{t_1t_1} f_{t_1}^{-2}\, e^{if(t_1, t_2)}\, dt_1\, dt_2\,, \tag{2.53}$$

and $I_1(\beta)$ is defined analogously to $I_1(\alpha)$.

We first estimate the integrated terms and consider, say, $I_1(\alpha)$. Suppose that $t_1 = \alpha(t_2)$ is a solution of $f_{t_1} = \sqrt{\lambda_1}$. Then

$$I_1(\alpha) = \frac{-i}{\sqrt{\lambda_1}} \int e^{if(\alpha(t), t)}\, dt\,.$$

We find

$$f'(\alpha(t), t) = f_{t_2} - f_{t_1} \frac{f_{t_1t_2}}{f_{t_1t_1}} = v\,,$$

$$\begin{aligned} f''(\alpha(t), t) &= \frac{1}{f_{t_1t_1}} \left\{ H(f) - \frac{f_{t_1}}{f_{t_1t_1}^2} \left(f_{t_1t_1}^2 f_{t_1t_2t_2} - 2 f_{t_1t_1} f_{t_1t_2} f_{t_1t_1t_2} + f_{t_1t_2}^2 f_{t_1t_1t_1} \right) \right\} \\ &= \frac{1}{f_{t_1t_1}} \frac{\partial(u, v)}{\partial(t_1, t_2)}\,. \end{aligned} \tag{2.54}$$

Because of $|f''(\alpha), t)| \gg \Lambda/\lambda_1$ Lemma 2.2 yields

$$I_1(\alpha) \ll 1/\sqrt{\Lambda}\,.$$

We now estimate $I_1(\alpha)$ under the assumption that $t_1 = \alpha(t_2)$ is part of the curve of boundary. Again, we use the notation $\alpha(t) = \varrho(t)$, and we divide $I_1(\alpha)$ into three parts:

$$I_1(\alpha) = I_1(\varrho) = I_{1,1}(\varrho) + I_{1,2}(\varrho) + I_{1,3}(\varrho)\,.$$

In $I_{1,1}(\varrho)$ let

$$|f_{t_1}\varrho' + f_{t_2}| \geqq \sqrt{\Lambda/\lambda_1}\,,$$

in $I_{1,2}(\varrho)$ let

$$|f_{t_1}\varrho' + f_{t_2}| < \sqrt{\Lambda/\lambda_1}\,, \qquad |f_{t_1t_1}\varrho' + f_{t_1t_2}| \geqq \sqrt{(1-\vartheta)\,\Lambda/2}\,,$$

in $I_{1,3}(\varrho)$ let

$$|f_{t_1}\varrho' + f_{t_2}| < \sqrt{\Lambda/\lambda_1}\,, \qquad |f_{t_1t_1}\varrho' + f_{t_1t_2}| < \sqrt{(1-\vartheta)\,\Lambda/2}$$

with suitable ϑ, $0 < \vartheta < 1$. As in the proof of Lemma 2.6 we obtain

$$I_{1,1}(\varrho) \ll 1/\sqrt{\Lambda}\,,$$

$$I_{1,2}(\varrho) \ll \frac{\log(\gamma_1/\lambda_1 + 1)}{\sqrt{\Lambda}} \ll \frac{1 + |\log\gamma_1| + |\log\lambda_1|}{\sqrt{\Lambda}}$$

In $I_{1,3}(\varrho)$ we use the substitution

$$y = f_{t_1}(\varrho(t), t)\,\varrho'(t) + f_{t_2}(\varrho(t), t)\,.$$

If $|\,f_{t_1t_1}f_{t_1}\varrho''| \leqq \vartheta|H(f)|$, we have

$$|y'| \geqq \frac{1}{|\,f_{t_1t_1}|}\{|H(f)| - |f_{t_1t_1}\varrho' + f_{t_1t_2}|^2 - |\,f_{t_1t_1}f_{t_1}\varrho''|\}$$
$$> \frac{1-\vartheta}{2}\,\frac{\Lambda}{|f_{t_1t_1}|} \gg \frac{\Lambda}{\lambda_1}$$

and therefore

$$I_{1,3}(\varrho) \ll \frac{1}{\sqrt{\lambda_1}}\int\limits_0^{\sqrt{\Lambda/\lambda_1}}\frac{dy}{|y'|} \ll \frac{1}{\sqrt{\Lambda}}\,.$$

If the above condition is satisfied throughout, we obtain $r = 0$. Otherwise we get from $|f_{t_1t_1}f_{t_1}\varrho''| > \vartheta|H(f)|$ the estimate

$|f_{t_1}| \gg \Lambda/\lambda_1 r_0$ and

$I_{1,3}(\varrho) \ll c_2 r\lambda_1/\Lambda$

with $r = r_0$.

Similar results hold if partly $t_1 = \alpha(t_2)$ is a solution of $f_{t_1} = \sqrt{\lambda_1}$ and curve of boundary.

We now consider the integral I_2 in the region D_1. We divide D_1 into two subregions. Let $D_{1,1}$ denote the part of D_1 lying between the curves

$$v = \pm\sqrt{\Lambda/\lambda_1}\,, \tag{2.55}$$

and let $D_{1,2}$ be the remainder of D_1. Let $\varphi(u, t_2)$ be defined by $f_{t_1}(\varphi, t_2) = u$. Then the function $\frac{\partial}{\partial t_2} f(\varphi, t_2) = v$ is monotonic for a fixed u in a bounded number of t_2-intervals. Consider the region $D_{1,2}$. Since $|\partial f/\partial t_2|$ there is no less than $\sqrt{\Lambda/\lambda_1}$, we have, by the second mean-value theorem,

$$\int e^{if(\varphi, t_2)}\, dt_2 \ll \sqrt{\lambda_1/\Lambda}\,.$$

Hence

$$\iint\limits_{D_{1,2}} f_{t_1 t_1} f_{t_1}^{-2}\, e^{if(t_1, t_2)}\, dt_1 dt_2 = \int \frac{du}{u^2} \int e^{if(\varphi, t_2)}\, dt_2$$

$$\ll \sqrt{\frac{\lambda_1}{\Lambda}} \int\limits_{\sqrt{\lambda_1}}^{\infty} \frac{du}{u^2} \ll \frac{1}{\sqrt{\Lambda}}\,.$$

For the remaining integral we have

$$\iint\limits_{D_{1,1}} f_{t_1 t_1} f_{t_1}^{-2}\, e^{if(t_1, t_2)}\, dt_1\, dt_2$$

$$\ll \lambda_1 \iint\limits_{D_{1,1}} f_{t_1}^{-2}\, dt_1\, dt_2 \ll \lambda_1 \iint \left|\frac{\partial(u, v)}{\partial(t_1, t_2)}\right|^{-1} \frac{du\, dv}{u^2} \ll \frac{\lambda_1}{\Lambda} \int\limits_{\sqrt{\lambda_1}}^{\infty} \frac{du}{u^2} \int\limits_{0}^{\sqrt{\Lambda/\lambda_1}} dv = \frac{1}{\sqrt{\Lambda}}\,.$$

Thus

$$I_2 \ll 1/\sqrt{\Lambda}\,.$$

So we have proved the required estimation (2.50) in the whole region D_1.

Now consider the region D_2. We divide D_2 into two subregions as above. Let $D_{2,1}$ be the strip between the curves (2.55) and D_2 be the remainder of D_2. Again, let $\varphi(u, t_2)$ be defined by $f_{t_1}(\varphi, t_2) = u$. Then, as in the preceding part of the proof,

$$\iint\limits_{D_{2,2}} e^{if(t_1, t_2)}\, dt_1\, dt_2 = \int du \int \frac{1}{f_{t_1 t_1}}\, e^{if(\varphi, t_2)}\, dt_2 \ll \int\limits_{0}^{\sqrt{\lambda_1}} \frac{du}{\sqrt{\lambda_1 \Lambda}} \ll \frac{1}{\sqrt{\Lambda}}\,.$$

Finally

$$\iint\limits_{D_{2,1}} e^{if(t_1,t_2)}\,dt_1\,dt_2 \ll \iint\limits_{D_{2,1}} dt_1\,dt_2 \ll \int\limits_0^{\sqrt{\lambda_1}} \int\limits_0^{\sqrt{\Lambda/\lambda_1}} \left|\frac{\partial(u,v)}{\partial(t_1,t_2)}\right|^{-1} du\,dv \ll 1/\sqrt{\Lambda}\,.$$

So we have also for the region D_2 the desired estimation (2.50).

A similar proof can be applied to the region D_3. This completes the proof of the Lemma.

2.2.1.2. Applications to double exponential sums and three-dimensional lattice point problems

We begin with the simplest result based on Lemma 2.6.

Theorem 2.16. *Suppose that*

$$|f_{t_1t_1}(t_1,t_2)| \asymp \lambda_1\,, \qquad |f_{t_2t_2}(t_1,t_2)| \asymp \lambda_2\,,$$

$$|f_{t_1t_2}(t_1,t_2)| \ll \sqrt{\lambda_1\lambda_2}\,, \qquad |H(f)| \gg \lambda_1\lambda_2$$

throughout the rectangle D'. For all parts of the curve of boundary let $t_2 =$ const or $t_1 = \varrho(t_2)$, where $\varrho(t)$ is partly twice differentiable and $|\varrho''(t)| \ll r$. If R is defined by

$$R = 1 + \log|D'| + |\log\lambda_1| + |\log\lambda_2| + c_2 r\sqrt{\lambda_1/\lambda_2}\,,$$

then

$$\sum_{(n_1,n_2)\in D} e^{2\pi i f(n_1,n_2)} \ll (c_1\lambda_1 + c_2\sqrt{\lambda_1\lambda_2} + 1)(c_2\lambda_2 + c_1\sqrt{\lambda_1\lambda_2} + 1)\frac{R}{\sqrt{\lambda_1\lambda_2}}\,. \tag{2.56}$$

Furthermore, if the additional condition

$$c_1\sqrt{\lambda_1} \asymp c_2\sqrt{\lambda_2}$$

is satisfied, then

$$\sum_{(n_1,n_2)\in D} \psi(f(n_1,n_2)) \ll \left(|D'|(\lambda_1\lambda_2)^{1/4} + (c_1+c_2)|\log\lambda_1\lambda_2| + \frac{1}{\sqrt{\lambda_1\lambda_2}}\right)R\,. \tag{2.57}$$

Proof. Clearly, the formulation of the conditions for $f_{t_1t_1}$ and $f_{t_2t_2}$ is substantially the same as in Lemma 2.6. We may assume that $\lambda_1\lambda_2 < 1$, since otherwise the theorem is trivial. Suppose that

$$\alpha_j \leqq f_{t_j}(t_1,t_2) \leqq \beta_j\,, \qquad \gamma_j = \beta_j - \alpha_j \qquad (j = 1, 2)\,.$$

The functions $f_{t_j}(t_1, t_2)$ are strictly monotonic with respect to t_j. Therefore, Lemma 2.3 can be twice applied to the exponential sum in (2.56). We first obtain

$$\sum_{(n_1, n_2) \in D} e^{2\pi i f(n_1, n_2)} = \sum_{n_2} \sum_{v_1} \int_{t_1} e^{2\pi i (f(t_1, n_2) - v_1 t_1)} dt_1 + O(c_2 \log (\gamma_1 + 2)).$$

Here the integral is taken over t_1 for a fixed n_2 such that $(t_1, n_2) \in D$. For v_1 there are the inequalities

$$f_{t_1}(\varrho_1(n_2), n_2) - \eta < v_1 < f_{t_1}(\varrho_2(n_2), n_2) + \eta,$$

where η is any positive constant less than 1, and ϱ_1, ϱ_2 denote the values for which f_{t_1} takes its lower and upper bounds depending on n_2. Changing the sums and the integral, so that the sum over n_2 is the inner sum, we obtain in a similar manner

$$\sum_{(n_1, n_2) \in D} e^{2\pi i f(n_1, n_2)} = \sum_{(v_1, v_2) \in D_2} \iint_{(t_1, t_2) \in D} e^{2\pi i (f(t_1, t_2) - v_1 t_1 - v_2 t_2)} dt_1 \, dt_2$$
$$+ O(c_1(\gamma_1 + 1) \log (\gamma_2 + 2)) + O(c_2 \log (\gamma_1 + 2)). \tag{2.58}$$

For v_1, v_2 we first of all find the inequalities

$$f_{t_1}(\varrho_1(t_2), t_2) - \eta < v_1 < f_{t_1}(\varrho_2(t_2), t_2) + \eta,$$

$$f_{t_2}(t_1, \varrho_3(t_1)) - \eta < v_2 < f_{t_2}(t_1, \varrho_4(t_1)) + \eta,$$

where ϱ_3, ϱ_4 denote the values for which f_{t_2} takes its lower and upper bounds depending on t_1. Clearly in both inequalities we can take the same value η. Let us suppose for a moment that $\eta = 0$. Then each (v_1, v_2) represents a lattice point of D_2, which is the image of D under the mapping

$$y_1 = f_{t_1}(t_1, t_2), \quad y_2 = f_{t_2}(t_1, t_2).$$

Since in reality $\eta > 0$, the (v_1, v_2) run over all lattice points of a somewhat bigger domain D_2^*. However, if we choose η sufficiently small, the number of lattice points of D_2 is at most of order $(\gamma_1 + 1)(\gamma_2 + 1)$. Hence, by (2.49),

$$\sum_{(n_1, n_2) \in D} e^{2\pi i f(n_1, n_2)} \ll \frac{(\gamma_1 + 1)(\gamma_2 + 1)}{\sqrt{\lambda_1 \lambda_2}} R + c_1(\gamma_1 + 1) \log (\gamma_2 + 2)$$
$$+ c_2 \log (\gamma_1 + 2).$$

Because of

$$\gamma_1 \ll c_1 \lambda_1 + c_2 \sqrt{\lambda_1 \lambda_2}, \qquad \gamma_2 \ll c_2 \lambda_2 + c_1 \sqrt{\lambda_1 \lambda_2}$$

the estimate (2.56) follows at once.

To prove (2.57) we use (1.18) and the estimate

$$\sum_{(n_1, n_2) \in D} e^{2\pi i f(n_1, n_2)} \ll \left(|D'| \sqrt{\lambda_1 \lambda_2} + c_1 + c_2 + \frac{1}{\sqrt{\lambda_1 \lambda_2}} \right) R ,$$

which holds under the additional condition $c_1 \sqrt{\lambda_1} \asymp c_2 \sqrt{\lambda_2}$. Then

$$\begin{aligned} \sum_{(n_1, n_2) \in D} \psi(f(n_1, n_2)) &\ll \frac{|D'|}{z} + \sum_{\nu=1}^{\infty} \min\left(\frac{z^2}{\nu^3}, \frac{1}{\nu}\right) \Big| \sum_{(n_1, n_2) \in D} e^{2\pi i \nu f(n_1, n_2)} \Big| \\ &\ll \frac{|D'|}{z} + \sum_{\nu \leqq z} \left(|D'| \sqrt{\lambda_1 \lambda_2} \, \frac{c_1 + c_2}{\nu} \right) (R + \log \nu) \\ &\quad + \sum_{\nu > z} \left(\frac{z}{\nu}\right)^2 \left(|D'| \sqrt{\lambda_1 \lambda_2} + \frac{c_1 c_2}{\nu} \right) (R + \log \nu) \\ &\quad + \sum_{\nu=1}^{\infty} \frac{1}{\nu^2} \frac{R + \log \nu}{\sqrt{\lambda_1 \lambda_2}} \\ &\ll \frac{|D'|}{z} + \left(z |D'| \sqrt{\lambda_1 \lambda_2} + (c_1 + c_2) \log z \right) (R + \log z) \\ &\quad + \frac{R}{\sqrt{\lambda_1 \lambda_2}} . \end{aligned}$$

Putting $z = (\lambda_1 \lambda_2)^{-1/4}$ the result (2.57) follows immediately.

We now derive from Lemma 2.7 a similar theorem. No doubt, the resulting theorem is much more complicated than Theorem 2.16, but it has the advantage that the Hessian may be small.

Theorem 2.17. *Suppose that*

$$\begin{aligned} &\alpha_j \leqq f_{t_j}(t_1, t_2) \leqq \beta_j , && \gamma_j = \beta_j - \alpha_j \qquad (j = 1, 2) , \\ &\alpha_1 \cdot \beta_1 > 0 , && \delta_1 = \beta_1 / \alpha_1 , \\ &\lambda_1 \leqq |f_{t_1 t_1}(t_1, t_2)| \ll \lambda_1 , && \Lambda \leqq |H(f)| \ll \Lambda , \\ &c_1(\gamma_1 + 1) \ll |D| \sqrt{\Lambda} . \end{aligned}$$

Moreover, with the notation

$$u = f_{t_1} , \qquad w = f_{t_1 t_2} / f_{t_1 t_1}$$

let

$$\left| f_{t_1} \frac{\partial(u, w)}{\partial(t_1 t_2)} \right| \leqq \vartheta \, |H(f)| ,$$

where ϑ *is a suitable constant with* $0 < \vartheta < 1$. *For all parts of the curve of boundary let* $t_2 = \text{const}$ *or* $t_1 = \varrho(t_2)$, *where* $\varrho(t)$ *is partly twice differentiable. Suppose that*

$$|f_{t_1} f_{t_1 t_1} \varrho''| \leqq \vartheta |H(f)| \qquad (0 < \vartheta < 1)$$

at the bound. Let

$$R_1 = 1 + |\log \gamma_1| + |\log \gamma_2| + |\log \lambda_1| ,$$

$$R_2 = R_1(1 + \log \delta_1) ,$$

$$R = (R_1 + |\log \Lambda|)\,(1 + \log \delta_1) .$$

Then

$$\sum_{(n_1, n_2) \in D} e^{2\pi i f(n_1, n_2)} \ll \left\{ |D| \sqrt{\Lambda} + \frac{\gamma_1 + \gamma_2 + 1}{\sqrt{\Lambda}} + c_2 \right\} R_2 , \tag{2.59}$$

$$\sum_{(n_1, n_2) \in D} \psi(f(n_1, n_2)) \ll \left\{ |D|\, \Lambda^{1/4} + \left(\frac{\gamma_1 + \gamma_2 + 1}{\sqrt{\Lambda}} + c_2 \right) |\log \Lambda| \right\} R . \tag{2.60}$$

Proof. We may assume that $\Lambda < 1$, since otherwise the theorem is trivial. We wish to use the representation (2.58) and to estimate the double integral by means of Lemma 2.7. There the function $f(t_1, t_2)$ is to replace by the function $f(t_1, t_2) - v_1 t_1 - v_2 t_2$. This is meaningless for the second derivatives. But the first derivative f_{t_1} is now replaced by $f_{t_1} - v_1$. Because of this property we divide the domain D into strips such that in each strip $|f_{t_1} - v_1| \leqq |f_{t_1}|$. This can be done by the curves $f_{t_1}(t_1, t_2) = 2^{-m}\beta_1$, $m = 0, 1, 2, \dots$, assuming without loss of generality that $\beta_1 > 0$. We now apply (2.58) and Lemma 2.7 to each strip. Hence, putting

$$u' = f_{t_1} - v_1 , \qquad v' = f_{t_2} - v_2 - (f_{t_1} - v_1) \frac{f_{t_1 t_2}}{f_{t_1 t_1}}$$

the functional determinant becomes

$$\left| \frac{\partial(u', v')}{\partial(t_1, t_2)} \right| = \left| H(f) - (f_{t_1} - v_1) \frac{\partial(u, w)}{\partial(t_1, t_2)} \right| \geqq |H(f)| - |f_{t_1} - v_1| \left| \frac{\partial(u, w)}{\partial(t_1, t_2)} \right|$$

$$\geqq |H(f)| - \left| f_{t_1} \frac{\partial(u, w)}{\partial(t_1, t_2)} \right| \geqq (1 - \vartheta)\, |H(f)| \gg \Lambda .$$

If the curve $f_{t_1}(t_1, t_2) = 2^{-m}\beta_1$ is a part of the bound, we put $t_1 = \varphi(t_2)$. Then

$$f_{t_1 t_1} \varphi' + f_{t_1 t_2} = 0 ,$$

$$f_{t_1 t_1} \varphi'' + f_{t_1 t_1 t_1} \varphi'^2 + 2 f_{t_1 t_1 t_2} \varphi' + f_{t_1 t_2 t_2} = 0 ,$$

$$f_{t_1 t_1} \varphi'' + f_{t_1 t_1 t_1} \frac{f_{t_1 t_2}}{f_{t_1 t_1}^2} - 2 f_{t_1 t_1 t_2} \frac{f_{t_1 t_2}}{f_{t_1 t_1}} + f_{t_1 t_2 t_2} = 0 .$$

Consequently, we obtain

$$f_{t_1 t_1}\varphi'' = -\frac{\partial(u, w)}{\partial(t_1, t_2)}$$

and

$$|f_{t_1} f_{t_1 t_1}\varphi''| = \left| f_{t_1} \frac{\partial(u, w)}{\partial(t_1, t_2)} \right| \leqq \vartheta\, |H(f)| .$$

Hence, all the conditions of Lemma 2.7 are satisfied, and we can use (2.58) and (2.50) with $r = 0$ to each strip. The number of strips is of order $1 + \log \delta_1$. Therefore we have

$$\sum_{(n_1, n_2) \in D} e^{2\pi i f(n_1, n_2)} \ll \left\{ \frac{G_2}{\sqrt{\Lambda}} + c_i(\gamma_1 + 1) + c_2 \right\} R_2 ,$$

where G_2 denotes the number of lattice points of D_2, the image of D under the mapping $y_1 = f_{t_1}, y_2 = f_{t_2}$. Clearly, G_2 is of order of the area of D_2 plus the order of length of the bound. Thus

$$G_2 \ll |D_2| + \gamma_1 + \gamma_2 + 1 \ll |D|\, \Lambda + \gamma_1 + \gamma_2 + 1 .$$

Now (2.59) follows immediately.

In order to prove (2.60), we use (1.18) and (2.59). Then

$$\begin{aligned}
\sum_{(n_1, n_2) \in D} \psi(f(n_1, n_2)) &\ll \frac{|D|}{z} + \sum_{\nu=1}^{\infty} \min\left(\frac{z^2}{\nu^3}, \frac{1}{\nu}\right) \Bigg| \sum_{(n_1, n_2) \in D} e^{2\pi i \nu f(n_1, n_2)} \Bigg| \\
&\ll \frac{|D|}{z} + \sum_{\nu=1}^{\infty} \min\left(\frac{z^2}{\nu^3}, \frac{1}{\nu}\right) \\
&\quad \times \left\{ \nu\, |D| \sqrt{\Lambda} + \frac{\gamma_1 + \gamma_2 + 1}{\sqrt{\Lambda}} + c_2 \right\} \\
&\quad \times (R_1 + \log \nu)(1 + \log \delta_1) \\
&\ll \frac{|D|}{z} + \left\{ z\, |D| \sqrt{\Lambda} + \left(\frac{\gamma_1 + \gamma_2 + 1}{\sqrt{\Lambda}} + c_2 \right) \log z \right\} \\
&\quad \times (R_1 + \log z)(1 + \log \delta_1) .
\end{aligned}$$

Putting $z = \Lambda^{-1/4}$, estimate (2.60) follows immediately.

2.2.2. *Estimation of double exponential sums by iterated application of the one-dimensional theory*

The gist of estimation of double exponential sums by iterated application of the one-dimensional theory consists in making use of the sharp transformation formula of

Theorem 2.11. In a double exponential sum

$$\sum_{(n_1, n_2)\in D} e^{2\pi i f(n_1, n_2)}$$

we apply this transformation formula once with respect to one of the variables n_1 or n_2. After transforming this sum we use van der Corput's theory for the other variable. This can be easily seen in the proofs of the following two theorems.

Theorem 2.18. *Suppose that*

$$\alpha_j \leqq f_{t_j}(t_1, t_2) \leqq \beta_j , \qquad \gamma_j = \beta_j - \alpha_j \quad (j = 1, 2) ,$$

$$|f_{t_1t_1}(t_1, t_2)| \asymp \lambda_1 , \qquad |H(f)| \asymp \Lambda$$

and either

$$0 < |f_{t_1t_1t_1}(t_1, t_2)| \ll \lambda' \ll \frac{\lambda_1}{c_1}$$

or

$$|f_{t_1t_1t_1}(t_1, t_2)| \asymp L(t_2) , \quad L(t_2) \ll \lambda' ,$$

where $L(t_2)$ may depend on t_2. Let the function $f_{t_1t_1}f_{t_1t_1t_2} - f_{t_1t_2}f_{t_1t_1t_1}$ be of constant sign in D. For each point $(a, b) \in D$ the function $F(t_1, a; f(t_1, b))$, defined by (2.37), *possesses only a bounded number of points of zero. Let $t_2 = const$ or $t_1 = \varrho(t_2)$ for all parts of the curve of boundary, where $\varrho(t)$ is twice continuously differentiable. Let*

$$T(a, b) = \begin{cases} 0 & \text{for } f_{t_1}(a, b) \in \mathbf{Z} , \\ \min\left(\dfrac{1}{\|f_{t_1}(a, b)\|}, \dfrac{1}{\sqrt{\lambda_1}}\right) & \text{for } f_{t_1}(a, b) \notin \mathbf{Z} , \end{cases} \tag{2.61}$$

where $\mathbf{Z}$ denotes the set of integers. Then

$$\sum_{(n_1, n_2)\in D} e^{2\pi i f(n_1, n_2)} \ll |D|\sqrt{\Lambda} + \textstyle\sum_1 T(\varrho(n_2), n_2) + \dfrac{\gamma_1 + \gamma_2 + 1}{\sqrt{\Lambda}}$$

$$+ \frac{\gamma_1 + 1}{\sqrt{\Lambda_1}} + c_2(\lambda' \lambda_1^{-2} + \log(c_1\lambda_1 + 2)) , \tag{2.62}$$

$SC(\sum_1)$: *n_2 runs over integers with $a_2 \leqq n_2 \leqq b_2$ such that the points (t_1, n_2) lie on the bound with $t_1 = \varrho(n_2)$.*

Proof. We apply Theorem 2.11 to the sum over n_1. Let $t_1 = \varrho_1(t_2)$ denote the lower part and $t_1 = \varrho_2(t)$ the upper part of the curve of boundary. Then

$$\sum_{(n_1, n_2)\in D} e^{2\pi i f(n_1, n_2)} = \sum_{a_2 \leqq n_2 \leqq b_2} \; \sum_{\varrho_1 \leqq n_1 \leqq \varrho_2} e^{2\pi i f(n_1, n_2)} ,$$

where $\varrho_1 = \varrho_1(n_2)$, $\varrho_2 = \varrho_2(n_2)$. Suppose that $f_{t_1t_1}(t_1, t_2) > 0$. Let D_1 be the image of D under the mapping

$$y_1 = f_{t_1}(t_1, t_2) , \quad y_2 = t_2 .$$

Let the function $\varphi(y_1, t_2)$ be defined by $f_{t_1}(\varphi, t_2) = y_1$ and put

$$f_1(y_1, t_2) = f(\varphi(y_1, t_2), t_2) - y_1\varphi(y_1, t_2).$$

Since $c_1\lambda' \ll \lambda_1$ or

$$(\varrho_2 - \varrho_1)\, L(n_2) \ll \left| \int_{\varrho_1}^{\varrho_2} f_{t_1t_1t_1}(t_1, n_2)\, dt_1 \right| = |f_{t_1t_1}(\varrho_2, n_2) - f_{t_1t_1}(\varrho_1, n_2)| \ll \lambda_1$$

and $L(n_2) \ll \lambda'$ we obtain from (2.38)

$$\begin{aligned}\sum_{\varrho_1 \leqq n_1 \leqq \varrho_2} e^{2\pi i f(n_1, n_2)} &= \varepsilon \sum{}'' \frac{1}{\sqrt{f_{t_1t_1}(\varphi, n_2)}}\, e^{2\pi i f_1(m_1, n_2)} \\ &\quad + O(T(\varrho_1(n_2), n_2)) + O(T(\varrho_2(n_2), n_2)) \\ &\quad + O(\lambda'\lambda_1^{-2}) + O(\log(c_1\lambda_1 + 2)),\end{aligned}$$

$$SC(\textstyle\sum''):\quad f_{t_1}(\varrho_1(n_2), n_2) \leqq m_1 \leqq f_{t_1}(\varrho_2(n_2), n_2)$$

and

$$\begin{aligned}\sum_{(n_1, n_2) \in D} e^{2\pi i f(n_1, n_2)} &= \varepsilon \sum_{\substack{m_1 \\ (m_1, n_2) \in D_1}} \sum_{n_2}{}'' \frac{1}{\sqrt{f_{t_1t_1}(\varphi, n_2)}}\, e^{2\pi i f_1(m_1, n_2)} \\ &\quad + O(\textstyle\sum_1 T(\varrho(n_2), n_2)) + O(c_2\lambda'\lambda_1^{-2}) \\ &\quad + O(c_2 \log(c_1\lambda_1 + 2)). \end{aligned} \tag{2.63}$$

We now apply Theorem 2.1 to the sum over n_2. We have to consider the function $f_1(m_1, t_2)$ and obtain

$$\frac{\partial}{\partial t_2} f_1(m_1, t_2) = f_{t_1}\varphi_{t_2} + f_{t_2} - m_1\varphi_{t_2} = f_{t_2},$$

$$\frac{\partial^2}{\partial t_2^2} f_1(m_1, t_2) = f_{t_1t_2}\varphi_{t_2} + f_{t_2t_2}$$

and, because of $f_{t_1t_1}\varphi_{t_2} + f_{t_1t_2} = 0$,

$$\left| \frac{\partial^2}{\partial t_2^2} f_1(m_1, t_2) \right| = \left| \frac{H(f)}{f_{t_1t_1}} \right| \asymp \frac{\Lambda}{\lambda_1}.$$

We assume $\partial^2 f_1/\partial t_2^2 > 0$ without loss of generality such that $\partial f_1/\partial t_2$ is strictly increasing. Let $\gamma = \gamma(m_1)$ and $\delta = \delta(m_1)$ denote the lower and upper bounds of D_1. Then we get from Theorem 2.1 with $\gamma < \tau \leqq \delta$

$$\sum_{\gamma \leqq n_2 \leqq \tau}{}'' e^{2\pi i f_1(m_1, n_2)} \ll \left(\frac{\partial}{\partial t_2} f_1(m_1, \delta) - \frac{\partial}{\partial t_2} f_1(m_1, \gamma) + 1 \right) \sqrt{\frac{\lambda_1}{\Lambda}} + 1.$$

Since the function

$$\frac{\partial}{\partial t_2}(f_{t_1t_1}(\varphi, t_2))^{-1/2} = \frac{1}{2}(f_{t_1t_1t_1}f_{t_1t_2} - f_{t_1t_1t_2}f_{t_1t_1})(f_{t_1t_1})^{-5/2}$$

keeps a constant sign, we obtain by partial summation

$$\sum_{\gamma \leqq n_2 \leqq \delta} \frac{1}{\sqrt{f_{t_1t_1}(\varphi, n_2)}} e^{2\pi i f_1(m_1, n_2)} \ll \left(\frac{\partial}{\partial t_2} f_1(m_1, \delta) - \frac{\partial}{\partial t_2} f_1(m_1, \gamma) + 1\right)\frac{1}{\sqrt{\Lambda}} + \frac{1}{\sqrt{\lambda_1}}.$$

Let D_2 denote the image of D under the mapping

$$y_1 = f_{t_1}(t_1, t_2), \quad y_2 = f_{t_2}(t_1, t_2).$$

Then

$$\begin{aligned}\sum_{(m_1, n_2) \in D_1} \frac{1}{\sqrt{f_{t_1t_1}(\varphi, n_2)}} e^{2\pi i f_1(m_1, n_2)} &\ll \Big(\sum_{(m_1, m_2) \in D_2} 1 + \gamma_1 + 1\Big)\frac{1}{\sqrt{\Lambda}} + \frac{\gamma_1 + 1}{\sqrt{\lambda_1}} \\ &\ll (|D_2| + \gamma_1 + \gamma_2 + 1)\frac{1}{\sqrt{\Lambda}} + \frac{\gamma_1 + 1}{\sqrt{\lambda_1}} \\ &\ll (|D|\Lambda + \gamma_1 + \gamma_2 + 1)\frac{1}{\sqrt{\Lambda}} + \frac{\gamma_1 + 1}{\sqrt{\lambda_1}}.\end{aligned}$$

Substituting this into (2.63), we get the result (2.62).

The theorem shows that it remains to estimate the sum over $T(\varrho(n_2), n_2)$. For this purpose the following lemma is useful.

Lemma 2.8. *Let $f(t)$ be a real, continuously differentiable function throughout the interval $[a, b]$, $a < b$, with $|f'(t)| \geqq q > 0$, and let z be a positive number. Then*

$$\sum_{a \leqq n \leqq b} \min\left(\frac{1}{\|f(n)\|}, z\right) \ll (|f(b) - f(a)| + 1)\left(z + \frac{1}{q}\log(b - a + 2)\right). \tag{2.64}$$

Proof. We may assume that $|f(b) - f(a)| < b - a$, since otherwise the lemma is trivial. We suppose that $f'(t) > 0$ without loss of generality, such that $f(t)$ is strictly increasing.

We first consider the case $[f(a)] = [f(b)]$. Then

$$f(n) - [f(n)] \geqq f(n) - f(a) = \int_a^n f'(t)\,dt \geqq q(n - a),$$

$$1 - (f(n) - [f(n)]) > f(b) - f(n) = \int_n^b f'(t)\,dt \geqq q(b - n).$$

Hence

$$S = \sum_{a \leqq n \leqq b} \min\left(\frac{1}{\|f(n)\|}, z\right) \leqq 2z + \frac{1}{q} \sum_{a<n<b} \frac{1}{\min(n-a, b-n)}$$

$$\ll z + \frac{1}{q} \log(b-a+2).$$

In the case $[f(a)] < [f(b)]$ we divide the sum S into

$$S = \sum_{k=[f(a)]}^{[f(b)]} S_k,$$

where S_k is defined by

$$S_k = \sum \min\left(\frac{1}{\|f(n)\|}, z\right), \qquad SC(\textstyle\sum): \; a \leqq n \leqq b, \qquad k \leqq f(n) < k+1.$$

Assume that either $f(n-1) < k$ or $k = [f(a)]$,

$$k \leqq f(n) < f(n+1) < \ldots < f(n+r) < k+1$$

and either $f(n+r+1) \geqq k+1$ or $k = [f(b)]$. Let $f^{-1}(t)$ denote the inverse function of $f(t)$. Then we obtain for $v = 0, 1, \ldots, r$

$$f(n+v) - [f(n+v)] = f(n+v) - k = \int_{f^{-1}(k)}^{n+v} f'(t)\,\mathrm{d}t$$

$$\geqq q(n+v-f^{-1}(k)) \geqq qv,$$

$$1 - (f(n+v) - [f(n+v)]) = k+1-f(n+v) = \int_{n+v}^{f^{-1}(k+1)} f'(t)\,\mathrm{d}t$$

$$\geqq q(f^{-1}(k+1) - n - v) > q(r-v).$$

Now

$$S_k \leqq 2z + \frac{1}{q} \sum_{v=1}^{r-1} \frac{1}{\min(v, r-v)} \ll z + \frac{1}{q} \log(b-a+2)$$

and (2.64) follows at once.

In the applications of Theorem 2.18 to three-dimensional lattice point problems in most cases the situation arises that at the bound of D partly either $f_{t_1}(\varrho(t), t)$ is an integer or a constant or $\left|\frac{\mathrm{d}}{\mathrm{d}t} f_{t_1}(\varrho(t), t)\right|$ has a fixed lower bound.

If T_1 denotes a part of the sum $\Sigma_1\, T(\varrho(n_2), n_2)$, where $f_{t_1}(\varrho(t), t)$ is an integer, then we get $T_1 = 0$.

When we say, $f_{t_1}(\varrho(t), t)$ is a constant, we mean that this constant does not depend on the parameters of the problem. If T_2 denotes such a part, we obtain

$$T_2 = \Sigma_2\, T(\varrho(n_2), n_2) \leqq \Sigma_2 \frac{1}{\|c\|} \ll c_2\,,$$

$$SC(\Sigma_2):\quad a_2 \leqq n_2 \leqq b_2,\ \ f_{t_1}(\varrho(n_2), n_2) = c \notin \mathbf{Z}\,.$$

We now consider a part T_3 with

$$\left|\frac{\mathrm{d}}{\mathrm{d}t} f_{t_1}(\varrho(t), t)\right| \gg q > 0\,.$$

We apply Lemma 2.8 with $z = 1/\sqrt{\lambda_1}$ and $a_2 \leqq a < b \leqq b_2$. The estimate (2.64) gives

$$\begin{aligned} T_3 &\ll \sum_{a \leqq n \leqq b} \min\left(\frac{1}{\|f_{t_1}(\varrho(n), n\|}, \frac{1}{\sqrt{\lambda_1}}\right) \\ &\ll (|f_{t_1}(\varrho(b), b) - f_{t_1}(\varrho(a), a)| + 1)\left(\frac{1}{\sqrt{\lambda_1}} + \frac{1}{q}\log(c_2 + 1)\right) \\ &\ll (\gamma_1 + 1)\left(\frac{1}{\sqrt{\lambda_1}} + \frac{1}{q}\log(c_2 + 1)\right). \end{aligned} \tag{2.65}$$

As we can take $q = \sqrt{\Lambda}$ in most cases, we are in a position to formulate the following theorem.

Theorem 2.19. *Suppose that*

$$\alpha_j \leqq f_{t_j}(t_1, t_2) \leqq \beta_j\,,\quad \gamma_j = \beta_j - \alpha_j \qquad (j = 1, 2)\,,$$

$$|f_{t_1t_1}(t_1, t_2)| \asymp \lambda_1\,, \qquad |H(f)| \asymp \Lambda$$

and either

$$0 < |f_{t_1t_1t_1}(t_1, t_2)| \ll \lambda' \ll \frac{\lambda_1}{c_1}$$

or

$$|f_{t_1t_1t_1}(t_1, t_2)| \asymp \mathbf{L}(t_2)\,, \qquad \mathbf{L}(t_2) \ll \lambda'\,,$$

where $\mathbf{L}(t_2)$ *may depend on* t_2. *Let the function* $f_{t_1t_1}f_{t_1t_1t_2} - f_{t_1t_2}f_{t_1t_1t_1}$ *be of constant sign in* D. *For each point* $(a, b) \in D$ *the function* $F(t_1, a; f(t_1, b))$ *possesses only a bounded number of points of zero. Let* $t_2 = \text{const}$ *or* $t_1 = \varrho(t_2)$ *for all parts of the curve of boundary, where* $\varrho(t)$ *is twice continuously differentiable. Assume that partly* $f_{t_1}(\varrho(t), t)$ *is an integer or a constant or*

$$\left|\frac{\mathrm{d}}{\mathrm{d}t} f_{t_1}(\varrho(t), t)\right| \gg \sqrt{\Lambda}\,.$$

Then

$$\sum_{(n_1,n_2)\in D} e^{2\pi i f(n_1,n_2)} \ll |D|\sqrt{\Lambda} + \frac{\gamma_1+\gamma_2+1}{\sqrt{\Lambda}}(1+\log c_2) + \frac{\gamma_1+1}{\sqrt{\lambda_1}}$$
$$+ c_2(\lambda'\lambda_1^{-2} + 1 + \log c_1 + |\log \lambda_1|), \tag{2.66}$$

$$\sum_{(n_1,n_2)\in D} \psi(f(n_1,n_2))$$
$$\ll |D|\Lambda^{1/4} + \frac{(\gamma_1+\gamma_2)|\log\Lambda|+1}{\sqrt{\Lambda}}(1+\log c_2) + \frac{1}{\sqrt{\lambda_1}}(\gamma_1\Lambda^{-1/8}+1)$$
$$+ c_2(\lambda'\lambda_1^{-2} + (1+\log c_1 + |\log\lambda_1| + |\log\Lambda|)|\log\Lambda|). \tag{2.67}$$

Proof. From (2.62) and the above remarks estimate (2.66) follows immediately. In order to prove (2.67), we use (2.66) and (1.18). Thus

$$\sum_{(n_1,n_2)\in D} \psi(f(n_1,n_2))$$
$$\ll \frac{|D|}{z} + \sum_{\nu=1}^{\infty} \min\left(\frac{z^2}{\nu^3}, \frac{1}{\nu}\right)\left|\sum_{(n_1,n_2)\in D} e^{2\pi i \nu f(n_1,n_2)}\right|$$
$$\ll \frac{|D|}{z} + \sum_{\nu=1}^{\infty} \min\left(\frac{z^2}{\nu^3}, \frac{1}{\nu}\right)\left\{\nu|D|\sqrt{\Lambda} + \frac{\gamma_1+\lambda_2+1/\nu}{\sqrt{\Lambda}}(1+\log c_2)\right.$$
$$\left. + \frac{1}{\sqrt{\lambda_1}}\left(\gamma_1\sqrt{\nu}+\frac{1}{\sqrt{\nu}}\right) + c_2\left(\frac{1}{\nu}\lambda'\lambda_1^{-2} + 1 + \log c_1 + |\log\lambda_1| + \log\nu\right)\right\}$$
$$\ll \frac{|D|}{z} + z|D|\sqrt{\Lambda} + \frac{(\gamma_1+\gamma_2)\log z + 1}{\sqrt{\Lambda}}(1+\log c_2) + \frac{1}{\sqrt{\lambda_1}}(\gamma_1\sqrt{z}+1)$$
$$+ c_2(\lambda'\lambda_1^{-2} + (1+\log c_1 + |\log\lambda_1| + \log z)\log z).$$

Putting $z = \Lambda^{-1/4}$, the result (2.67) follows at once.

Comparing the results (2.60) and (2.67), we see that they are essentially the same. (2.67) is slightly better because its leading term omits the logarithm factor, but this is rather unimportant. In the applications the other terms are of smaller order.

In the preceding theorem we used the simplest result of van der Corput's theory. Considering the further developments of his theory, we can also prove sharper results. Being an important example, we formulate the following result which is based on the method of exponent pairs.

Theorem 2.20. *Let* u_1, u_2 *be fixed positive numbers and* $1 \leqq a_1 < b_1 \leqq u_1 a_1$, $1 \leqq a_2 < b_2 \leqq u_2 a_2$. *Suppose that*

$$\left|\frac{\partial^\nu}{\partial t_1^\nu} f(t_1, t_2)\right| \asymp \frac{\lambda}{a_1^\nu} \qquad (\nu = 1, 2, 3).$$

Let the function $f_{t_1 t_1} f_{t_1 t_1 t_2} - f_{t_1 t_2} f_{t_1 t_1 t_1}$ *be of constant sign in* D. *For each point* $(a, b) \in D$ *the function* $F(t_1, a; f(t_1, b))$ *possesses only a bounded number of points of zero. Let* $t_2 = \text{const}$ *or* $t_1 = \varrho(t_2)$ *for all parts of the curve of boundary, where* $\varrho(t)$ *is twice continuously differentiable. Assume that partly* $f_{t_1}(\varrho(t), t)$ *is an integer or a constant or*

$$\left|\frac{\mathrm{d}}{\mathrm{d}t} f_{t_1}(\varrho(t), t)\right| \gg \frac{\lambda}{a_1 a_2}.$$

Let the function $\varphi(y_1, t_2)$ *be defined by* $f_{t_1}(\varphi, t_2) = y_1$. *For each possible integer* m_1 *let the function*

$$f_1(m_1, t_2) = f(\varphi(m_1, t_2), t_2) - m_1 \varphi(m_1, t_2)$$

satisfy the conditions of the method of exponent pairs with respect to t_2, *where in the notation of Definition* 2.1 $z = \lambda/a_2$, $a = a_2$. *If* (k, l) *denotes any exponent pair and if* $\lambda \gg a_1, a_2$, *then*

$$\sum_{(n_1, n_2) \in D} \mathrm{e}^{2\pi i f(n_1, n_2)} \ll a_2^{l-k} \lambda^{k+1/2} \log \lambda \tag{2.68}$$

and

$$\sum_{(n_1, n_2) \in D} \psi(f(n_1, n_2)) \ll (a_1^{2k+1} a_2^{2l+1} \lambda^{2k+1})^{\frac{1}{2k+3}} \log \lambda. \tag{2.69}$$

Proof. We can apply the proof of Theorem 2.18 up to equation (2.63). There we have $c_1 \ll a_1$, $c_2 \ll a_2$, $\lambda_1 = \lambda/a_1^2$, $\lambda' = \lambda/a_1^3$. Because of (2.65) and $q = \lambda/a_1 a_2$ and $\gamma_1 \ll \lambda/a_1$ we get for the sum over $T(\varrho(n_2), n_2)$ that

$$\Sigma_1 \, T(\varrho(n_2), n_2) \ll \sqrt{\lambda} + a_2 \log a_2 \ll \sqrt{\lambda} + a_2 \log \lambda.$$

Then equation (2.63) becomes

$$\sum_{(n_1, n_2) \in D} \mathrm{e}^{2\pi i f(n_1, n_2)} = \varepsilon \sum_{(m_1, n_2) \in D_1}'' \frac{1}{\sqrt{f_{t_1 t_1}(\varphi, n_2)}} \mathrm{e}^{2\pi i f_1(m_1, n_2)} + O(\sqrt{\lambda}) + O(a_2 \log \lambda).$$

Further on we proceed similarly to the proof of Theorem 2.18. From the definition of exponent pairs we obtain

$$\sum_{n_2}'' \mathrm{e}^{2\pi i f_1(m_1, n_2)} \ll a_2^{l-k} \lambda^k,$$

where n_2 runs over any subinterval of $[a_2, b_2]$. Now we obtain (2.68) in the same way as in the proof of Theorem 2.18.

We get the result (2.69) from (2.68) by means of (1.18):

$$\sum_{(n_1, n_2) \in D} \psi(f(n_1, n_2)) \ll \frac{a_1 a_2}{z} + \sum_{\nu=1}^{\infty} \min\left(\frac{z^2}{\nu^3}, \frac{1}{\nu}\right) \left| \sum_{(n_1, n_2) \in D} e^{2\pi i \nu f(n_1, n_2)} \right|$$

$$\ll \frac{a_1 a_2}{z} + \sum_{\nu=1}^{\infty} \min\left(\frac{z^2}{\nu^3}, \frac{1}{\nu}\right) a_2^{l-k} (\lambda\nu)^{k+1/2} \log(\lambda\nu)$$

$$\ll \frac{a_1 a_2}{z} + a_2^{l-k} (\lambda z)^{k+1/2} \log(\lambda z) .$$

Putting

$$z = (a_1^2 a_2^{2(k-l+1)} \lambda^{-2k-1})^{\frac{1}{2k+3}} ,$$

we obtain the estimate (2.69).

2.2.3. *Applications of Weyl's steps*

Generalizing the one-dimensional case we formulate the basic theorem for applying Weyl's steps on double exponential sums.

Theorem 2.21. *Let $f(t_1, t_2)$ be a real function in D', and let H_1, H_2 be integers with $1 \leqq H_1 \leqq c_1$, $1 \leqq H_2 \leqq c_2$. Let*

$$S = \sum_{(n_1, n_2) \in D} e^{2\pi i f(n_1, n_2)} ,$$

$$S_1 = \sum\nolimits_1 e^{2\pi i (f(n_1 + h_1, n_2 + h_2) - f(n_1, n_2))} ,$$

$$SC(\textstyle\sum_1): (n_1, n_2) \in D , \qquad (n_1 + h_1, n_2 + h_2) \in D ,$$

$$S_2 = \sum\nolimits_2 e^{2\pi i (f(n_1 + h_1, n_2 - h_2) - f(n_1, n_2))} ,$$

$$SC(\textstyle\sum_2): (n_1, n_2) \in D , \qquad (n_1 + h_1, n_2 - h_2) \in D .$$

Then

$$S \ll \frac{|D'|}{\sqrt{H_1 H_2}} + \left\{ \frac{|D'|}{H_1 H_2} \sum_{h_1=1}^{H_1-1} \sum_{h_2=0}^{H_2-1} |S_1| \right\}^{1/2} + \left\{ \frac{|D'|}{H_1 H_2} \sum_{h_1=0}^{H_1-1} \sum_{h_2=1}^{H_2-1} |S_2| \right\}^{1/2} . \tag{2.70}$$

In particular, if $H_1 = H$, $H_2 = 1$, we have

$$S \ll \frac{|D'|}{\sqrt{H}} + \left\{ \frac{|D'|}{H} \sum_{h=1}^{H-1} \left| \sum\nolimits_3 e^{2\pi i (f(n_1 + h, n_2) - f(n_1, n_2))} \right| \right\}^{1/2} , \tag{2.71}$$

$$SC(\textstyle\sum_3): (n_1, n_2) \in D , \qquad (n_1 + h, n_2) \in D .$$

Proof. Clearly, (2.71) is an immediate consequence of (2.70). Therefore, we now prove (2.70). We have

$$(H_1H_2\,|S|)^2 = \left|\sum_{m_1=1}^{H_1}\sum_{m_2=1}^{H_2}\sum_{(n_1+m_1,\,n_2+m_2)\in D} e^{2\pi i f(n_1+m_1,\,n_2+m_2)}\right|^2$$

$$= \left|\sum_{n_1}\sum_{n_2}\sum\nolimits_4 e^{2\pi i f(n_1+m_1,\,n_2+m_2)}\right|^2,$$

$$SC(\Sigma_4):\quad (n_1+m_1, n_2+m_2)\in D\,,\quad 1\leqq m_1\leqq H_1\,,\quad 1\leqq m_2\leqq H_2\,.$$

Applying Schwarz's inequality, we obtain

$$(H_1H_2\,|S|)^2 \leqq \sum_{n_1'}\sum_{n_2'} 1 \sum_{n_1}\sum_{n_2}\left|\sum\nolimits_4 e^{2\pi i f(n_1+m_1,\,n_2+m_2)}\right|^2$$

$$\ll |D'|\left|\sum\nolimits_5 e^{2\pi i(f(n_1+m_1,\,n_2+m_2)-f(n_1+m_1',\,n_2+m_2'))}\right|,$$

$$SC(\Sigma_5):\quad (n_1+m_1, n_2+m_2)\in D\,,\quad (n_1+m_1', n_2+m_2')\in D\,,$$
$$1\leqq m_1, m_1'\leqq H_1\,,\quad 1\leqq m_2, m_2'\leqq H_2\,.$$

$$(H_1H_2\,|S|)^2 \ll |D'|^2 H_1H_2 + |D'|\left|\sum\nolimits_6 e^{2\pi i(f(n_1+m_1,\,n_2+m_2)-f(n_1+m_1',\,n_2+m_2'))}\right| + |D'|\left|\sum\nolimits_7 e^{2\pi i(f(n_1+m_1,\,n_2+m_2)-f(n_1+m_1',\,n_2+m_2'))}\right|,$$

$$SC(\Sigma_6):\quad (n_1+m_1, n_2+m_2)\in D\,,\quad (n_1+m_1', n_2+m_2')\in D\,,$$
$$1\leqq m_1' < m_1\leqq H_1\,,\quad 1\leqq m_2'\leqq m_2\leqq H_2\,,$$

$$SC(\Sigma_7):\quad (n_1+m_1, n_2+m_2)\in D\,,\quad (n_1+m_1', n_2+m_2')\in D\,,$$
$$1\leqq m_1'\leqq m_1\leqq H_1\,,\quad 1\leqq m_2 < m_2'\leqq H_2\,.$$

$$(H_1H_2\,|S|)^2 \ll |D'|^2 H_1H_2 + |D'|\left|\sum\nolimits_8 e^{2\pi i(f(n_1+m_1-m_1',\,n_2+m_2-m_2')-f(n_1,n_2))}\right| + |D'|\left|\sum\nolimits_9 e^{2\pi i(f(n_1+m_1-m_1',\,n_2+m_2-m_2')-f(n_1,n_2))}\right|,$$

$$SC(\Sigma_8):\quad (n_1+m_1-m_1', n_2+m_2-m_2')\in D\,,\quad (n_1, n_2)\in D\,,$$
$$1\leqq m_1' < m_1\leqq H_1\,,\quad 1\leqq m_2'\leqq m_2\leqq H_2\,,$$

$$SC(\Sigma_9):\quad (n_1+m_1-m_1', n_2+m_2-m_2')\in D\,,\quad (n_1, n_2)\in D\,,$$
$$1\leqq m_1'\leqq m_1\leqq H_1\,,\quad 1\leqq m_2 < m_2'\leqq H_2\,.$$

Putting $m_1 - m_1' = h_1, m_2 - m_2' = h_2$ in sum 8 and $m_1 - m_1' = h_1, m_2 - m_2' = -h_2$ in sum 9, we obtain

$$(H_1H_2\,|S|)^2 \ll |D'|^2 H_1H_2 + |D'|\,H_1H_2\sum_{h_1=1}^{H_1-1}\sum_{h_2=0}^{H_2-1}|S_1| + |D'|\,H_1H_2\sum_{h_1=0}^{H_1-1}\sum_{h_2=0}^{H_2-1}|S_2|\,.$$

Therefore result (2.70) follows.

In applying Theorem 2.21, (2.70) provides sharper estimates than those by using (2.71) in most cases. But great difficulties arise depending on the special properties of the function f, so that each problem has to be considered on its own merits. Therefore in the forthcoming general theorems we use only (2.71). We begin with a theorem being useful for three-dimensional lattice point problems.

Theorem 2.22. *Let u_1, u_2 be fixed positive numbers and $1 \leqq a_1 < b_1 \leqq a_1 u_1$, $1 \leqq a_2 < b_2 \leqq a_2 u_2$, $k \leqq 2$, $K = 2^k$. Suppose that*

$$\left|\frac{\partial^\nu f(t_1, t_2)}{\partial t_1^\nu}\right| \asymp \frac{\lambda}{a_1^\nu} \qquad (\nu = 1, 2, \dots, k),$$

$$\left|\frac{\partial^k f(t_1, t_2)}{\partial t_1^{k-2}\, \partial t_2^2}\right| \asymp \frac{\lambda}{a_1^{k-2} a_2^2}, \qquad \left|\frac{\partial^k f(t_1, t_2)}{\partial t^{k-1}\, \partial t_2}\right| \ll \frac{\lambda}{a^{k-1} a_2},$$

$$\left| H\left(\int_0^1 \dots \int_0^1 \frac{\partial^{k-2} f(t_1', t_2)}{\partial t_1^{k-2}}\, \mathrm{d}\tau_1 \dots \mathrm{d}\tau_{k-2}\right)\right| \gg \frac{\lambda^2}{a_1^{2k-2} a_2^2} \prod_{\nu=1}^{k-2} \frac{1}{h_\nu^2}$$

with $t_1' = t_1$ for $k = 2$ and

$$t_1' = t_1 + h_1\tau_1 + \dots + h_{k-2}\tau_{k-2}, \quad 1 \leqq h_1, h_2, \dots, h_{k-2} \leqq c_1,$$

for $k > 2$. Let $t_2 =$ const or $t_1 = \varrho(t_2)$ for all parts of the curve of boundary, where $\varrho(t_2)$ is twice differentiable with $|\varrho''(t_2)| \ll a_1/a_2^2$. If $\lambda \gg a_1, a_2$, then

$$\sum_{(n_1, n_2) \in D} \mathrm{e}^{2\pi i f(n_1, n_2)} \ll \left(a_1^{3K-4k-4} a_2^{3K-12} \lambda^4\right)^{\frac{1}{3K-8}} \log \lambda \tag{2.72}$$

and

$$\sum_{(n_1, n_2) \in D} \psi(f(n_1, n_2)) \ll \left(a_1^{3K-4k} a_2^{3K-8} \lambda^4\right)^{\frac{1}{3K-4}} \log \lambda\,. \tag{2.73}$$

Proof. We begin with the proof of (2.72). The case $k = 2$ is contained in Theorem 2.16. Putting there

$$\lambda_1 = \frac{\lambda}{a_1^2}, \quad \lambda_2 = \frac{\lambda}{a_2^2}, \qquad r = \frac{a_1}{a_2^2},$$

then (2.56) becomes

$$S = \sum_{(n_1, n_2) \in D} \mathrm{e}^{2\pi i f(n_1, n_2)} \ll \left(\lambda + a_1 + a_2 + \frac{a_1 a_2}{\lambda}\right)(1 + \log(a_1 a_2) + \log \lambda)\,. \tag{2.74}$$

Clearly, this is correct without the condition that $\lambda \gg a_1, a_2$. Only $\lambda \geqq 1$ is required. But if this additional condition is satisfied, (2.72) follows for $k = 2$ at once.

Now auppose that (2.72) is true for $k-1$. We apply Theorem 2.21 with $H_1 = H$, $H_2 = 1$. In applying (2.72) with $k-1$ instead of k, we have to replace the number λ by $h\lambda/a_1$ because of

$$f(t_1+h, t_2) - f(t_1, t_2) = h \int_0^1 f_{t_1}(t_1 + h\tau, t_2)\, d\tau .$$

From (2.71) we obtain

$$S \ll \frac{a_1 a_2}{\sqrt{H}} + \left\{ \frac{a_1 a_2}{H} \sum_{h=1}^{H-1} \left| \sum e^{2\pi i(f(n_1+h,\, n_2) - f(n_1,\, n_2))} \right| \right\}^{1/2},$$

$$SC(\Sigma):\quad (n_1, n_2) \in D\,, \quad (n_1 + h, n_2) \in D\,.$$

We divide the sum over h into three parts such that

$$S \ll \frac{a_1 a_2}{\sqrt{H}} + S_1 + S_2 + S_3\,,$$

and the terms S_j ($j = 1, 2, 3$) are defined by

$$S_j = \left\{ \frac{a_1 a_2}{H} \sum\nolimits_j \left| \sum e^{2\pi i(f(n_1+h,\, n_2) - f(n_1,\, n_2))} \right| \right\}^{1/2},$$

$$SC(\Sigma_1):\quad \lambda h/a_1 \leqq a_1, a_2\,,$$

$$SC(\Sigma_2):\quad \lambda h/a_1 > a_1, a_2\,,$$

$$SC(\Sigma_3):\quad \lambda h/a_1 \text{ between } a_1 \text{ and } a_2\,.$$

Using (2.74) in the first sum, then

$$S_1 \ll \left\{ \frac{a_1 a_2}{H} \sum\nolimits_1 \frac{a_1^2 a_2}{h} \log \lambda \right\}^{1/2} \ll \frac{a_1 a_2}{\sqrt{H}} \log \lambda\,.$$

We apply (2.74) in the third sum as well. If $a_1 \leqq a_2$, then

$$S_3 \ll \left\{ \frac{a_1 a_2}{H} \sum\nolimits_3 a_2 \log \lambda \right\}^{1/2} \ll \left\{ \frac{(a_1 a_2)^2}{H} \frac{a_2}{\lambda} \log \lambda \right\}^{1/2} \ll \frac{a_1 a_2}{\sqrt{H}} \log \lambda\,.$$

If $a_2 \leqq a_1$, then

$$S_3 \ll \left\{ \frac{a_1 a_2}{H} \sum\nolimits_3 a_1 \log \lambda \right\}^{1/2} \ll \left\{ \frac{(a_1 a_2)^2}{H} \frac{a_1^2}{a_2 \lambda} \log \lambda \right\}^{1/2}$$

$$\ll \frac{a_1 a_2}{\sqrt{H}} \frac{a_1}{\sqrt{a_2 \lambda}} \log \lambda\,.$$

In the second sum we apply (2.72) with $k-1$ for k:

$$S_2 \ll \left\{ \frac{(a_1 a_2)^2}{H} \sum_{h=1}^{H} \left(\frac{\lambda h}{a_1^{k-1} a_2} \right)^{\frac{8}{3K-16}} \right\}^{1/2} \ll a_1 a_2 \left(\frac{\lambda H}{a_1^{k-1} a_2} \right)^{\frac{4}{3K-16}}.$$

Altogether we get

$$S \ll a_1 a_2 \left\{ \frac{1}{\sqrt{H}} \left(1 + \frac{a_1}{\sqrt{a_2 \lambda}}\right) + \left(\frac{\lambda H}{a_1^{k-1} a_2}\right)^{\frac{4}{3K-16}} \right\} \log \lambda .$$

If $a_2 \lambda \gg a_1^2$, the terms are of the same order by putting

$$H = \left[\left(\frac{a_1^{k-1} a_2}{\lambda} \right)^{\frac{8}{3K-8}} \right].$$

Thus (2.72) follows, provided that $1 \leqq H \leqq c_1$. Clearly, for $\lambda > a_1^{k-1} a_2$ the estimate (2.72) is trivial. Also, if $H > c_1$, by a trivial estimate we find

$$S \ll c_1 c_2 \ll H a_2 \ll (a_1^{8k-8} a_2^{3K} \lambda^{-8})^{\frac{1}{3K-8}} \ll (a_1^{3K-4K-4} a_2^{3K-12} \lambda^4)^{\frac{1}{3K-8}},$$

provided that $k \geqq 4$. If $k = 3$, we use Theorem 2.7. Putting $k = 2$ (this is another k) there, we obtain from (2.15)

$$S = \sum_{\substack{n_1 \, n_2 \\ (n_1, n_2) \in D}} \sum e^{2\pi i f(n_1, n_2)} \ll \sum_{a_1 \leqq n_1 \leqq b_1} \sqrt{\lambda} \ll c_1 \sqrt{\lambda} \ll H \sqrt{\lambda} \ll (a_1^2 a_2)^{1/2}$$
$$\ll (a_1^2 a_2^3 \lambda)^{1/4},$$

since $a_1^2 \ll a_2 \lambda$. Hence, (2.75) again follows.

In case $\lambda \ll a_1^2 / a_2$ and $a_2 \leqq a_1$, we apply Theorem 2.7 with $k - 1$ for k in (2.15) and $k \geqq 3$. Using the theorem with respect to the variable n_1 in S, we obtain

$$S \ll \sum_{a_2 \leqq n_2 \leqq b_2} a_1 \left(\frac{\lambda}{a_1^{k-1}} \right)^{\frac{2}{K-4}} \ll a_1 a_2 \left(\frac{\lambda}{a_1^{k-1}} \right)^{\frac{2}{K-4}}.$$

Because of $a_2^{K-8} \ll a_1^{(k-3)K}$, we get $\lambda^K \ll (a_1^2 a_2^{-1})^K \ll a_1^{(k-1)K} a_2^{8-2K}$ and therefore

$$\left(\frac{\lambda}{a_1^{k-1}} \right)^{3K-8} \ll \left(\frac{\lambda}{a_1^{k-1} a_2} \right)^{2K-8},$$

which leads to

$$S \ll a_1 a_2 \left(\frac{\lambda}{a_1^{k-1} a_2} \right)^{\frac{4}{3K-8}}.$$

This completes the proof of (2.72).

In order to prove (2.73), we use (1.18) and (2.72). Then

$$\sum_{(n_1, n_2) \in D} \psi(f(n_1, n_2)) \ll \frac{a_1 a_2}{z} + \sum_{\nu=1}^{\infty} \min\left(\frac{z^2}{\nu^3}, \frac{1}{\nu} \right) a_1 a_2 \left(\frac{\lambda \nu}{a_1^{k-1} a_2} \right)^{\frac{4}{3K-8}} \log(\lambda \nu)$$

$$\ll \frac{a_1 a_2}{z} + a_1 a_2 \left(\frac{\lambda z}{a_1^{k-1} a_2} \right)^{\frac{4}{3K-8}} \log(\lambda z).$$

Putting

$$z = \left(\frac{a_1^{k-1} a_2}{\lambda}\right)^{\frac{4}{3K-4}}$$

estimate (2.73) follows at once.

In the last theorem the main point was the proof of estimate (2.72) of the exponential sum. The easily deduced result (2.73) enables us to estimate the error terms of three-dimensional lattice point problems. But it is also possible to use (2.72) for two-dimensional problems. For this end we need only the special case $k = 5$. This leads to the following theorem.

Theorem 2.23. *Let u be a fixed positive number and $1 \leqq a < b \leqq au$. Let $f(t)$ be real in $[a, b]$ with continuous derivatives up to the fifth order. Suppose that*

$$|f^{(\nu)}(t)| \asymp \frac{\lambda}{a^\nu} \qquad (\nu = 1, 2, 3, 4).$$

Let the function $\varphi(\tau)$ be defined by $f'(\varphi(\tau)) = \tau$. Suppose that

$$|\varphi^{(\nu)}(\tau)| \asymp \lambda \left(\frac{a}{\lambda}\right)^{\nu+1} \qquad (\nu = 3, 4).$$

Let $\alpha = \min f'(t)$, $\beta = \max f'(t)$, $t_2 \geqq 1$, $\alpha t_2 \leqq t_1 \leqq \beta t_2$. Suppose that

$$t_2^4 \left| \frac{\partial^2}{\partial t_2^2} \frac{1}{t_2^2} \varphi''\left(\frac{t_1}{t_2}\right) \right| \asymp \frac{a^3}{\lambda^2}$$

and

$$t_2^8 \left| H\left(\int_0^1\int_0^1\int_0^1 t_2^{-2} \varphi''\left(\frac{t_1'}{t_2}\right) d\tau_1\, d\tau_2\, d\tau_3 \right)\right| \gg \frac{a^8}{\lambda^6}$$

with

$$t_1' = t_1 + h_1\tau_1 + h_2\tau_2 + h_3\tau_3\,, \quad 1 \leqq h_1, h_2, h_3 \leqq (\beta - \alpha)\, t_2\,.$$

If $\lambda \gg a$, then

$$\sum_{a<n\leqq b} \psi(f(n)) \ll (a^{11}\lambda^8)^{1/29} \log \lambda\,. \tag{2.75}$$

Proof. In view of Theorem 1.8 we consider the sum

$$S = \sum_{n_2=1}^{\infty} c_{n_2} \sum_{a<n\leqq b} e^{2\pi i n_2 f(n)}\,,$$

where

$$c_t = \frac{z^2}{8\pi^3\, i t^3} (e^{-2\pi i t/z} - 1)^2 \ll \min\left(\frac{z^2}{t^3}, \frac{1}{t}\right).$$

For the inner sum we apply Theorem 2.9. Because of the conditions of our theorem we can use the error term (2.29). Let $f_1(t_1, t_2)$ denote the function

$$f_1(t_1, t_2) = t_2 f(\varphi(t_1/t_2)) - t_1\varphi(t_1/t_2)\,.$$

Then we obtain from (2.27)

$$S = \sum_{n_2=1}^{\infty} c_{n_2}\left\{\varepsilon \sum_{\alpha n_2 < n_1 \leqq \beta n_2} \frac{1}{\sqrt{n_2\,|f''(\varphi(n_1/n_2))|}}\, e^{2\pi i f_1(n_1, n_2)} + O\left(\frac{a}{\sqrt{n_2\lambda}}\right) + O((n_2\lambda)^{1/3})\right\}$$

$$= \varepsilon \sum_{n_2=1}^{\infty} \sum_{\alpha n_2 \leqq n_1 \leqq \beta n_2} \frac{c_{n_2}}{\sqrt{n_2\,|f''(\varphi(n_1/n_2))|}}\, e^{2\pi i f_1(n_1, n_2)} + O\left(\frac{a}{\sqrt{\lambda}}\right) + O((z\lambda)^{1/3})\,. \tag{2.76}$$

Let D denote the domain

$$D = \{(t_1, t_2)\colon \alpha t_2 \leqq t_1 \leqq \beta t_2,\ a_2 \leqq t_2 \leqq 2a_2\}\,.$$

Then we consider the sum

$$R = \sum_{(n_1, n_2) \in D} e^{2\pi i f_1(n_1, n_2)}\,,$$

and we write R_j, if the restriction $n_j \leqq t_j$ $(j = 1, 2)$ is given, and R_{12}, if $n_1 \leqq t_1$, $n_2 \leqq t_2$. We estimate the sum R_{12} with arbitrary, but suitable values t_1, t_2 by means of Theorem 2.22. There we take $k = 5$ and because of our domain D we have $a_1 \asymp \lambda a_2/a$. Since

$$\frac{\partial}{\partial t_1} f_1(t_1, t_2) = -\varphi(t_1/t_2)$$

and since $\varphi(t_1/t_2) \asymp a$, we see that in Theorem 2.22 λ has to be replaced by λa_2. The condition

$$\left|\frac{\partial^\nu f_1(t_1, t_2)}{\partial t_1^\nu}\right| \asymp \lambda a_2 \left(\frac{a}{\lambda a_2}\right)^\nu$$

is satisfied for $\nu = 1$ by the preceding remark, for $\nu = 2, 3$ because of

$$\varphi'(\tau) = \frac{1}{f''(\varphi)}, \quad \varphi''(\tau) = -\frac{f'''(\varphi)}{f''^3(\varphi)},$$

and for $v = 4, 5$ because of the assumption of our theorem. In the same way it can be seen that the other conditions of Theorem 2.22 are also fulfilled. For the curve of boundary $t_1 = \varrho(t_2)$ we apparently have $\varrho''(t_2) = 0$. Then (2.72) gives

$$R_{12} \ll \left(\left(\frac{\lambda a_2}{a}\right)^{72} a_2^{84}(\lambda a_2)^4\right)^{1/88} \log(\lambda a_2) = (a^{-18} a_2^{40} \lambda^{19})^{1/22} \log(\lambda a_2). \quad (2.77)$$

In order to estimate the sum

$$H = \sum_{(n_1, n_2) \in D} h(n_1, n_2)\, e^{2\pi i f_1(n_1, n_2)}$$

with

$$h(n_1, n_2) = \frac{c_{n_2}}{\sqrt{n_2 |f''(\varphi(n_1/n_2))|}},$$

we use partial summation iterating the one-dimensional formula (1.5). Then, by summing over n_1,

$$H = \sum_{(n_1, n_2) \in D} h(\beta n_2, n_2)\, e^{2\pi i f_1(n_1, n_2)} - \int \Sigma_1\, h_{t_1}(t_1, n_2)\, e^{2\pi i f_1(n_1, n_2)}\, dt_1,$$

$$SC(\textstyle\int \Sigma_1):\quad (n_1, n_2) \in D, \quad n_1 \leqq t_1, \qquad (t_1, n_2) \in D.$$

Now, by summing over n_2, we obtain

$$H = h(2\beta a_2, 2a_2)\, R - \int_{a_2}^{2a_2} \left(\frac{d}{dt_2} h(\beta t_2, t_2)\right) R_2\, dt_2$$

$$- \int_{\alpha a_2}^{2\beta a_2} h_{t_1}(t_1, N_2)\, R_1\, dt_1 + \iint_{(t_1, t_2) \in D} h_{t_1 t_2}(t_1, t_2)\, R_{12}\, dt_1\, dt_2$$

with $N_2 = \min(t_1/\alpha, 2a_2)$. Clearly, for R, R_1, R_2, R_{12} we have the estimate (2.77). It is easily seen that

$$h(2\beta a_2, 2a_2) \ll \frac{a}{\sqrt{a_2 \lambda}} \min\left(\frac{z^2}{a_2^3}, \frac{1}{a_2}\right),$$

$$h_{t_1}(t_1, N_2) \ll \frac{a}{a_2 \lambda} \cdot \frac{a}{\sqrt{a_2 \lambda}} \min\left(\frac{z^2}{a_2^3}, \frac{1}{a_2}\right),$$

$$h_{t_1 t_2}(t_1, t_2) \ll \frac{a}{a_2^2 \lambda} \cdot \frac{a}{\sqrt{a_2 \lambda}} \min\left(\frac{z^2}{a_2^3}, \frac{1}{a_2}\right).$$

In order to estimate $\mathrm{d}h(\beta t_2, t_2)/\mathrm{d}t_2$, we remark that

$$c_t' = \frac{-3z^2}{8\pi^3 \mathrm{i} t^4} (\mathrm{e}^{-2\pi \mathrm{i}t/z} - 1)^2 - \frac{z}{2\pi^2 t^3} (\mathrm{e}^{-2\pi \mathrm{i}t/z} - 1)\, \mathrm{e}^{-2\pi \mathrm{i}t/z}$$

$$\ll \min\left(\frac{z^2}{t^4}, \frac{1}{t^2}\right) + \min\left(\frac{z}{t^3}, \frac{1}{t^2}\right) \ll \min\left(\frac{z}{t^3}, \frac{1}{t^2}\right).$$

Thus

$$\frac{\mathrm{d}}{\mathrm{d}t_2} h(\beta t_2, t_2) \ll \frac{1}{a_2} \cdot \frac{a}{\sqrt{a_2 \lambda}} \min\left(\frac{z}{a_2^2}, \frac{1}{a_2}\right).$$

Hence

$$H \ll \frac{a}{\sqrt{a_2 \lambda}} \min\left(\frac{z}{a_2^2}, \frac{1}{a_2}\right) (a^{-18} a_2^{40} \lambda^{19})^{1/22} \log(\lambda a_2)$$

$$\ll \min\left(\frac{z}{a_2}, 1\right) (a^4 a_2^7 \lambda^8)^{1/22} \log(\lambda a_2).$$

Dividing the sum over n_2 in (2.76) into parts $2^{r-1} \leqq n_2 < 2^r$, $r = 1, 2, \ldots$, we obtain

$$S \ll \sum_{r=1}^{\infty} \min(z2^{-r}, 1)(a^4 2^{7r} \lambda^8)^{1/22} (r + \log \lambda) + \frac{a}{\sqrt{\lambda}} + (z\lambda)^{1/3}$$

$$\ll (a^4 z^7 \lambda^8)^{1/22} (\log z + \log \lambda) + \frac{a}{\sqrt{\lambda}} + (z\lambda)^{1/3}.$$

Therefore, by (1.17),

$$\sum_{a<n\leqq b} \psi(f(n)) \ll \frac{a}{z} + (a^4 z^7 \lambda^8)^{1/22} (\log z + \log \lambda) + \frac{a}{\sqrt{\lambda}} + (z\lambda)^{1/3}.$$

The first two terms are of the same order apart from the logarithm if we put $z = (a^9 \lambda^{-4})^{2/29}$. Then

$$\sum_{a<n\leqq b} \psi(f(n)) \ll (a^{11} \lambda^8)^{1/29} \log \lambda + \frac{a}{\sqrt{\lambda}} + (a^6 \lambda^7)^{1/29},$$

and estimate (2.75) follows.

2.2.4. *Transformation of double exponential sums*

The proof of the following theorem concerning the transformation of double exponential sums is quite similar to the proof of Theorem 2.18. Having used there the sharp transformation formula of Theorem 2.11 once and then van der Corput's Theorem 2.1, we now apply the transformation formula twice.

Theorem 2.24. *Suppose that*

$$\alpha_j \leqq f_{t_j}(t_1, t_2) \leqq \beta_j, \quad \gamma_j = \beta_j - \alpha_j \qquad (j = 1, 2),$$

$$|f_{t_1 t_1}(t_1, t_2)| \asymp \lambda_1, \quad |H(f)| \asymp \Lambda$$

and either

$$0 < |f_{t_1 t_1 t_1}| \ll \lambda' \ll \frac{\lambda_1}{c_1}$$

or

$$|f_{t_1 t_1 t_1}(t_1, t_2)| \asymp L_1(t_2), \quad L_1(t_2) \ll \lambda',$$

where $L_1(t_2)$ may depend on t_2, and

$$|g(t_1, t_2)| \ll G, \quad |g_{t_1}(t_1, t_2) \ll G_1.$$

For each point $(a, b) \in D$ the function $F(t_1, a; f(t_1, b))$, defined by (2.37), possesses only a bounded number of points of zero. Assume the function $g_{t_1}(t_1, t_2)$ to be monotonic with respect to t_1. Let $t_2 = \text{const}$ or $t_1 = \varrho(t_2)$ for all parts of the curve of boundary, where $\varrho(t)$ is twice continuously differentiable, and let $T(a, b)$ be defined by (2.61).

We denote the image of D under the mapping

$$y_1 = f_{t_1}(t_1, t_2), \quad y_2 = t_2$$

by D_1. Let the functions $\varphi_1(y_1, y_2), f_1(y_1, y_2)$ be defined by

$$f_{t_1}(\varphi_1, y_2) = y_1, \quad f_1(y_1, y_2) = f(\varphi_1(y_1, y_2), y_2) - y_1\varphi_1(y_1, y_2).$$

Suppose that either

$$0 < \left|\frac{\partial^3}{\partial y_2^3} f_1(y_1, y_2)\right| \ll \lambda'' \ll \frac{\Lambda}{c_2\lambda_1}$$

or

$$\left|\frac{\partial^3}{\partial y_2^3} f_1(y_1, y_2)\right| \asymp L_2(y_1), \qquad L_2(y_1) \ll \lambda'',$$

where $L_2(y_1)$ may depend on y_1, and

$$\frac{\partial}{\partial y_2} \frac{g(\varphi_1, y_2)}{\sqrt{|f_{t_1 t_1}(\varphi_1, y_2)|}} \ll G_2.$$

For each point $(a, b) \in D$ the function $F(y_2, b; f_1(a, y_2))$ possesses only a bounded number of points of zero. Let the function

$$\frac{\partial}{\partial y_2} \frac{g(\varphi_1, y_2)}{\sqrt{|f_{t_1 t_1}(\varphi_1, y_2)|}}$$

be monotonic with respect to y_2. *For all parts of the curve of boundary of* D_1 *we set* $y_2 = \text{const}$ *or* $y_1 = \sigma(y_2)$, *where* $\sigma(y)$ *is twice continuously differentiable. Suppose that* D_1 *satisfies condition* (C) *in the beginning of Section* 2.2.

We denote the image of D under the mapping

$$y_1 = f_{t_1}(t_1, t_2)\,, \quad y_2 = f_{t_2}(t_1, t_2)$$

by D_2. *Let the functions* $\varphi_j(y_1, y_2)$ $(j = 1, 2)$, $f_2(y_1, y_2)$ *be defined by*

$$f_{t_1}(\varphi_1, \varphi_2) = y_1\,, \quad f_{t_2}(\varphi_1, \varphi_2) = y_2\,,$$
$$f_2(y_1, y_2) = f(\varphi_1, \varphi_2) - y_1\varphi_1(y_1, y_2) - y_2\varphi_2(y_1, y_2)\,.$$

Finally let

$$\varepsilon = \begin{cases} \mathrm{i} & \text{for} \quad H(f) > 0\,, \quad f_{t_1t_1} > 0\,, \\ -\mathrm{i} & \text{for} \quad H(f) > 0\,, \quad f_{t_1t_1} < 0\,, \\ 1 & \text{for} \quad H(f) < 0\,. \end{cases}$$

Then

$$\sum_{(n_1, n_2) \in D} g(n_1, n_2)\, \mathrm{e}^{2\pi \mathrm{i} f(n_1, n_2)}$$
$$= \varepsilon \sum_{(m_1, m_2) \in D_2} \frac{g(\varphi_1, \varphi_2)}{\sqrt{|H(f(\varphi_1, \varphi_2))|}}\, \mathrm{e}^{2\pi \mathrm{i} f_2(m_1, m_2)}$$
$$+ O(G\,\Sigma_1\, T(\varrho(n_2), n_2)) + O(G\Delta) + O(G_1\Delta_1) + O(G_2\Delta_2)\,, \tag{2.78}$$

$SC(\Sigma_1)$: n_2 *runs over integers with* $a_2 \leqq n_2 \leqq b_2$ *such that the points* (t_1, n_2) *lie on the bound with* $t_1 = \varrho(n_2)$,

$$\Delta = \frac{\gamma_1 + \gamma_2 + 1}{\sqrt{\Lambda}} + \frac{\gamma_1 + 1}{\sqrt{\lambda_1}}\{\lambda''\lambda_1^2\Lambda^{-2} + \log(c_2\Lambda\lambda_1^{-1} + 2)\}$$
$$+ c_2\{\lambda'\lambda_1^{-2} + \log(c_1\lambda_1 + 2)\}\,,$$

$$\Delta_1 = |D| + c_2\lambda_1^{-1}\,,$$
$$\Delta_2 = |D_1| + \gamma_1 + c_2 + (\gamma_1 + 1)\,\lambda_1\Lambda^{-1}\,.$$

Remark. If we assume that $f_{t_1}(\varrho(t), t)$ is partly an integer or a constant or

$$\left|\frac{\mathrm{d}}{\mathrm{d}t} f_{t_1}(\varrho(t), t)\right| \gg \sqrt{\Lambda}\,,$$

we obtain by (2.65) and the preceding remarks

$$\Sigma_1\, T(\varrho(n_2), n_2) \ll \frac{\gamma_1 + 1}{\sqrt{\lambda_1}} + \left(c_2 + \frac{1}{\sqrt{\Lambda}}\right) \log(c_2 + 1)\,. \tag{2.79}$$

Proof. We proceed similarly to the proof of Theorem 2.18 up to equation (2.63). Denoting the left-hand side of (2.78) by S, we obtain by means of (2.38)

$$S = e^{\pm\pi i/4} \sum_{\substack{m_1 \; n_2 \\ (m_1, n_2) \in D_1}} \sum{}'' \frac{g(\varphi_1, n_2)}{\sqrt{|f_{t_1 t_1}(\varphi_1, n_2)|}} e^{2\pi i f(m_1, n_2)}$$

$$+ G\{O(\Sigma_1\, T(\varrho(n_2), n_2)) + O(c_2(\lambda' \lambda_1^{-2} + \log(c_1\lambda_1 + 2)))\} + O(G_1 \Delta_1'). \tag{2.80}$$

If $t_1 = \varrho_1(t_2)$ denotes the lower and $t_1 = \varrho_2(t_2)$, the upper part of the curve of boundary, (2.38) shows that the remainder Δ_1' is given by

$$\Delta_1' = \sum_{n_2} \{\varrho_2(n_2) - \varrho_1(n_2) + 1/\lambda_1\} \ll \sum_{(n_1, n_2) \in D_1} 1 + c_2/\lambda_1 + c_2 \ll |D|$$

$$+ c_2/\lambda_1 = \Delta_1 .$$

Now we again apply the transformation formula (2.38) to the inner sum on the right-hand side of (2.80). Since we assume that the domain D_1 satisfies condition (C), its curve of boundary $y_1 = \sigma(y_2)$ is partly monotonic, and $y_1 =$ const is only possible at the lower or upper bounds of D_1. We omit the primes at the sum of the non-constant part and get an error term of order $G/\sqrt{\lambda_1} + G\gamma_2/\sqrt{\Lambda}$. Applying Theorem 2.11, we have to consider the function $f_1(y_1, y_2)$. We know that

$$\left| \frac{\partial^2}{\partial y_2^2} f_1 \right| = \left| \frac{H(f)}{f_{t_1 t_1}} \right| \asymp \frac{\Lambda}{\lambda_1} .$$

In (2.38) we may trivially estimate the values T by $\sqrt{\lambda_1/\Lambda}$. Let $\delta_1 = \delta_1(m_1)$ and $\delta_2 = \delta_2(m_1)$ denote the left-hand and right-hand bounds of D_1. Then

$$\sum_{\delta_1 \leqq n_2 \leqq \delta_2}'' \frac{g(\varphi_1, n_2)}{\sqrt{|f_{t_1 t_1}(\varphi_1, n_2)|}} e^{2\pi i f_1(m_1, n_2)}$$

$$= e^{\pm\pi i/4} \sum_{m_2} \frac{g(\varphi_1, \varphi_2)}{\sqrt{|H(f(\varphi_1, \varphi_2))|}} e^{2\pi i f_2(m_1, m_2)}$$

$$+ O\left(\frac{G}{\sqrt{\lambda_1}} \{\lambda'' \lambda_1^2 \Lambda^{-2} + \log(c_2 \Lambda \lambda_1^{-1} + 2)\} \right)$$

$$+ O(G_2(\delta_2 - \delta_1 + \lambda_1 \Lambda^{-1})) + O\left(\frac{G}{\sqrt{\Lambda}} \right). \tag{2.81}$$

If we omit the primes at the sum over n_2 on such parts having the constant bound $y_1 = m_1$, we represent the sum as above. However, on the right-hand side of (2.81) there occurs an additional term which is minus the half of the sum over m_2. Estimating this sum trivially, we get an error term of order $(\gamma_2 + 1)\, G/\sqrt{\Lambda}$. Hence

$$\sum_{\substack{m_1\\(m_1,n_2)\in D_1}} \sum_{n_2}'' \frac{g(\varphi_1, n_2)}{\sqrt{|f_{t_1t_1}(\varphi_1, n_2)|}} e^{2\pi i f_1(m_1, n_2)}$$

$$= \varepsilon \sum_{(m_1, m_2)\in D_2} \frac{g(\varphi_1, \varphi_2)}{\sqrt{|H(f(\varphi_1, \varphi_2))|}} e^{2\pi i f_2(m_1, m_2)}$$

$$+ O\left(\frac{G(\gamma_1 + 1)}{\sqrt{\lambda_1}} \{\lambda'' \lambda_1^2 \Lambda^{-2} + \log (c_2 \Lambda \lambda_1^{-1} + 2)\}\right)$$

$$+ O(G_2 \Delta_2') + O\left(G \frac{\gamma_1 + \gamma_2 + 1}{\sqrt{\Lambda}}\right).$$

(2.38) shows that the remainder Δ_2' is given by

$$\Delta_2' = \sum_{m_1} \{\delta_2(m_1) - \delta_1(m_1) + \lambda_1 \Lambda^{-1}\}$$

$$\ll \sum_{(m_1, n_2)\in D_1} 1 + (\gamma_1 + 1)(\lambda_1 \Lambda^{-1} + 1)$$

$$\ll |D_1| + \gamma_1 + c_2 + (\gamma_1 + 1)\, \lambda_1 \Lambda^{-1} = \Delta_2 .$$

If we substitute the results into (2.80), we obtain formula (2.78) at once.

We now consider the estimation of exponential sums in the following way: We transform the exponential sum, and then we estimate the transformed sum by applying Weyl's steps. For the sake of simplicity we consider somewhat special domains.

Theorem 2.25. *Let u_1, u_2 be fixed positive numbers and $1 \leqq a_1 < b_1 \leqq u_1 a_1$, $1 \leqq a_2 < b_2 \leqq u_2 a_2$. Let $f(t_1, t_2)$ satisfy all the conditions of Theorem 2.24 with $c_1 \asymp a_1$, $c_2 \asymp a_2$, $\Lambda = \lambda_1 \lambda_2$, $\lambda_1 = \lambda a_1^{-2}$, $\lambda_2 = \lambda a_2^{-2}$, $\lambda' = \lambda a_1^{-3}$, $\lambda'' = \lambda a_2^{-3}$. Put $g(t_1, t_2) = 1$ such that $G = 1$, $G_1 = 0$ and suppose that $G_2 = a_1/a_2 \sqrt{\lambda}$. Let either $f_{t_1}(\varrho(t), t)$ be partly an integer or a constant or*

$$\left|\frac{d}{dt} f_{t_1}(\varrho(t), t)\right| \gg \frac{\lambda}{a_1 a_2}$$

on the bound of D. Let the function $f_2(y_1, y_2)$ satisfy all the conditions of Theorem 2.22 for $k = 3$ with the same λ and with λ/a_1, λ/a_2 instead of a_1, a_2 respectively in D_2'. Suppose that the functions $H_{y_1}(f(\varphi_1, \varphi_2))$, $H_{y_2}(f(\varphi_1, \varphi_2))$, $H_{y_1 y_2}(f(\varphi_1, \varphi_2))$ hold the sign fixed. If $\lambda \gg a_1, a_2$, then

$$\sum_{(n_1, n_2)\in D} e^{2\pi i f(n_1, n_2)} \ll \left\{\lambda (a_1^{k-1} a_2 \lambda^{1-k})^{\frac{4}{3K-8}} + a_2\right\} \log \lambda . \tag{2.82}$$

Proof. We first apply transformation formula (2.78). It can be seen that $\gamma_1 \ll \lambda/a_1$, $\gamma_2 \ll \lambda/a_2$. Hence

$$\begin{aligned}\Delta &\ll a_1 + a_2 + \sqrt{\lambda}(a_2\lambda^{-1} + \log\lambda) + a_2(a_1\lambda^{-1} + \log\lambda)\\ &\ll (a_1 + a_2 + \sqrt{\lambda})\log\lambda\,,\\ \Delta_2 &\ll a_2\lambda a_1^{-1} + a_2^2 a_1^{-1} \ll a_2\lambda a_1^{-1}\end{aligned}$$

and, by means of (2.65),

$$\Sigma_1\, T(\varrho(n_2), n_2) \ll (a_2 + \sqrt{\lambda})\log\lambda\,.$$

Then we obtain by (2.78)

$$\sum_{(n_1,n_2)\in D} e^{2\pi i f(n_1,n_2)} = \varepsilon \sum_{(m_1,m_2)\in D_2} \frac{1}{\sqrt{|H(f(\varphi_1,\varphi_2))|}}\, e^{2\pi i f_2(m_1,m_2)} + O((a_1 + a_2 + \sqrt{\lambda})\log\lambda)\,.$$

We consider the sum

$$S = \Sigma\, e^{2\pi\, i f_2(m_1,m_2)}$$

in any subdomain of D_2 with $m_1 \leqq t_1$, $m_2 \leqq t_2$. Such a region lies in a rectangle D_2' whose lengths of sides are of order λ/a_1 and λ/a_2. We can also say that (m_1, m_2) runs over the lattice points of D_2', where $e^{2\pi i f_2(m_1,m_2)}$ must be replaced by 0 if $(m_1, m_2) \notin D_2$. Then we obtain by (2.72)

$$S \ll \frac{\lambda^2}{a_1 a_2}(a_1^{k-1} a_2 \lambda^{1-k})^{\frac{4}{3K-8}} \log\lambda$$

and, by partial summation applying (1.16),

$$\sum_{(n_1,n_2)\in D} e^{2\pi i f(n_1,n_2)} \ll \left\{\lambda(a_1^{k-1}a_2\lambda^{1-k})^{\frac{4}{3K-8}} + a_1 + a_2 + \sqrt{\lambda}\right\}\log\lambda\,.$$

Obviously, the second term is less than the first. The last term is less than the first because of $k \geqq 3$. This gives the result (2.82).

Note that in (2.82) the second term is less than the first if we additionally assume that $a_1 \geqq a_2$. However, we apply Theorem 2.25 to three-dimensional lattice point problems only in the cases $k = 3, 4$. Here we may omit the second term without this restriction.

Theorem 2.26. *Let all the conditions of Theorem* 2.25 *be satisfied for* $k = 3$. *Then*

$$\sum_{(n_1,n_2)\in D} \psi(f(n_1,n_2)) \ll (a_1^4 a_2^3 \lambda^2)^{1/6} \log\lambda\,. \tag{2.83}$$

Proof. (2.82) shows that

$$\sum_{(n_1,n_2)\in D} e^{2\pi i f(n_1,n_2)} \ll \{(a_1^2 a_2 \lambda^2)^{1/4} + a_2\}\log\lambda \ll (a_1^2 a_2\lambda^2)^{1/4}\log\lambda\,,$$

provided that $a_1^2\lambda^2 \gg a_2^3$. Otherwise we use van der Corput's Theorem 2.1 with respect to n_2. Then

$$\sum_{(n_1, n_2)\in D} e^{2\pi i f(n_1, n_2)} \ll a_1 \sqrt{\lambda} \ll (a_1^2 a_2 \lambda^2)^{1/4} \log \lambda ,$$

since we deduce $a_1^2 \ll a_2$ from $a_1^2\lambda^2 \ll a_2^3$ and $a_2 \ll \lambda$. Thus, the estimate

$$\sum_{(n_1, n_2)\in D} e^{2\pi i f(n_1, n_2)} \ll (a_1^2 a_2 \lambda^2)^{1/4} \log \lambda$$

holds without restriction. Now (1.18) leads to

$$\sum_{(n_1, n_2)\in D} \psi(f(n_1, n_2)) \ll \frac{a_1 a_2}{z} + \sum_{\nu=1}^{\infty} \min\left(\frac{z}{\nu^2}, \frac{1}{\nu}\right) (a_1^2 a_2 \lambda^2 \nu^2)^{1/4} \log (\lambda\nu)$$

$$\ll \frac{a_1 a_2}{z} + (a_1^2 a_2 \lambda^2 z^2)^{1/4} \log (\lambda z) .$$

Choosing z such that $z = (a_1^2 a_2^3 \lambda^{-2})^{1/6}$, we obtain (2.83) at once.

Theorem 2.27. *Let all the conditions of Theorem 2.25 be satisfied for $k = 4$. Then*

$$\sum_{(n_1, n_2)\in D} \psi(f(n_1, n_2)) \ll (a_1^{10} a_2^8 \lambda^7)^{1/17} \log \lambda . \tag{2.84}$$

Proof. We get by (2.82) for $k = 4$

$$\sum_{(n_1, n_2)\in D} e^{2\pi i f(n_1, n_2)} \ll \{(a_1^3 a_2 \lambda^7)^{1/10} + a_2\} \log \lambda$$

and, by means of (1.18),

$$\sum_{(n_1, n_2)\in D} \psi(f(n_1, n_2))$$

$$\ll \frac{a_1 a_2}{z} + \sum_{\nu=1}^{\infty} \min\left(\frac{z}{\nu^2}, \frac{1}{\nu}\right) \{(a_1^3 a_2 \lambda^7 \nu^7)^{1/10} + a_2\} \log (\lambda\nu)$$

$$\ll \frac{a_1 a_2}{z} + \{(a_1^3 a_2 \lambda^7 z^7)^{1/10} + a_2 \log z\} \log (\lambda z) .$$

If we put $z = (a_1^7 a_2^9 \lambda^{-7})^{1/17}$, then

$$\sum_{(n_1, n_2)\in D} \psi(f(n_1, n_2)) \ll \{(a_1^{10} a_2^8 \lambda^7)^{1/17} + a_2 \log \lambda\} \log \lambda$$

$$\ll (a_1^{10} a_2^8 \lambda^7)^{1/17} \log \lambda ,$$

provided that $a_1^{10}\lambda^7 \gg a_2^9 \log^{17} \lambda$. Otherwise we apply van der Corput's Theorem 2.2 with respect to n_2. Then

$$\sum_{(n_1, n_2)\in D} \psi(f(n_1, n_2)) \ll a_1 (a_2 \lambda)^{1/3} .$$

Because of $a_1^{10}\lambda^7 \ll a_2^9 \log^{17} \lambda$ and $a_2 \ll \lambda$ we find

$$a_1(a_2\lambda)^{1/3} = (a_1^{17}a_2^{17/3}\lambda^{17/3})^{1/17} \ll (a_1^{10}a_2^{17/3+63/10}\lambda^{17/3-49/10})^{1/17} \log \lambda$$
$$\ll (a_1^{10}a_1^{172/30}\lambda^7)^{1/17} \log \lambda \ll (a_1^{10}a_2^8\lambda^7)^{1/17} \log \lambda .$$

Therefore (2.84) also follows in this case.

2.3. MULTIPLE EXPONENTIAL SUMS

The extension of the method of E. C. Titchmarsh to multiple exponential sums leads to enormous difficulties. But it is possible to proceed in the sense of Sections 2.2.2 and 2.2.4. We can transform the exponential sums and trivial estimates of the new sums or estimates after several Weyl's steps lead to non-trivial estimates of the original sums. G. Kolesnik developed a general transformation formula in a somewhat special case, which is applicable only for multiplicative problems. Here we give a very general form of this transform which can also be used for additive problems. But we do not develop a full theory of multiple exponential sums. For, the quality of the estimates decreases with increasing dimension. There it is often better to use methods depending on special functions and complex function theory.

Throughout Section 2.3 the following conditions are always assumed to be true.

(A) Let D be a bounded p-dimensional domain ($p \geqq 2$) with a volume $|D|$, where the number of lattice points are of order $|D|$.

(B) Suppose that D is a subset of the hyper-rectangle

$$D' = \{\boldsymbol{t}: \boldsymbol{t} = (t_1, t_2, \dots, t_p),\ a_j \leqq t_j \leqq b_j\ (j = 1, 2, \dots, p)\}$$

with $c_j = b_j - a_j \geqq 1$ $(j = 1, 2, \dots, p)$, $|D'| = c_1 c_2 \cdot \dots \cdot c_p$.

(C) Any straight line parallel to any of the coordinate axes intersects D in a bounded number of line segments. For the sake of simplicity we only consider such domains D where these straight lines intersect the boundary of D in at most two points or in one line segment. We can do this without loss of generality, because each such general domain can be divided into a finite number of these special domains.

(D) Let $f(\boldsymbol{t})$ and $g(\boldsymbol{t})$ be real functions in D' with continuous partial derivatives of as many orders as may be required. Suppose that the functions $f_{t_j}(\boldsymbol{t})$ are monotonic with respect to t_j $(j = 1, 2, \dots, p)$.

(E) Intersections of D with domains of the type $f_{t_j}(\boldsymbol{t}) \leqq c$ or $f_{t_j}(\boldsymbol{t}) \geqq c$ $(j = 1, 2, \dots, p)$ are to satisfy condition (C) as well.

(F) The boundary of D can be divided into a bounded number of parts. In each part the boundary is given by $t_j = r_j(t_{j+1}, \dots, t_p)$ $(j = 2, \dots, p-1)$ or $t_p = \text{const}$ or $t_1 = \varrho(t_2, t_3, \dots, t_p)$ with continuous partial derivatives of as many orders as may be required.

(G) Let D_j ($j = 1, 2, \ldots, p$) be the image of D under the mapping

$$y_\nu = f_{t_\nu}(\boldsymbol{t}) \quad \text{for} \quad \nu = 1, 2, \ldots, j,$$
$$y_\nu = t_\nu \quad \text{for} \quad \nu = j+1, j+2, \ldots, p.$$

Suppose that the number of lattice points of D_j is of order of the volume $|D_j|$. Furthermore, we put $D_0 = D$.

(H) Let E_μ ($\mu = 1, 2, \ldots, p$) denote the hyper-plane $y_\mu = c$. We consider the intersections $D_j \cap E_\mu$ ($j = 0, 1, \ldots, p$) for each constant c. Suppose that the greatest intersection is contained in a domain $R_{j,\mu}$ with

$$1 \ll |R_{j,\mu}| \ll |D_j|.$$

(I) Let the functions $\varphi_\nu(\boldsymbol{y})$ and $f_j(\boldsymbol{y})$, $\boldsymbol{y} = (y_1, y_2, \ldots, y_p)$, be defined by

$$f_{t_\nu}(\varphi_1, \ldots, \varphi_j, y_{j+1}, \ldots, y_p) = y_\nu \qquad (\nu = 1, 2, \ldots, j),$$

$$f_j(\boldsymbol{y}) = f(\varphi_1, \ldots, \varphi_j, y_{j+1}, \ldots, y_p) - \sum_{\nu=1}^{j} y_\nu \varphi_\nu(\boldsymbol{y}) \qquad (j = 1, 2, \ldots, p).$$

Furthermore, we put $f_0(\boldsymbol{y}) = f(\boldsymbol{y})$.

(K) Suppose that for each point $\gamma \in D_j$, $\gamma = (\gamma_1, \gamma_2, \ldots, \gamma_p)$, the functions

$$\left(\frac{\partial}{\partial y_{j+1}} f_j(\boldsymbol{y}_j) - \frac{\partial}{\partial y_{j+1}} f_j(\gamma)\right)^6$$
$$-8 \frac{\partial^2}{\partial y_{j+1}^2} f_j(\gamma) \left(\frac{\partial^2}{\partial y_{j+1}^2} f_j(\boldsymbol{y}_j)\right)^2 \left(f_j(\boldsymbol{y}_j) - f_j(\gamma) - (y_{j+1} - \gamma_{j+1}) \frac{\partial}{\partial y_{j+1}} f_j(\gamma)\right)^6,$$

where $\boldsymbol{y}_j = (\gamma_1, \ldots, \gamma_j, y_{j+1}, \gamma_{j+2}, \ldots, \gamma_p)$ and $j = 0, 1, \ldots, p-1$, have only a bounded number of points of zero. In particular, this condition is satisfied when $f(\boldsymbol{t})$ is an algebraic function.

(L) Let the Hessian $H_j(f)$ ($j = 1, 2, \ldots, p$) be defined by the functional determinant

$$H_j(f) = \frac{\partial(f_{t_1}, f_{t_2}, \ldots, f_{t_j})}{\partial(t_1, t_2, \ldots, t_j)} = \begin{vmatrix} f_{t_1t_1} & f_{t_1t_2} & \cdots & f_{t_1t_j} \\ f_{t_2t_1} & f_{t_2t_2} & \cdots & f_{t_2t_j} \\ \vdots & \vdots & & \vdots \\ f_{t_jt_1} & f_{t_jt_2} & \cdots & f_{t_jt_j} \end{vmatrix}.$$

Furthermore, we put $H_0(f) = 1$.

2.3.1. *Transformation of multiple exponential sums*

The following theorem concerning the transformation of multiple exponential sums is a direct generalization of Theorem 2.24 with some sharper, but unimportant conditions being assumed. Therefore, starting from (2.78), we prove the transformation formula (2.85) by induction, applying Theorem 2.11.

Theorem 2.28. *Let* $j = 1, 2, \ldots, p$. *Suppose that*

$$\Lambda_0 = 1\,, \quad |H_j(f)| \asymp \Lambda_j\,,$$

$$\left|\frac{\partial^3}{\partial y_j^3} f_{j-1}(\boldsymbol{y})\right| \asymp L_j\,, \qquad L_j \ll \lambda_j\,,$$

where L_j *may depend on* $y_1, \ldots, y_{j-1}, y_{j+1}, \ldots, y_p$. *Assume the functions*

$$g_j(\boldsymbol{y}) = \frac{\partial}{\partial y_j} \frac{g(\varphi_1, \ldots, \varphi_{j-1}, y_j, \ldots, y_p)}{\sqrt{|H_{j-1}(f(\varphi_1, \ldots, \varphi_{j-1}, y_j, \ldots, y_p))|}}$$

to be monotonic with respect to y_j, *and let*

$$|g(\boldsymbol{t})| \ll G\,, \quad |g_j(\boldsymbol{y})| \asymp G_j\,.$$

Furthermore, let $\boldsymbol{n} = (n_1, n_2, \ldots, n_p)$, $\boldsymbol{m} = (m_1, m_2, \ldots, m_p)$, $\boldsymbol{\varphi} = (\varphi_1, \varphi_2, \ldots, \varphi_p)$, $\varphi_j = \varphi_j(\boldsymbol{m})$, *and let* r_p *denote the number of changes of signs in the sequence* 1, $H_1(f), H_2(f), \ldots, H_p(f)$. *Then*

$$\sum_{\boldsymbol{n} \in D} g(\boldsymbol{n})\, \mathrm{e}^{2\pi i f(\boldsymbol{n})} = \mathrm{e}^{\frac{\pi i}{4}(p-2r_p)} \sum_{\boldsymbol{m} \in D_p} \frac{g(\boldsymbol{\varphi})}{\sqrt{|H_p(f(\boldsymbol{\varphi}))|}}\, \mathrm{e}^{2\pi i f_p(\boldsymbol{m})}$$
$$+ O(G \textstyle\sum_1 T(\varrho, n_2, \ldots, n_p)) + O(G\Delta)\,, \tag{2.85}$$

where

$$T(\varrho, n_2, \ldots, n_p) = \begin{cases} 0 & \text{for } f_{t_1} \in \mathbf{Z}\,, \\ \min\left(\dfrac{1}{\| f_{t_1}(\varrho, n_2, \ldots, n_p)\|}, \dfrac{1}{\sqrt{\Lambda_1}}\right) & \text{for } f_{t_1} \notin \mathbf{Z} \end{cases}$$

$SC(\Sigma_1)$: *Summation over all integers* $n_2, n_3, \ldots, n_p$ *such that the points* $(t_1, n_2, n_3, \ldots, n_p)$ *lie on the bound of* D *with* $t_1 = \varrho(n_2, n_3, \ldots, n_p)$,

$$\Delta = \frac{|R_{2,1}|}{\sqrt{\Lambda_2}} + \sum_{j=2}^{p} \frac{|R_{j-1,j}|}{\sqrt{\Lambda_j}} + \sum_{j=1}^{p} \frac{|R_{j-1,j}|}{\sqrt{\Lambda_{j-1}}} \left\{\frac{\lambda_j \Lambda_{j-1}^2}{\Lambda_j^2} + \log\left(\frac{c_j \Lambda_j}{\Lambda_{j-1}} + 2\right)\right\}.$$

Proof. For $p = 2$ we consider (2.78) under the sharper conditions (G), (H) and for $g_j(y)$. Then

$$G_1 \Delta_1 = G_1(|D| + c_2 \Lambda_1^{-1}) \ll G_1 |D| (1 + c_2 |D_1|^{-1})$$
$$\ll \iint_{(t_1, t_2) \in D} g_{t_1}(t_1, t_2)\, \mathrm{d}t_1\, \mathrm{d}t_2 \ll Gc_2 \ll G\Delta$$

and

$$G_2 \Delta_2 = G_2(|D_1| + \gamma_1 + c_2 + (\gamma_1 + 1)\, \Lambda_1 \Lambda_2^{-1}) \ll G_2 |D_1| (1 + |R_{2,2}|\, |D_2|^{-1})$$
$$\ll \iint_{(y_1, y_2) \in D_1} g_2(y_1, y_2)\, \mathrm{d}y_1\, \mathrm{d}y_2 \ll \frac{G|R_{1,2}|}{\sqrt{\Lambda_1}} \ll G\Delta\,.$$

This proves (2.85) in case of $p = 2$.

Now we use induction from $p - 1$ to p. If we write

$$\boldsymbol{n}' = (n_1, n_2, \dots, n_{p-1}), \qquad \boldsymbol{m}' = (m_1, m_2, \dots, m_{p-1}),$$
$$\boldsymbol{\varphi}' = (\varphi_1, \varphi_2, \dots, \varphi_{p-1}, n_p), \qquad \varphi_j = \varphi_j(\boldsymbol{m}', n_p),$$

we obtain

$$\sum_{\boldsymbol{n}=D} g(\boldsymbol{n})\, \mathrm{e}^{2\pi i f(\boldsymbol{n})} = \sum_{\substack{n_p \quad \boldsymbol{n}' \\ (\boldsymbol{n}', n_p) \in D}} \sum g(\boldsymbol{n}', n_p)\, \mathrm{e}^{2\pi i f(\boldsymbol{n}', n_p)}$$
$$= \mathrm{e}^{\frac{\pi i}{4}(p-1-2r_{p-1})} \sum_{\substack{n_p \quad \boldsymbol{m}' \\ (\boldsymbol{m}', n_p) \in D_{p-1}}} \sum \frac{g(\boldsymbol{\varphi}')}{\sqrt{|H_{p-1}(f(\boldsymbol{\varphi}'))|}}\, \mathrm{e}^{2\pi i f_{p-1}(\boldsymbol{m}', n_p)}$$
$$+ O(G \textstyle\sum_1 T(\varrho, n_1, \dots, n_2)) + O\Big(G \sum_{n_p} \Delta'\Big).$$

The remainder Δ' is given by the above value Δ with $p - 1$ for p. Here the $(p - 2)$-dimensional regions $R_{j-1,j}$ depend on n_p. Summing over n_p, we obtain under the conditions (G) and (H)

$$\sum_{\boldsymbol{n} \in D} g(\boldsymbol{n})\, \mathrm{e}^{2\pi i f(\boldsymbol{n})} = \mathrm{e}^{\frac{\pi i}{4}(p-1-2r_{p-1})} \sum_{\substack{\boldsymbol{m}' \quad n_p \\ (\boldsymbol{m}', n_p) \in D_{p-1}}} \sum \frac{g(\boldsymbol{\varphi}')}{\sqrt{|H_{p-1}(f(\boldsymbol{\varphi}'))|}}\, \mathrm{e}^{2\pi i f_{p-1}(\boldsymbol{m}', n_p)}$$
$$+ O\Big(G \sum_1 T(n)\Big) + O(G\Delta). \tag{2.86}$$

We now apply Theorem 2.11 with respect to the sum over n_p. Now writing $\boldsymbol{\varphi}' = (\varphi_1, \dots, \varphi_{p-1}, y_p)$, $\varphi_j = \varphi_j(\boldsymbol{m}', y_p)$, we have to consider the function

$$f_{p-1}(\boldsymbol{m}', y_p) = f(\boldsymbol{\varphi}') - \sum_{j=1}^{p-1} m_j \varphi_j.$$

The notation $f_{t_j}(\boldsymbol{\varphi}')$, in turn, means $\partial f(t)/\partial t_j$ at the point $t = \boldsymbol{\varphi}'$. Then we obtain

$$\frac{\partial}{\partial y_p} f_{p-1}(\boldsymbol{m}', y_p) = \sum_{j=1}^{p-1} f_{t_j}(\boldsymbol{\varphi}') \frac{\partial \varphi_j}{\partial y_p} + f_{t_p}(\boldsymbol{\varphi}') - \sum_{j=1}^{p-1} m_j \frac{\partial \varphi_j}{\partial y_p} = f_{t_p}(\boldsymbol{\varphi}').$$

From (I) it can be seen that $f_{t_j}(\boldsymbol{\varphi}') = y_j$ for $j = 1, 2, \dots, p - 1$. Hence

$$\sum_{\nu=1}^{p-1} f_{t_j t_\nu}(\boldsymbol{\varphi}') \frac{\partial \varphi_\nu}{\partial y_p} + f_{t_j t_p}(\boldsymbol{\varphi}') = 0.$$

This is a system of $p - 1$ linear equations for the unknowns $\partial \varphi_\nu / \partial y_p$. Therefore, by Cramer's rule,

$$\frac{\partial \varphi_\nu}{\partial y_p} = \frac{F_\nu}{H_{p-1}(f)}.$$

The determinant F_ν is obtained from the coefficient determinant $H_{p-1}(f)$ by replacing the ν-th column by the elements $-f_{t_j t_p}(\varphi')$. Thus

$$\frac{\partial^2}{\partial y_p^2} f_{p-1}(\boldsymbol{m}', y_p) = \sum_{j=1}^{p-1} f_{t_j t_p}(\varphi') \frac{\partial \varphi_j}{\partial y_p} + f_{t_p t_p}(\varphi')$$

$$= \frac{1}{H_{p-1}(f)} \left\{ \sum_{j=1}^{p-1} f_{t_j t_p}(\varphi') F_j + f_{t_p t_p}(\varphi') H_{p-1}(f) \right\} = \frac{H_p(f)}{H_{p-1}(f)}.$$

Applying Theorem 2.11 with respect to the sum over n_p and estimating the corresponding terms T trivially by $\sqrt{\Lambda_{p-1}/\Lambda_p}$, we obtain

$$\mathrm{e}^{\frac{\pi \mathrm{i}}{4}(p-1-2r_{p-1})} \sum_{\substack{m'}} \sum_{\substack{n_p \\ (m', n_p) \in D_{p-1}}} \frac{g(\varphi')}{\sqrt{|H_{p-1}(f(\varphi'))|}} \mathrm{e}^{2\pi \mathrm{i} f_{p-1}(m', n_p)}$$

$$= \mathrm{e}^{\frac{\pi \mathrm{i}}{4}(p-2r_p)} \sum_{\boldsymbol{m} \in D_p} \frac{g(\varphi)}{\sqrt{|H_p(f(\varphi))|}} \mathrm{e}^{2\pi \mathrm{i} f_p(\boldsymbol{m})} + O\left(G \frac{|R_{p-1,p}|}{\sqrt{\Lambda_p}}\right)$$

$$+ O\left(G \frac{|R_{p-1,p}|}{\sqrt{\Lambda_{p-1}}} \left\{ \frac{\lambda_p \Lambda_{p-1}^2}{\Lambda_p^2} + \log\left(\frac{c_p \Lambda_p}{\Lambda_{p-1}} + 2\right)\right\}\right)$$

$$+ O\left(G_p \left\{ |D_{p-1}| + |R_{p-1,p}| \frac{\Lambda_{p-1}}{\Lambda_p} \right\}\right).$$

We get for the last error term

$$G_p \left\{ |D_{p-1}| + |R_{p-1,p}| \frac{\Lambda_{p-1}}{\Lambda_p} \right\}$$

$$\ll G_p |D_{p-1}| \left\{ 1 + \frac{|R_{p-1,p}|}{|D_p|} \right\} = G_p |D_{p-1}| \left\{ 1 + \frac{|R_{p,p}|}{|D_p|} \right\}$$

$$\ll G_p |D_{p-1}| \ll \int_{y \in D_{p-1}} g_p(y)\, \mathrm{d}y \ll \frac{G |R_{p-1,p}|}{\sqrt{\Lambda_{p-1}}}.$$

Substituting the results into (2.86), formula (2.85) follows at once.

In the applications of Theorem 2.28 to lattice point problems in most cases the situation arises that at the boundary of D partly either $f_{t_1}(\varrho(t_2, \dots, t_p), t_2, \dots, t_p)$ is an integer or a constant not depending on the parameters of the problem or the first derivative with respect to t_2, has a fixed lower bound. We obtain the following results in the same way as in Section 2.2.2: If T_1 denotes a part of the sum

$$\Sigma_1\ T(\varrho(n_2, \dots, n_p), n_2, \dots, n_p),$$

where $f_{t_1}(\varrho(t_2, \dots, t_p), t_2, \dots, t_p)$ is an integer, then $T_1 = 0$. If t_2 denotes a part, where $f_{t_1}(\varrho(t_2, \dots, t_p), t_2, \dots, t_p) = \text{const}$, then $T_2 \ll |R_{0,1}|$. For a part T_3 with

$$\left| \frac{\partial}{\partial t_2} f_{t_1}(\varrho(t_2, \dots, t_p), t_2, \dots, t_p) \right| \gg q > 0$$

we apply Lemma 2.10 with $z = 1/\sqrt{\Lambda_1}$. Corresponding to (2.65), we get from (2.64)

$$T_3 \ll |R_{1,2}| \left(\frac{1}{\sqrt{\Lambda_1}} + \frac{\log (c_2 + 1)}{q} \right). \tag{2.87}$$

Note that in most cases we can take $q = \sqrt{\Lambda_2}$.

Finally, we consider the special case, where D is a subset of the hyper-rectangle D' with $1 \leqq a_j < b_j \leqq a_j u_j$ $(j = 1, 2, \dots, p)$, where the values u_j denote some fixed constants. We suppose that

$$\left| \frac{\partial^k f(t)}{\partial t_1^{\nu_1} \partial t_2^{\nu_2} \cdot \dots \cdot \partial t_p^{\nu_p}} \right| \asymp \frac{\lambda}{a_1^{\nu_1} a_2^{\nu_2} \cdot \dots \cdot a_p^{\nu_p}}$$

with $\nu_1 + \nu_2 + \dots + \nu_p = k$ $(k = 1, 2, 3)$ and

$$\Lambda_j = \frac{\lambda^j}{a_1^2 a_2^2 \cdot \dots \cdot a_j^2}, \quad \lambda_j = \frac{\lambda}{a_j^3}.$$

Further we assume that at the boundary of D partly either $f_{t_1}(\varrho, t_2, \dots, t_p)$ is an integer or a constant not depending on the parameters of the problem or

$$\left| \frac{\partial}{\partial t_2} f_{t_1}(\varrho, t_2, \dots, t_p) \right| \gg \frac{\lambda}{a_1 a_2}.$$

Let $\lambda \gg a_1, a_2, \dots, a_p$. Under these conditions we have in Theorem 2.28 that

$$|R_{2,1}| \asymp \frac{\lambda}{a_2} a_3 \cdot \dots \cdot a_p, \qquad |R_{j-1,}| \asymp \frac{a_{j+1} \cdot \dots \cdot a_p}{a_1 \cdot \dots \cdot a_{j-1}} \lambda^{j-1}$$

and by means of (2.87)

$$\sum\nolimits_1 T(\varrho, n_2, \dots, n_p) \ll |R_{0,1}| + |R_{1,2}| \left(\frac{1}{\sqrt{\Lambda_1}} + \frac{\log \lambda}{\sqrt{\Lambda_2}} \right)$$

$$\ll a_3 \cdot \dots \cdot a_p (\sqrt{\lambda} + a_2) \log \lambda .$$

We get for the remainder Δ

$$\Delta \ll a_1 a_3 \cdot \dots \cdot a_p + \sum_{j=2}^{p} a_j \cdot \dots \cdot a_p \lambda^{\frac{j}{2} - 1} + \sum_{j=1}^{p} a_{j+1} \cdot \dots \cdot a_p \lambda^{\frac{j-1}{2}} \left(\frac{a_j}{\lambda} + \log \lambda \right)$$

$$\ll a_1 a_3 \cdot \dots \cdot a_p + \sum_{j=1}^{p} a_{j+1} \cdot \dots \cdot a_p \lambda^{\frac{j-1}{2}} \log \lambda .$$

Consequently, we obtain the transformation formula

$$\sum_{n \in D} g(n)\, e^{2\pi i f(n)} = e^{\frac{\pi i}{4}(p-2r_p)} \sum_{m \in D_p} \frac{g(\varphi)}{\sqrt{|H_p(f(\varphi))|}}\, e^{2\pi i f_p(m)} + O(G\acute{a}_1 a_3 \cdot \ldots \cdot a_p)$$

$$+ \sum_{j=1}^{p} O\left(G a_{j+1} \cdot \ldots \cdot a_p \lambda^{\frac{j-1}{2}} \log \lambda\right). \tag{2.88}$$

2.3.2. *The basic estimates*

Estimating the sum on the right-hand side of the transformation formula (2.85) trivially, we obtain a good non-trivial estimate of the exponential sum on the left-hand side. In the most important applications to lattice point theory we can use the assumptions on the boundary of D mentioned in the preceding section. Hence, we formulate the following theorem.

Theorem 2.29. *Let $j = 1, 2, \ldots, p$. Suppose that*

$$\Lambda_0 = 1\,, \quad |H_j(f)| \asymp \Lambda_j\,,$$

$$\left|\frac{\partial^3}{\partial y_j^3} f_{j-1}(y)\right| \asymp L_j\,, \quad L_j \ll \lambda_j \ll \Lambda_j^2 \Lambda_{j-1}^{-2}\,,$$

where L_j may depend on $y_1, \ldots, y_{j-1}, y_{j+1}, \ldots, y_p$. Assume the functions

$$g_j(y) = \frac{\partial}{\partial y_j} |H_{j-1}(f(\varphi_1, \ldots, \varphi_{j-1}, y_j, \ldots, y_p))|^{-1/2}$$

to be monotonic with respect the y_j, and let

$$|g_j(y)| \asymp G_j\,.$$

Suppose that at the boundary of D partly either $f_{t_1}(\varrho(t_2, \ldots, t_p), t_2, \ldots, t_p)$ is an integer or a constant not depending on the parameters of the problem or

$$\left|\frac{\partial}{\partial t_2} f_{t_1}(\varrho(t_2, \ldots, t_p), t_2, \ldots, t_p)\right| \gg \sqrt{\Lambda_2}\,.$$

Then

$$\sum_{n \in D} e^{2\pi i f(n)} \ll |D| \sqrt{\Lambda_p} + \Delta_1\,, \tag{2.89}$$

$$\sum_{n \in D} \psi(f(n)) \ll |D|\, \Lambda_p^{\frac{1}{p+2}} + \Delta_2 \tag{2.90}$$

with

$$\Delta_1 = \frac{|R_{2,1}| + |R_{1,2}| \log(c_2+1)}{\sqrt{\Lambda_2}} + \sum_{j=3}^{p} \frac{|R_{j-1,j}|}{\sqrt{\Lambda_j}} + \sum_{j=1}^{p} \frac{|R_{j-1,j}|}{\sqrt{\Lambda_{j-1}}} \log\left(\frac{c_j\Lambda_j}{\Lambda_{j-1}} + 2\right),$$

$$\Delta_2 = \frac{|R_{2,1}| + |R_{1,2}| \log(c_2+1)}{\sqrt{\Lambda_2}} |\log \Lambda_p| + \sum_{j=3}^{p} \Lambda_p^{\frac{2-j}{2p+4}} \frac{|R_{j-1,j}|}{\sqrt{\Lambda_j}} + \sum_{j=1}^{p} \Lambda_p^{\frac{1-j}{2p+4}} \frac{|R_{j-1,j}|}{\sqrt{\Lambda_{j-1}}} \log\left(\Lambda_p^{-\frac{1}{p+2}} \frac{c_j\Lambda_j}{\Lambda_{j-1}} + 2\right) |\log \Lambda_p| .$$

Proof. Estimating the exponential sum on the right-hand side of (2.85) trivially, then we find

$$\sum_{m\in D_p} \frac{1}{\sqrt{|H_p(f(\varphi))|}} e^{2\pi i f_p(m)} \ll \frac{1}{\sqrt{\Lambda_p}} \sum_{m\in D_p} 1 \ll \frac{|D_p|}{\sqrt{\Lambda_p}} \ll |D| \sqrt{\Lambda_p} .$$

We have, by means of (2.87),

$$\textstyle\sum_1 T(\varrho, n_2, \ldots, n_p) \ll |R_{0,1}| + |R_{1,2}| \left(\frac{1}{\sqrt{\Lambda_1}} + \frac{\log(c_2+1)}{\sqrt{\Lambda_2}}\right).$$

Because of $\lambda_j \ll \Lambda_j^2 \Lambda_{j-1}^{-2}$ result (2.89) now follows from (2.85) at once.

In order to prove (2.90), we apply (1.18)

$$\sum_{n\in D} \psi(f(n)) \ll \frac{|D|}{z} + \sum_{\nu=1}^{\infty} \min\left(\frac{z^s}{\nu^{s+1}}, \frac{1}{\nu}\right) \left|\sum_{n\in D} e^{2\pi i \nu f(n)}\right|$$

with sufficiently large s. Applying (2.89), we have to replace $|R_{2,1}|$ by $\nu|R_{2,1}|$, $|R_{j-1,j}|$ by $\nu^{j-1}|R_{j-1,j}|$, and Λ_j by $\nu^j\Lambda_j$. Then

$$\sum_{n\in D} \psi(f(n)) \ll \frac{|D|}{z} + \sum_{\nu=1}^{\infty} \min\left(\frac{z^s}{\nu^{s+1}}, \frac{1}{\nu}\right) \times \left\{ |D| \sqrt{\nu^p \Lambda_p} + \frac{|R_{2,1}| + |R_{1,2}| \log(c_2+1)}{\sqrt{\Lambda_2}} + \sum_{j=3}^{p} \nu^{\frac{j}{2}-1} \frac{|R_{j-1,j}|}{\sqrt{\Lambda_j}} + \sum_{j=1}^{p} \nu^{\frac{j-1}{2}} \frac{|R_{j-1,j}|}{\sqrt{\Lambda_{j-1}}} \times \log\left(\frac{\nu c_j \Lambda_j}{\Lambda_{j-1}} + 2\right) \right\}$$

with $s > p/2$. Now

$$\sum_{n \in D} \psi(f(\mathbf{n})) \ll \frac{|D|}{z} + |D| \sqrt{z^p \Lambda_p} + \frac{|R_{2,1}| + |R_{1,2}| \log (c_2 + 1)}{\sqrt{\Lambda_2}} \log z$$

$$+ \sum_{j=3}^{p} z^{\frac{j}{2}-1} \frac{R_{j-1,j}}{\sqrt{\Lambda_j}} + \sum_{j=1}^{p} z^{\frac{j-1}{2}} \frac{|R_{j-1,j}|}{\sqrt{\Lambda_{j-1}}} \log \left(\frac{z c_j \Lambda_j}{\Lambda_{j-1}} + 2 \right) \log z \, .$$

If we put $z = \Lambda_p^{-1/(p+2)}$, the first two terms are of the same order and (2.91) follows.

Finally, we reformulate Theorem 2.29 for the case of multiplicative problems.

Theorem 2.30. *Let D be a subset of the hyper-rectangle D' with $1 \leqq a_j < b_j \leqq a_j u_j$ $(j = 1, 2, \ldots, p)$, where u_j are positive, fixed constants. Suppose that*

$$\left| \frac{\partial^k f(t)}{\partial t_1^{\nu_1} \partial t_2^{\nu_2} \cdot \ldots \cdot \partial t_p^{\nu_p}} \right| \asymp \frac{\lambda}{a_1^{\nu_1} a_2^{\nu_2} \cdot \ldots \cdot a_p^{\nu_p}}$$

with $\nu_1 + \nu_2 + \ldots + \nu_p = k$ $(k = 1, 2, 3)$ and

$$|H_j(f)| \asymp \Lambda_j = \frac{\lambda^j}{a_1^2 a_2^2 \cdot \ldots \cdot a_j^2}$$

for $j = 1, 2, \ldots, p$ throughout the hyper-rectangle D'. Furthermore, let

$$\left| \frac{\partial^3}{\partial y_j^3} f_{j-1}(\mathbf{y}) \right| \asymp L_j \, , \qquad L_j \ll \lambda_j = \frac{\lambda}{a_j^3} \, .$$

Assume the functions $g_j(\mathbf{y})$, defined in Theorem 2.29, to be monotonic with respect to y_j, and let $|g_j(\mathbf{y})| \asymp G_j$. Suppose that at the boundary of D partly either $f_{t_1}(\varrho(t_2, \ldots, t_p), t_2, \ldots, t_p)$ is an integer or a constant or

$$\left| \frac{\partial}{\partial t_2} f_{t_1}(\varrho(t_2, \ldots, t_p), t_2, \ldots, t_p) \right| \gg \frac{\lambda}{a_1 a_2} \, .$$

If $\lambda \gg a_1 \geqq a_2 \geqq \ldots \geqq a_p$, then

$$\sum_{n \in D} \psi(f(\mathbf{n})) \ll (a_1 a_2 \cdot \ldots \cdot a_p \lambda)^{\frac{p}{p+2}} \varepsilon_p \tag{2.91}$$

with $\varepsilon_2 = \log \lambda$ and $\varepsilon_p = 1$ for $p > 2$.

Proof. Since $\lambda_j \ll \Lambda_j^2 \Lambda_{j-1}^{-2}$ and

$$|R_{2,1}| \asymp \frac{\lambda}{a_2} a_3 \cdot \ldots \cdot a_p \, , \quad |R_{j-1,j}| \asymp \frac{a_{j+1} \cdot \ldots \cdot a_p}{a_1 \cdot \ldots \cdot a_{j-1}} \lambda^{j-1} \, ,$$

(2.90) becomes

$$\sum_{n \in D} \psi(f(\mathbf{n})) \ll (a_1 a_2 \cdot \ldots \cdot a_p \lambda)^{\frac{p}{p+2}} + a_1 a_3 \cdot \ldots \cdot a_p \log \lambda$$

$$+ \sum_{j=1}^{p} a_{j+1} \cdot \ldots \cdot a_p (a_1 \cdot \ldots \cdot a_p \lambda)^{\frac{j-1}{p+2}} \log^2 \lambda \, . \tag{2.92}$$

We first consider the case $p = 2$. Here we obtain

$$\sum_{n \in D} \psi(f(n)) \ll \sqrt{a_1 a_2 \lambda} + a_1 \log \lambda + a_2 \log^2 \lambda + (a_1 a_2 \lambda)^{1/4} \log^2 \lambda$$
$$\ll \sqrt{a_1 a_2 \lambda} \log \lambda ,$$

which proves (2.91) for $p = 2$.

If $p > 2$, the inequality

$$(a_1 a_2 \cdot \ldots \cdot a_p \lambda)^p \gg a_1^{p+2} (a_2 \cdot \ldots \cdot a_p)^p \lambda^{p-2}$$
$$\gg (a_1 a_3 \cdot \ldots \cdot a_p)^{p+2} \lambda \gg (a_1 a_3 \cdot \ldots \cdot a_p \log \lambda)^{p+2}$$

shows that the second term in (2.92) is less than the first one. Further, we obtain the inequality

$$(a_1 \cdot \ldots \cdot a_j \lambda)^{p-j+1} \geq a_j^{j(p-j)+j} \lambda^{p-j+1}$$
$$\geq (a_{j+1} \cdot \ldots \cdot a_p)^j \lambda^{p-j+1} \gg (a_{j+1} \cdot \ldots \cdot a_p)^{j+1} (\log \lambda)^{2p+4} .$$

Hence

$$(a_1 \cdot \ldots \cdot a_p \lambda)^p \gg (a_{j+1} \cdot \ldots \cdot a_p)^{p+2} (a_1 \cdot \ldots \cdot a_p \lambda)^{j-1} (\log \lambda)^{2p+4} ,$$

which shows that the other terms in (2.92) are also less than the first ones. This proves (2.91).

NOTES ON CHAPTER 2

Section 2.1

2.1.1. The basic estimates for the number of lattice points in plane domains, presented in Theorems 2.1—2.3, were proved by J. G. van der Corput [3], [7] in 1921 and 1923. V. Jarník remarked that Theorem 2.2 is the best of its kind (see E. Landau [18]). In 1917 I. M. Vinogradov developed an elementary method which gives the same results apart from a logarithmic factor. The method is described in I. M. Vinogradov [2], A. O. Gelfond and Y. V. Linnik [1], E. Krätzel [11].

2.1.2. Theorem 2.5 was obtain by H. Weyl [1] in 1916. Theorem 2.6 represents an inequality proved by E. C. Titchmarsh [5]. A similar inequality was given by J. G. van der Corput [10] in 1929. In general, Theorems 2.6 and 2.7 will be applied in connection with the van der Corput transform of Section 2.1.3. However, Theorem 2.8 may be used advantageously, if the lattice point problem is unsymmetric.

2.1.3. In Theorem 2.9 the approximate functional equation (2.27) represents the so-called van der Corput transform. It is due to J. G. van der Corput [5] in 1922 with the error term (2.28). The sharper error term (2.29) was obtained by E. Phillips [1] in 1933 and was important in his simplification of van der Corput's theory of exponent pairs. Lemma 2.5 and Theorem 2.11 are based on an idea of I. M. Vinogradov [2], [7] and were used in his investigations of the sphere problem. A somewhat weaker form of Theorem 2.11 is due to G. Kolesnik [4] in 1981, in this form it was stated by E. Krätzel [17] in 1982. The very sharp error term in the transformation formula (2.38) is required, since this transformation will be used in iterated manner in the investigations of multiple exponential sums.

2.1.4. The explanation of this section are based on the work of E. Phillips [1], 1933, and R. A. Rankin [1], 1955. See also E. C. Titchmarsh [1]. Substantially, Theorems 2.12 and 2.13 go back to J. G. van der Corput [5].

Recently M. N. Huxley and N. Watt [1] proved that $(9/56 + \varepsilon, 37/56 + \varepsilon)$ is an exponent pair. Then proof is based on a method of E. Bombieri and H. Iwaniec. Then

$$A\left(\frac{9}{56} + \varepsilon, \frac{37}{56} + \varepsilon\right) = \left(\frac{9}{130} + \varepsilon_1, \frac{102}{130} + \varepsilon_1\right),$$

$$ABA\left(\frac{9}{56} + \varepsilon, \frac{37}{56} + \varepsilon\right) = \left(\frac{37}{334} + \varepsilon_2, \frac{241}{334} + \varepsilon_2\right)$$

are also important exponent pairs.

Section 2.2

2.2.1. The argument of this section follows the lines of E. C. Titchmarsh [2], [3] and S.-H. Min [1]. Whereas E. C. Titchmarsh proved Lemma 2.6 for rectangles only the idea of considering the curve of boundary goes back to H.-E. Richert [1]. Lemma 2.6 in this form and Theorem 2.16 were stated by E. Krätzel [17]. A similar result was also obtained by W.-G. Nowak [6]. Lemma 2.7 is a refinement of a result of E. C. Titchmarsh [3]. This and Theorem 2.17 were proved by E. Krätzel [18].

2.2.2. The method of proof in this section is basically due to I. M. Vinogradov, as previously mentioned in the notes to Section 2.1.3. Similar results were obtained by G. Kolesnik [4], [6]. Theorems 2.18 and 2.19 in this form were proved by E. Krätzel [17].

2.2.3. Theorem 2.21 is a result of E. C. Titchmarsh [3], 1934. Theorem 2.22 is due to E. Krätzel [17], and Theorem 2.23 represents an unpublished result of the same authors.

2.2.4. The basic approach to the problem of transforming double as well as multiple exponential sums has been developed by G. Kolesnik [4], [6]. Whereas G. Kolesnik's transformation formula can be applied only to special problems, Theorem 2.24 represents a very universal result. All results of this section are based on a paper of E. Krätzel [17].

Section 2.3

All theorems of this section represent unpublished results of E. Krätzel. A special case of the transformation formula (2.85) was obtained by G. Kolesnik as mentioned in the notes to Section 2.2.4.

Finally we remark that B. R. Srinivasan [2] has considered a generalization of van der Corput's and Titchmarsh's method to multiple exponential sums. But the proofs are not entirely satisfactory in all details.

Chapter 3

Plane additive problems

In this chapter we are concerned with counting the number of lattice points whose curves of boundary are given by functions of the type $f(\xi) + f(\eta) = x$. Historically, the first curve which was investigated in this connection was the circle, as already mentioned in the introductory chapter. Let

$$R(x) = \# \{(n, m) : n^2 + m^2 \leq x\}$$

be the number of lattice points which lie on or inside the circle $\xi^2 + \eta^2 = x$. We known that $R(x)$ is represented by the area πx of the circle in the first approximation. Hence, if we denote the remainder by $\Delta(x)$, we write

$$R(x) = \pi x + \Delta(x) .$$

Then there arise three principal problems:

1. *In the first place there is the problem of finding the best possible estimate of the remainder* $\Delta(x)$. *This is called the "O-problem".*
2. *Secondly, there is the problem of proving that such an estimate of the remainder really is the best and that no more precise relation can be true. This is called the "*Ω*-problem".*
3. *Finally, there is the problem of finding an exact formula for the remainder* $\Delta(x)$.

Clearly, the first problem and the second one are the most important and doubtless the most difficult problems. As regards the circle there are very difficult open questions. The third problem is of little importance and can be completely solved for the circle.

New interesting questions arise when we consider more general curves such as Lamé's curves $|\xi|^k + |\eta|^k = x$, where the parameter k is not too small. For the number of latticepoints on and inside such curves we are in a position to prove such precise estimates that the O- and Ω-estimates are identical. So far the problem is completely solved. Moreover, we can give an exact representation of a second approximation of the number of lattice points with the new remainder. Thus there is a new problem of finding an estimate for this remainder.

3.1. DOMAINS OF THE TYPE $f(|\xi|) + f(|\eta|) \leqq x$

3.1.1. *Trivial estimates*

Suppose that $f(t)$ is a strictly increasing and continuous function defined for $t \geqq 0$. For the sake of simplicity we assume that $f(0) = 0$. We are interested in counting the number of lattice points on and inside the closed curve $f(|\xi|) + f(|\eta|) = x$ (see Fig. 5). Because of the symmetry of the problem we may count the lattice points with non-negative coordinates four times, where each lattice point on any of the coordinate axes gets the factor 1/2, the point (0,0) therefore gets the factor 1/4.

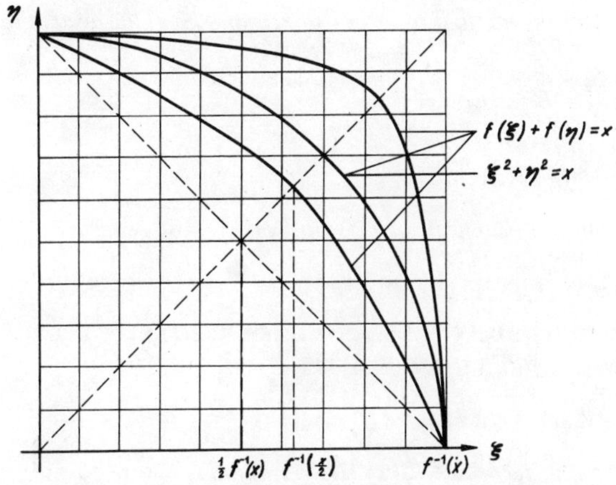

Fig. 5

We shall only consider convex curves such that we always assume that

$$\frac{1}{2} f^{-1}(t) < f^{-1}\left(\frac{t}{2}\right) \qquad (t > 0),$$

where $f^{-1}(t)$ denotes the inverse function of $f(t)$.

If

$$R(x) = \# \{(n, m): f(|n|) + f(|m|) \leqq x\}, \tag{3.1}$$

then the equations

$$R(x) = 4 \sum_{\substack{f(n)+f(m) \leqq x \\ n,\, m \geqq 0}}'' 1 = 4 \sum_{\substack{f(n) \leqq x \\ n \geqq 0}}' \left\{[f^{-1}(x - f(n))] + \frac{1}{2}\right\}$$

$$= 4 \sum_{\substack{f(n) \leqq x \\ n \geqq 0}} f^{-1}(x - f(n)) + O(f^{-1}(x))$$

hold. From Theorem 1.1 we have immediately

$$R(x) = H(x) + O(f^{-1}(x)),$$

where

$$H(x) = 4\int_0^{f^{-1}(x)} f^{-1}(x - f(t))\,dt = \iint\limits_{f(|\xi|)+f(|\eta|)\leqq x} d\xi\,d\eta$$

represents the area of our domain. It is easily seen that

$$(f^{-1}(x))^2 < 4\left(f^{-1}\left(\frac{x}{2}\right)\right)^2 < H(x) < 4(f^{-1}(x))^2 .$$

So we have our first, but trivial theorem.

Theorem 3.1. *Let $f(t)$ be a strictly increasing and continuous function, defined for $t \geqq 0$, with $f(0) = 0$. If $R(x)$ denotes the number of lattice points and $H(x)$ the area of the domain*

$$f(|\xi|) + f(|\eta|) \leqq x ,$$

then

$$R(x) = H(x) + O(f^{-1}(x)) , \tag{3.2}$$

where $f^{-1}(t)$ denotes the inverse function of $f(t)$ and $H(x)$ is of order

$$H(x) \asymp (f^{-1}(x))^2 .$$

Now we make some remarks on the Ω-problems. Let $R(x)$ be defined by (3.1). Suppose that there is a representation

$$R(x) = H(x) + \Delta(x) , \quad \Delta(x) = o(H(x)) , \tag{3.3}$$

where $H(x)$ is called the *main term* and $\Delta(x)$ the *remainder*. We need only some few *assumptions on the functions $f(n)$ and $H(x)$*:

1. Let $f(n)$ be a non-negative, non-decreasing function defined for $n = 0, 1, 2, \ldots$
2. Let $H(x)$ be a non-negative, non-decreasing function defined for $x \geqq 0$. Suppose that $H(x) \to \infty$ for $x \to \infty$.

We begin with a simple theorem, which is most applicable when $H(x) \gg x$.

Theorem 3.2. *If* (3.3) *holds and if* $\log H(x) \neq o(\log x)$, *then it is impossible to have both*

$$R(x) = H(x) + o\left(\frac{H(x)}{x}\right), \qquad R(x-1) = H(x) + o\left(\frac{H(x)}{x}\right).$$

Proof. Suppose that both the relations hold, then we get

$$R(x) - R(x-1) = o\left(\frac{H(x)}{x}\right) = o\left(\frac{R(x)}{x}\right).$$

Hence

$$R(x-1) = \left(1 + o\left(\frac{1}{x}\right)\right) R(x), \qquad \log R(x-1) = \log R(x) + o\left(\frac{1}{x}\right).$$

Let x_0 be sufficiently large, and let $x > x_0$ such that $x - x_0$ is an integer. Then

$$\log R(x) - \log R(x_0) = \sum_{n=0}^{x-x_0+1} (\log R(x-n) - \log R(x-n-1))$$

$$= \sum_{n=0}^{x-x_0+1} o\left(\frac{1}{x-n}\right) = o(\log x)$$

and we have

$$\log R(x) = o(\log x) .$$

This forms a contradiction to our hypothesis on $H(x)$, since $H(x) \sim R(x)$.

Theorem 3.3. *Let* $H(x) - H(x-1) = o(1)$. *If* (3.3) *holds, then* $\Delta x \neq O(1)$.

Proof. Suppose, on the contrary, that there is a positive value N such that $|\Delta(x)| < N$ for all $x \geqq 0$. Now we choose two values a and b so that

$$2N + 1 \leqq \sum_{a < f(|n|) \leqq b} 1 .$$

Let $f(|m_0|) = M$, where M can be chosen arbitrarily large. Then

$$2N + 1 \leqq \sum_{a < f(|n|) \leqq b} 1 \leqq \sum_{a+M < f(|n|) + f(|m|) \leqq b+M} 1 = R(b+M) - R(a+M)$$

$$= H(b+M) - H(a+M) + \Delta(b+M) - \Delta(a+M)$$

$$\leqq H(b+M) - H(a+M) + 2N$$

and therefore

$$H(b+M) - H(a+M) \geqq 1 .$$

On the other hand, because of our assumption $H(x) - H(x-1) = o(1)$, we can choose M so large that

$$H(b+M) - H(a+M) < 1 .$$

This gives a contradiction.

If we know a little more about the exact order of $H(x)$, we can find a sharper result.

Theorem 3.4. *Let* $0 < \alpha < 1$. *If* (3.3) *holds with* $x^\alpha \ll H(x) = o(x)$ *and if*

$$H(x) - 2H(x-1) + H(x-2) \leqq 0$$

for sufficiently large x, *then we have for every* $\varepsilon > 0$

$$\Delta(x) = \Omega\left(\left(\frac{x}{H(x)}\right)^{\frac{\alpha}{2-\alpha} - \varepsilon}\right) .$$

Proof. Let x_0 be sufficiently large, and let $x > x_0$ such that $x - x_0$ is an integer. Then

$$(x - x_0 + 2)(H(x) - H(x-1)) \leq \sum_{n=0}^{x-x_0+1} (H(x-n) - H(x-n-1))$$
$$= H(x) - H(x_0)$$

and hence

$$H(x) - H(x-1) \ll \frac{H(x)}{x}.$$

Let $f(n_0) = x + 1$, where x can be chosen arbitrarily large. Choose y such that $2 < y < x$. Then we have

$$R(x+y) - R(x) = \sum_{x < f(|n|) + f(|m|) \leq x+y} 1 \geq \sum_{f(|m|) \leq y-1} 1$$
$$\geq \sqrt{R(y-1)} \gg \sqrt{H(y-1)} \gg y^{\alpha/2}.$$

If we assume that

$$R(x) = H(x) + o\left(\left(\frac{x}{H(x)}\right)^{\frac{\alpha}{2-\alpha} - \varepsilon}\right).$$

then we get

$$R(x+y) - R(x) = H(x+y) - H(x) + o\left(\left(\frac{x}{H(x)}\right)^{\frac{\alpha}{2-\alpha} - \varepsilon}\right)$$
$$\ll y(H(x) - H(x-1)) + o\left(\left(\frac{x}{H(x)}\right)^{\frac{\alpha}{2-\alpha} - \varepsilon}\right)$$

and therefore

$$y^{\frac{\alpha}{2}} \ll \frac{y}{x} H(x) + o\left(\left(\frac{x}{H(x)}\right)^{\frac{\alpha}{2-\alpha} - \varepsilon}\right).$$

This leads to a contradiction, if we choose y such that

$$y = \left(\frac{x}{H(x)}\right)^{\frac{2}{2-\alpha} - \varepsilon}.$$

Clearly, $y < x$ and, when ε is small, y becomes large with x.

3.1.2. Representation and estimation of the number of lattice points

Again, let $R(x)$ denote the number of lattice points inside and on the closed curve $f(|\xi|) + f(|\eta|) = x$. We establish a particular representation of $R(x)$ and, by means of van der Corput's Theorem 2.3, we estimate the remainder. We assume that *the function $f(t)$ possesses the following properties*:

1. Let $f(t)$ for $t \geqq 0$ once and for $t > 0$ be twice continuously differentiable.
2. $f(0) = f'(0) = 0$.
3. For every $t > 0$ let $f(t), f'(t), f''(t) > 0$ and $f(t) \geqq t$.

Hence, we see that the function $f(t)$ is strictly increasing and has a strictly increasing inverse function $f^{-1}(t)$. We put

$$\eta = \eta(\xi) = f^{-1}(x - f(\xi)) .$$

Then we have

$$-\eta'(\xi) = \frac{f'(\xi)}{f'(\eta)},$$

$$-\eta''(\xi) = \frac{f''(\xi) f'^2(\eta) + f''(\eta) f'^2(\xi)}{f'^3(\eta)} > 0 . \tag{3.4}$$

Theorem 3.5. *Let $R(x)$ denote the number of lattice points and $H(x)$ the area of the domain $f(|\xi|) + f(|\eta|) \leqq x$. Then $R(x)$ is represented by*

$$R(x) = H(x) + \psi(f; x) + \Delta(f; x) , \tag{3.5}$$

where $\psi(f; x)$ and $\Delta(f; x)$ are given by

$$\psi(f; x) = 8 \int_0^{f^{-1}(x)} \eta'(\xi)\, \psi(\xi)\, d\xi , \tag{3.6}$$

$$\begin{aligned} \Delta(f; x) = &-8 \sum_{\substack{x/2 < f(n) \leqq x \\ n > 0}} \psi(\eta(n)) \\ &- 8 \int_0^{f^{-1}(x/2)} \eta'(\xi)\, \psi(\xi)\, d\xi + 4\psi^2\left(f^{-1}\left(\frac{x}{2}\right)\right). \end{aligned} \tag{3.7}$$

Proof.

$$\begin{aligned} R(x) &= 4 \sideset{}{''}\sum_{\substack{f(n) + f(m) \leqq x \\ n, m \geqq 0}} 1 = 8 \sideset{}{''}\sum_{\substack{f(n) + f(m) \leqq x \\ f(n) > x/2}} 1 + 4 \Bigg(\sideset{}{'}\sum_{f(n) \leqq x/2} 1 \Bigg)^2 \\ &= 8 \sum_{x/2 < f(n) \leqq x} \left([\eta(n)] + \frac{1}{2} \right) + 4 \left(\left[f^{-1}\left(\frac{x}{2}\right)\right] + \frac{1}{2} \right)^2 \\ &= 8 \sum_{x/2 < f(n) \leqq x} \eta(n) + 4\left(f^{-1}\left(\frac{x}{2}\right)\right)^2 - 8 f^{-1}\left(\frac{x}{2}\right) \psi\left(f^{-1}\left(\frac{x}{2}\right)\right) \\ &\quad - 8 \sum_{x/2 < f(n) \leqq x} \psi(\eta(n)) + 4\psi^2\left(f^{-1}\left(\frac{x}{2}\right)\right). \end{aligned}$$

Applying the Euler-Maclaurin sum formula to the first sum, we obtain

$$R(x) = 8 \int_{f^{-1}(x/2)}^{f^{-1}(x)} \eta(t)\,\mathrm{d}t + 4\left(f^{-1}\left(\frac{x}{2}\right)\right)^2 + 8 \int_{f^{-1}(x/2)}^{f^{-1}(x)} \eta'(\xi)\,\psi(\xi)\,\mathrm{d}\xi$$

$$-8 \sum_{x/2 < f(n) \leq x} \psi(\eta(n)) + 4\psi^2\left(f^{-1}\left(\frac{x}{2}\right)\right).$$

Using the notations (3.6) and (3.7), the representation (3.5) now follows at once.

By means of Theorem 3.1 it is seen in (3.5) that $H(x)$ is of order $(f^{-1}(x))^2$. In order to estimate the functions $\psi(f; x)$ and $\Delta(f; x)$, we need some additional, but quite harmless assumptions.

Lemma 3.1. *If* $f'(t) \ll f(t)/t$ *for* $t \geqq 1$, *then*

$$\psi(f; x) \ll f^{-1}\left(\frac{x}{f^{-1}(x)}\right). \tag{3.8}$$

Proof. We obtain from (3.6)

$$\psi(f; x) = 8\left(\int_0^{f^{-1}(z)} + \int_{f^{-1}(z)}^{f^{-1}(x)}\right) \eta'(\xi)\,\psi(\xi)\,\mathrm{d}\xi\,,$$

where z is any positive number less than x. We use the notation

$$\psi_1(t) = \int_0^t \psi(\tau)\,\mathrm{d}\tau\,.$$

Then we get by partial integration of the first integral

$$\psi(f; x) = 8\eta'(f^{-1}(z))\,\psi_1(f^{-1}(z)) - 8 \int_0^{f^{-1}(z)} \eta''(\xi)\,\psi_1(\xi)\,\mathrm{d}\xi$$

$$+8 \int_{f^{-1}(z)}^{f^{-1}(x)} \eta'(\xi)\,\psi(\xi)\,\mathrm{d}\xi\,.$$

Now we use the property $\psi_1(t) = O(1)$, and we estimate both integrals trivially. Then

$$\psi(f; x) \ll -\eta'(f^{-1}(z)) + \eta(f^{-1}(z))$$

$$= \frac{f'(f^{-1}(z))}{f'(f^{-1}(x-z))} + f^{-1}(x-z) \ll \frac{f'(f^{-1}(x))}{f'(f^{-1}(x-z))} + f^{-1}(x-z)\,.$$

Since $f'(t)$ is a strictly increasing function and since $f(0) = 0$, we have for $t > 0$

$$f(t) = \int_0^t f'(\tau)\,\mathrm{d}\tau < tf'(t)\,,$$

Thus we have the inequality $f(t) < tf'(t) \ll f(t)$, which leads to

$$\psi(f; x) \ll \frac{x}{f^{-1}(x)} \cdot \frac{f^{-1}(x-z)}{x-z} + f^{-1}(x-z)\,.$$

Choosing $z = x - \dfrac{x}{f^{-1}(x)}$, estimate (3.8) follows. Note that $x - z \geqq 1$ if $x \geqq 1$.

Lemma 3.2. *If $f'(t) \ll f(t)/t$, $f''(t) \asymp f(t)/t^2$ for $t \geqq 1$, and if $-\eta''(\xi)$ is monotonically increasing for $\xi \geqq f^{-1}(x/2)$, then*

$$\Delta(f; x) \ll (f^{-1}(x))^{2/3}\,. \tag{3.9}$$

Proof. Clearly, the last two terms in (3.7) are of order 1. Hence

$$\Delta(f; x) = -8 \sum_{x/2 < f(n) \leqq x} \psi(\eta(n)) + O(1)\,.$$

Let z denote the value

$$z = x(f^{-1}(x))^{-1/2}\,,$$

where $z < x/2$, if x is sufficiently large. Then, by means of the property $f'(t) > f(t)/t$, we obtain

$$f^{-1}(x) - f^{-1}(x-z) = \int_{f^{-1}(x-z)}^{f^{-1}(x)} \mathrm{d}\xi = \int_{x-z}^{x} \frac{\mathrm{d}t}{f'(f^{-1}(t))} \ll \int_{x-z}^{x} \frac{f^{-1}(t)}{t}\,\mathrm{d}t$$
$$\ll \frac{z}{x} f^{-1}(x) \ll (f^{-1}(x))^{1/2}\,.$$

Therefore we have

$$\Delta(f; x) = -8 \sum_{x/2 < f(n) \leqq x-z} \psi(\eta(n)) + O((f^{-1}(x))^{1/2})$$

and by Theorem 2.3

$$\Delta(f; x) \ll \int_{f^{-1}(x/2)}^{f^{-1}(x-z)} (-\eta''(\xi))^{1/3}\,\mathrm{d}\xi + \left(-\eta''\left(f^{-1}\left(\frac{x}{2}\right)\right)\right)^{-1/2} + (f^{-1}(x))^{1/2}\,.$$

Because of our assumptions we get from (3.4)

$$-\eta''(\xi) \asymp \frac{\eta f(\xi)}{\xi^2 f^2(\eta)}\,(f(\xi) + f(\eta)) = \frac{x\eta f(\xi)}{\xi^2 (x - f(\xi))^2}\,,$$

$$-\eta''\left(f^{-1}\left(\frac{x}{2}\right)\right) \gg \frac{1}{f^{-1}\left(\dfrac{x}{2}\right)} \gg \frac{1}{f^{-1}(x)}\,.$$

Hence

$$\Delta(f;x) \ll \int\limits_{f^{-1}(x/2)}^{f^{-1}(x)} \left(\frac{x\eta f(\xi)}{\xi^2(x-f(\xi))^2}\right)^{1/3} d\xi + (f^{-1}(x))^{1/2} .$$

Using the substitution $f(\xi) = t$ and the inequality

$$1/f'(\xi) < \xi/f(\xi) = f^{-1}(t)/t ,$$

we obtain

$$\Delta(f;x) \ll \int\limits_{x/2}^{x} \left(\frac{xf^{-1}(t)\,f^{-1}(x-t)}{t^2(x-t)^2}\right)^{1/3} dt + (f^{-1}(x))^{1/2} \ll (f^{-1}(x))^{2/3} ,$$

which proves (3.9).

When we form a parallel between the estimates (3.8) and (3.9), we can find two possibilities. It is either

$$f^{-1}\left(\frac{x}{f^{-1}(x)}\right) \ll (f^{-1}(x))^{2/3}$$

or

$$(f^{-1}(x))^{2/3} = o\left(f^{-1}\left(\frac{x}{f^{-1}(x)}\right)\right).$$

In the second case the problem is completely solved in general, since under some assumptions on $f(t)$ it can be proved that

$$\psi(f;x) = \Omega\left(f^{-1}\left(\frac{x}{f^{-1}(x)}\right)\right).$$

But then we can write down not only the first approximation $H(x)$ but also a second approximation $\psi(f; x)$ in the asymptotic representation of $R(x)$. Thus the problem of estimating the remainder $\Delta(f; x)$ arises. Altogether we formulate the following theorem.

Theorem 3. 6. *Let $R(x)$ denote the number of lattice points and $H(x)$ the area of the domain $f(|\xi|) + f(|\eta|) \leqq x$. If $f'(t) \ll f(t)/t$, $f''(t) \asymp f(t)/t^2$ and if $-\eta''(\xi)$ is monotonically increasing for $\xi \geqq f^{-1}(x/2)$, then*

$$R(x) = H(x) + \psi(f;x) + O(f^{-1}(x))^{2/3} , \tag{3.10}$$

where

$$H(x) \asymp (f^{-1}(x))^2 , \qquad \psi(f;x) \ll f^{-1}\left(\frac{x}{f^{-1}(x)}\right).$$

3.1.3. The Erdös-Fuchs Theorem

We consider an integer-valued, arithmetical function with $0 \leqq f(0) \leqq f(1) \leqq \ldots$ Let $R(x)$ denote the number of solutions of the inequality $f(n) + f(m) \leqq x$. In 1941 P. Erdös and P. Turán [1] conjectured that

$$R(x) = cx + O(1) \qquad (c > 0)$$

cannot hold. This remained unproved until 1954 when P. Erdös and W. H. J. Fuchs [1] obtained the much more precise result

$$R(x) = cx + \Omega(x^{1/4} \log^{-1/2} x)\,,$$

which is called the Erdös-Fuchs Theorem. Proofs by means of the theory of Fourier series one can find in the books of H. Halberstam and K. F. Roth [2] and F. Fricker [1]. A very short proof of the somewhat weaker result

$$R(x) = cx + \Omega(x^{1/4-\varepsilon}) \qquad (\varepsilon > 0)$$

without Fourier theory was given by D. J. Newman [1]. P. T. Bateman, E. E. Kohlbecker and J. P. Tull [1] generalized the Erdös-Fuchs Theorem to the case where the main term has the form $xL(x)$, $L(x)$ being a slowly oscillating function which is convex or concave on some interval of the form $[a, \infty)$. As a refinement of this work P. T. Bateman [1] showed that similar results can be obtained when the main term is a convex or concave function. R. C. Vaughan [1] investigated the case that the main term is a linear combination of powers of x and generalized the results to arbitrary dimensions. Combining the ideas of R. C. Vaughan and P. T. Bateman, E. K. Hayashi [1], [3] extended the theorems to main terms whose fractional differences are of one sign for large x. W. Jurkat developed a method which allows to sharpen the result of P. Erdös and W. H. J. Fuchs to

$$R(x) = cx + \Omega(x^{1/4})\,.$$

For the sake of simplicity we start with the method of D. J. Newman. Then we apply Jurkat's method, in order to investigate main terms which are powers of x.

Theorem 3.7. *Let $f(n)$ be an integer-valued arithmetical function such that $0 \leqq f(0) \leqq f(1) \leqq \ldots$ Let $R(x)$ be the number of solutions of $f(n) + f(m) \leqq x$. Then*

$$R(x) = cx + \Omega(x^{1/4-\varepsilon}) \tag{3.11}$$

for $c > 0$ and every $\varepsilon > 0$.

Proof with Newman's method. We suppose that

$$R(x) = cx + O(x^{1/4-\varepsilon}) \tag{3.12}$$

holds with some $\varepsilon > 0$ and force a contradiction. Since equation (3.12) implies that $R(x)$ is finite, we may assume that there cannot be an infinity of successive equalities in the sequence $f(0), f(1), \ldots$ Moreover, the number of successive equalities is bounded. We write

$$R(x) = c(x+1) + \Delta(x)\,, \quad \Delta(x) = O(x^{1/4-\varepsilon})\,.$$

Now we introduce the function

$$F(z) = \sum_{n=0}^{\infty} z^{f(n)}\,,$$

where z is a complex variable with $|z| < 1$. Then

$$F^2(z) = \frac{c}{1-z} + (1-z) \sum_{n=0}^{\infty} \Delta(n)\, z^n\,. \tag{3.13}$$

Let $h \geq 0$ be an integer. We consider the integral

$$I = \frac{1}{2\pi} \int_{-\pi}^{+\pi} \left| \frac{1-z^h}{1-z} F(z) \right|^2 d\varphi\,,$$

where

$$z = r\,e^{i\varphi}\,, \quad r = e^{-1/s}\,, \quad s > s_0 > 0\,.$$

Because of

$$\frac{1-z^h}{1-z} F(z) = \sum_{n=0}^{\infty} c_n z^n$$

with non-negative integers c_n we obtain

$$I = \sum_{n=0}^{\infty} c_n^2 r^{2n} \geqq \sum_{n=0}^{\infty} c_n r^{2n} = \frac{1-r^{2h}}{1-r^2} F(r^2) \gg hF(r^2)\,.$$

From (3.13) we see that

$$F^2(r^2) \sim \frac{c}{1-r^2} \quad (r \to 1) \quad \text{and} \quad F(r^2) \gg \frac{1}{\sqrt{1-r}} \gg \sqrt{s}\,.$$

Hence

$$I \gg h\sqrt{s}\,. \tag{3.14}$$

Otherwise

$$I = \frac{1}{2\pi} \int_{-\pi}^{+\pi} \left| \frac{1-z^h}{1-z} \right|^2 \left| \frac{c}{1-z} + (1-z) \sum_{n=0}^{\infty} \Delta(n)\, z^n \right| d\varphi$$

$$\ll h^2 \int_{-\pi}^{+\pi} \frac{d\varphi}{|1-z|} + \int_{-\pi}^{+\pi} \left| \frac{1-z^h}{1-z} \sum_{n=0}^{\infty} \Delta(n)\, z^n \right| d\varphi\,.$$

We obtain, by the subsequent Lemma 3.3,

$$\int_{-\pi}^{+\pi} \frac{d\varphi}{|1-z|} \ll \log s$$

and for the second integral, by Schwarz's inequality,

$$\left(\int_{-\pi}^{+\pi} \left| \frac{1-z^h}{1-z} \sum_{n=0}^{\infty} \Delta(n)\, z^n \right| d\varphi \right)^2 \leqq \int_{-\pi}^{+\pi} \left| \frac{1-z^h}{1-z} \right|^2 d\varphi \int_{-\pi}^{+\pi} \left| \sum_{n=0}^{\infty} \Delta(n)\, z^n \right|^2 d\varphi$$

$$= 4\pi^2 \frac{1-r^{2h}}{1-r^2} \sum_{n=0}^{\infty} \Delta^2(n)\, r^{2n} \ll h \sum_{n=0}^{\infty} n^{1/2-2\varepsilon} r^{2n} \ll h s^{3/2-2\varepsilon} .$$

Thus we get $I \ll h^2 \log s + \sqrt{h}\, s^{3/4-\varepsilon}$ and by (3.14)

$$h\sqrt{s} \ll h^2 \log s + \sqrt{h}\, s^{3/4-\varepsilon}, \qquad 1 \ll \frac{h \log s}{\sqrt{s}} + \frac{s^{1/4-\varepsilon}}{\sqrt{h}} .$$

Choosing $h = [s^{1/2-\varepsilon}]$, it can be seen that this inequality is impossible. This means that our assumption (3.12) is false and therefore (3.11) must hold.

For the next theorem we need the following lemmas.

Lemma 3.3. *Let* $z = r\, e^{i\varphi}$, $r = e^{-1/s}$, $s > s_0 > 0$. *Then*

$$\int_{-\pi}^{+\pi} |1-z|^{-\alpha}\, d\varphi \begin{cases} s^{\alpha-1} & \textit{for} \quad \alpha > 1 , \\ \log s & \textit{for} \quad \alpha = 1 . \end{cases}$$

Proof. From $\sin^2 \dfrac{\varphi}{2} \geqq \left(\dfrac{\varphi}{\pi}\right)^2$ for $|\varphi| \leqq \pi$ we find

$$|1-z|^2 = (1 - r\, e^{i\varphi})(1 - r\, e^{-i\varphi}) = (1-r)^2 + 4r \sin^2 \frac{\varphi}{2} \gg s^{-2} + \varphi^2 .$$

Hence

$$\int_{-\pi}^{+\pi} |1-z|^{-\alpha}\, d\varphi \ll \int_{-\pi}^{+\pi} \frac{d\varphi}{\max(s^{-\alpha}, |\varphi|^{\alpha})} = 2 \int_0^{1/s} s^{\alpha}\, d\varphi + 2 \int_{1/s}^{\pi} \varphi^{-\alpha}\, d\varphi$$

$$\ll \begin{cases} s^{\alpha-1} & \text{for} \quad \alpha > 1 , \\ \log s & \text{for} \quad \alpha = 1 . \end{cases}$$

Lemma 3.4. *Let* $\alpha > -1$, $\beta \geqq 0$, $s > 4$. *Then*

$$\sum_{n=2}^{\infty} \frac{n^{\alpha}}{\log^{\beta} n}\, e^{-n/s} \ll \frac{s^{\alpha+1}}{\log^{\beta} s} .$$

Proof. We divide the sum into two parts:

$$\sum_{n=2}^{[s]} \frac{n^\alpha}{\log^\beta n} e^{-n/s} \ll \sum_{n \leqq \sqrt{s}} n^\alpha + \frac{1}{\log^\beta s} \sum_{\sqrt{s} < n \leqq s} n^\alpha \ll \frac{s^{\alpha+1}}{\log^\beta s},$$

$$\sum_{n=[s]+1}^{\infty} \frac{n^\alpha}{\log^\beta n} e^{-n/s} \ll \int_s^\infty \frac{t^\alpha}{\log^\beta t} e^{-t/s} \, dt \ll \frac{s^{\alpha+1}}{\log^\beta s}.$$

This proves the lemma.

Lemma 3.5. *Let $\alpha > 0$, $\beta \geqq 0$, $s > 4$, and let (c_n) be a sequence of real numbers with*

$$|c_n| \leqq \varepsilon(n) \frac{n^\alpha}{\log^\beta n},$$

where $\varepsilon(n) \to 0$ as $n \to \infty$. If

$$\eta > \sup_{n > \sqrt{s}} \varepsilon(n), \qquad \eta \gg s^{-1/2},$$

then

$$\sum_{n=2}^{\infty} c_n e^{-n/s} \ll \eta \frac{s^{\alpha+1}}{\log^\beta s}.$$

Proof. Applying Lemma 3.4, we get

$$\sum_{n=2}^{\infty} c_n e^{-n/s} \ll \sum_{n \leqq \sqrt{s}} n^\alpha + \eta \sum_{n > \sqrt{s}} \frac{n^\alpha}{\log^\beta n} e^{-n/s} \ll s^{(\alpha+1)/2} + \eta \frac{s^{\alpha+1}}{\log^\beta s}$$

$$\ll \frac{s^{\alpha+1}}{\log^\beta s} (s^{-1/2} + \eta) \ll \eta \frac{s^{\alpha+1}}{\log^\beta s}.$$

Theorem 3.8. *Let $f(n)$ be an integer-valued, arithmetical function such that $0 \leqq f(0) \leqq f(1) \leqq \ldots$ Let $R(x)$ be the number of solutions of $f(n) + f(m) \leqq x$, and let $c, c_1, c_2, \ldots, c_r, \lambda, \lambda_1, \lambda_2, \ldots, \lambda_r$ be real numbers with $c > 0$, $0 < \lambda < 2$, $0 < \lambda_1 < \lambda_2 < \ldots < \lambda_r < \lambda$. Then*

$$R(x) = cx^\lambda + \sum_{\nu=1}^{r} c_\nu x^{\lambda_\nu} + \Omega(x^{\lambda/4}). \tag{3.15}$$

Remark. We may assume that x is an integer. Then we know that

$$(-1)^x \binom{-1-\alpha}{x} = \frac{x^\alpha}{\Gamma(\alpha+1)} + bx^{\alpha-1} + O(x^{\alpha-2}).$$

Assuming that

$$R(x) = cx^\lambda + \sum_{\nu=1}^{r} c_\nu x^{\lambda_\nu} + o(x^{\lambda/4}),$$

we can find numbers $c', c_1', c_2', \dots, c_s', \lambda_1', \lambda_2', \dots, \lambda_s'$ with $c' > 0$, $0 < \lambda_1' < \lambda_2' < \dots < \lambda_s' < \lambda$ and

$$R(x) = c'(-1)^x \binom{-1-\lambda}{x} + \sum_{\nu=1}^{s} c_\nu'(-1)^x \binom{-1-\lambda_\nu'}{x} + o(x^{\lambda/4}) .$$

This enables us to replace (3.15) by

$$R(x) = c(-1)^x \binom{-1-\lambda}{x} + \sum_{\nu=1}^{r} c_\nu(-1)^x \binom{-1-\lambda_\nu}{x} + \Omega(x^{\lambda/4}), \tag{3.16}$$

where again the numbers c, c_ν, λ, λ_ν have the above meaning, and x is an integer.

Proof of (3.16) *with Jurkat's method.* We write

$$R(x) = c(-1)^x \binom{-1-\lambda}{x} + \sum_{\nu=1}^{r} c_\nu(-1)^x \binom{-1-\lambda_\nu}{x} + \Delta(x)$$

and suppose that

$$\Delta(x) = o(x^{\lambda/4}) .$$

We consider the function

$$F(z) = \sum_{n=0}^{\infty} z^{f(n)}$$

for $|z| < 1$. Then

$$F^2(z) = \frac{c}{(1-z)^\lambda} + \sum_{\nu=1}^{r} \frac{c_\nu}{(1-z)^{\lambda_\nu}} + (1-z) \sum_{n=0}^{\infty} \Delta(n)\, z^n ,$$

$$2F(z)\, F'(z) = \frac{c\lambda}{(1-z)^{\lambda+1}} + \sum_{\nu=1}^{r} \frac{c_\nu}{(1-z)^{\lambda_\nu+1}}$$

$$- \sum_{n=0}^{\infty} \Delta(n)\, z^n + (1-z) \sum_{n=1}^{\infty} n\, \Delta(n)\, z^{n-1} ,$$

$$F'(z) \sim \frac{\sqrt{c\lambda}}{2} (1-z)^{-1-\lambda/2} \quad (z \to 1). \tag{3.17}$$

Let $h > 0$ be an integer and

$$F_1(z) = \frac{1-z^h}{1-z} F(z) = \sum_{n=0}^{\infty} a_n z^n , \qquad F_2(z) = z\,\frac{1-z^h}{1-z} F'(z) = \sum_{n=0}^{\infty} b_n z^n .$$

The numbers a_n and b_n are non-negative integers, and if $b_n \neq 0$, then so is $a_n \neq 0$. We now consider the integral

$$I = \frac{1}{2\pi} \int_{-\pi}^{+\pi} F_1(z)\, F_2(\bar{z})\, d\varphi ,$$

where

$$z = r\,\mathrm{e}^{\mathrm{i}\varphi}, \quad \bar{z} = r\,\mathrm{e}^{-\mathrm{i}\varphi}, \quad r = \mathrm{e}^{-1/s}, \quad s > 8\,.$$

Because of (3.17) we have

$$I = \sum_{n=0}^{\infty} a_n b_n r^{2n} \geqq \sum_{n=0}^{\infty} b_n r^{2n} = F_2(r^2) = r^2 \frac{1 - r^{2h}}{1 - r^2} F'(r^2) \gg h s^{1-\lambda/2}. \tag{3.18}$$

Otherwise

$$I \leqq \frac{1}{2\pi} \int_{-\pi}^{+\pi} \left| \frac{1 - z^h}{1 - z} \right|^2 |zF(z)\,F'(z)|\,\mathrm{d}\varphi$$

$$\ll \int_{-\pi}^{+\pi} \left| \frac{1 - z^h}{1 - z} \right|^2 \left\{ \frac{1}{|1 - z|^{\lambda+1}} + \sum_{\nu=1}^{r} \frac{1}{|1 - z|^{\lambda_\nu + 1}} \right.$$

$$\left. + \left| \sum_{n=0}^{\infty} \Delta(n)\,z^n \right| + \left| (1 - z) \sum_{n=1}^{\infty} n\,\Delta(n)\,z^n \right| \right\} \mathrm{d}\varphi\,.$$

To the first two integrals we apply Lemma 3.3 and obtain

$$\int_{-\pi}^{+\pi} \left| \frac{1 - z^h}{1 - z} \right|^2 \left\{ \frac{1}{|1 - z|^{\lambda+1}} + \sum_{\nu=1}^{r} \frac{1}{|1 - z|^{\lambda\nu+1}} \right\} \mathrm{d}\varphi \ll h^2 \int_{-\pi}^{+\pi} \frac{\mathrm{d}\varphi}{|1 - z|^{\lambda+1}} \ll h^2 s^{\lambda}\,.$$

If we use Schwarz's inequality and apply Lemma 3.5 with respect to $c_n = \Delta^2(n)$, $\alpha = \lambda/2$, $\beta = 0$, we get

$$\left(\int_{-\pi}^{+\pi} \left| \frac{1 - z^h}{1 - z} \right|^2 \left| \sum_{n=0}^{\infty} \Delta(n)\,z^n \right| \mathrm{d}\varphi \right)^2 \ll h^2 \left(\int_{-\pi}^{+\pi} \left| \frac{1 - z^h}{1 - z} \right| \left| \sum_{n=0}^{\infty} \Delta(n)\,z^n \right| \mathrm{d}\varphi \right)^2$$

$$\ll h^2 \int_{-\pi}^{+\pi} \left| \frac{1 - z^h}{1 - z} \right|^2 \mathrm{d}\varphi \int_{-\pi}^{+\pi} \left| \sum_{n=0}^{\infty} \Delta(n)\,z^n \right|^2 \mathrm{d}\varphi$$

$$= 4\pi^2 h^2 \frac{1 - r^{2h}}{1 - r^2} \sum_{n=0}^{\infty} \Delta^2(n)\,r^{2n} \ll \eta h^3 s^{1+\lambda/2}$$

and similarly

$$\left(\int_{-\pi}^{+\pi} \left| \frac{1 - z^h}{1 - z} \right|^2 \left| (1 - z) \sum_{n=0}^{\infty} n\,\Delta(n)\,z^n \right| \mathrm{d}\varphi \right)^2$$

$$\ll \int_{-\pi}^{+\pi} \left| \frac{1 - z^h}{1 - z} \right|^2 \mathrm{d}\varphi \int_{-\pi}^{+\pi} \left| \sum_{n=0}^{\infty} n\,\Delta(n)\,z^n \right|^2 \mathrm{d}\varphi$$

$$= 4\pi^2 \frac{1 - r^{2h}}{1 - r^2} \sum_{n=0}^{\infty} n^2\,\Delta^2(n)\,r^{2n} \ll \eta h s^{3+\lambda/2}\,.$$

Hence

$$I \ll h^2 s^\lambda + \eta^{1/2} h^{3/2} s^{1/2+\lambda/4} + \eta^{1/2} h^{1/2} s^{3/2+\lambda/4}$$

and by (3.18)

$$1 \ll h s^{\lambda/2-1} + \eta^{1/2} h^{1/2} s^{-1/2-\lambda/4} + \eta^{1/2} h^{-1/2} s^{1/2-\lambda/4} . \tag{3.19}$$

For $s \to \infty$ we may suppose that $s^{\lambda/2-1} < \eta^{1/2} = o(1)$. Putting $h = [\eta^{1/2} s^{1-\lambda/2}]$, (3.19) gives $1 \ll o(1)$ which forms a contradiction.

3.2. THE CIRCLE PROBLEM

3.2.1. The basic estimates

In what follows let $r(n)$ denote the number of solutions of the Diophantine equation $n_1^2 + n_2^2 = n$ in positive, negative, or zero integers. Further let

$$R(x) = \sum_{n \leqq x} r(n) = \# \{(n_1, n_2): n_1, n_2 \in \mathbf{Z}, n_1^2 + n_2^2 \leqq x\}$$

such that $R(x)$ denotes the number of lattice points which lie on or inside the circle $\xi^2 + \eta^2 = x$. As already mentioned in the introductory chapter it is known, after C. F. Gauss, that

$$R(x) = \pi x + O(\sqrt{x}) .$$

On the other side it is easily seen that

$$R(x) = \pi x + \Omega(1) .$$

Namely, assuming on the contrary that

$$R(x) = \pi x + o(1) ,$$

we obtain for integers x

$$0 = R(x + \tfrac{1}{2}) - R(x) = \frac{\pi}{2} + o(1)$$

which forms a contradiction.

For a long time there was no progress in the development of the problem. It was first proved by W. Sierpiński [1] in 1906 that more than this trivial O-estimate is true. Applying a method of G. Voronoi [1] used in his treatment of Dirichlet's divisor problem, he proved the sharper estimate (3.20). However, the proof is very long and difficult. In the following years simple proofs were discovered by E. Landau, J. G. van der Corput and others. Here the result is a special case of Theorem 3.6. The interesting method of E. Landau will be shown in Section 3.2.3.

Theorem 3.9.

$$R(x) = \pi x + O(x^{1/3}) . \tag{3.20}$$

Proof. We apply Theorem 3.6 with $f(t) = t^2$, $f^{-1}(t) = \sqrt{t}$. It follows from $\xi^2 + \eta^2 = x$ that $-\eta''(\xi) = x(x - \xi^2)^{-3/2}$ which is monotonically increasing. Of course, $H(x) = \pi x$ and $\psi(f; x) \ll x^{1/4}$. Hence, (3.20) follows from (3.10).

In the other direction it follows that Sierpiński's constant 1/3 cannot be replaced by any number less than 1/4. E. Landau [3] proved in 1915 that

$$R(x) = \pi x + \Omega(x^{1/4-\varepsilon})$$

for every positive ε. In the same year G. H. Hardy [1] proved the somewhat better result

$$R(x) = \pi x + \Omega_{\pm}(x^{1/4}) .$$

Hardy's method will be represented in Section 3.2.5. Later E. Landau developed a very simple method in order to prove that

$$R(x) = \pi x + \Omega(x^{1/4})$$

which will be demonstrated in Section 3.2.3. Here we show that the Ω-estimate in a special case of the Erdös-Fuchs Theorem.

Theorem 3.10.

$$R(x) = \pi x + \Omega(x^{1/4}) . \tag{3.21}$$

Proof. We write

$$R(x) = \sum_{m^2+n^2 \leq x} 1 = 4 \sum_{\substack{m^2+n^2 \leq x \\ m,\, n \geq 0}}'' 1 = 4 \sum_{\substack{m^2+n^2 \leq x \\ m,\, n \geq 0}} 1 - 4\sqrt{x} + O(1)$$

and apply Theorem 3.8 with $f(n) = n^2$, $c = \pi/4$, $\lambda = 1$, $r = 1$, $c_1 = -1$, $\lambda_1 = 1/2$. From (3.15) result (3.21) now follows at once.

3.2.2. *The Hardy Identity*

In this section we are concerned with an identity which expresses $R(x)$ as an infinite series of Bessel functions. Then the identity

$$R(x) = \pi x + \sqrt{x} \sum_{n=1}^{\infty} \frac{r(n)}{\sqrt{n}} J_1(2\pi\sqrt{nx}) , \tag{3.22}$$

where x is not an integer, was first stated by G. Voronoi [2] who expressively disclaimed possessing an accurate proof. The first exact proof was obtained by G. H. Hardy [1] in 1915. Therefore, equation (3.22) is called the *Hardy Identity*. Hardy's proof depends on the theory of analytic functions. The second proof given by E. Landau [9] involves real analysis only. However, both the proofs are very difficult in detail.

Later G. H. Hardy and E. Landau gave simpler proofs. Here we use a method which goes back to G. H. Hardy and E. Landau [1] and implies real analysis only.

The difficulty lies in the following: $R(x)$ is not a continuous function. If x is an integer representable by a sum of two squares, then x is a discontinuity of $R(x)$. Therefore, the infinite series in (3.22) cannot be uniformly convergent. But we shall prove:

1. The series is convergent for every value of x.
2. It is uniformly convergent throughout any interval free from integral values of x.
3. For non-integral values of x the series represents $R(x)$ and for integers x, the mean value $(R(x+0) + R(x-0))/2$.

We begin with the much simpler problem in order to prove an identity for the continuous function $\int_0^x R(t)\,dt$. Then we obtain the Hardy Identity by differentiation.

In what follows a star denotes the convolution of two functions:

$$f(x) * g(x) = \int_0^x f(t)\, g(x-t)\,dt\,.$$

Lemma 3.6.

$$1 * R(x) = \frac{\pi}{2} x^2 - 8\sqrt{x} * \psi(\sqrt{x}) + 4\psi(\sqrt{x}) * \psi(\sqrt{x})\,. \tag{3.23}$$

Proof.

$$1 * R(x) = \int_0^x R(t)\,dt = 4\int_0^x \sum_{\substack{n^2+m^2\leqq t\\ n,m\geqq 0}}'' 1\,dt = 4\sum_{\substack{n^2+m^2\leqq x\\ n,m\geqq 0}}'' (x-n^2-m^2)$$

$$= 4\int_0^x \sum_{0\leqq n^2\leqq t}' 1 \sum_{0\leqq m^2\leqq x-t}' 1\,dt = 4\left([\sqrt{x}]+\frac{1}{2}\right) * \left([\sqrt{x}]+\frac{1}{2}\right)$$

$$= 4(\sqrt{x}-\psi(\sqrt{x})) * (\sqrt{x}-\psi(\sqrt{x}))\,.$$

From this the equation (3.23) follows at once.

Theorem 3.11. *We have*

$$1 * R(x) = \frac{\pi}{2} x^2 + \frac{x}{\pi}\sum_{n=1}^{\infty} \frac{r(n)}{n} J_2(2\pi\sqrt{nx})\,, \tag{3.24}$$

where the series is absolutely convergent.

Proof. We apply Lemma 3.6 and use the fact that the Fourier expansion (1.10) of the function $\psi(t)$ uniformly converges in every closed interval which does not contain an integer, and that the partical sums of the Fourier series are uniformly bounded. Hence, in what follows we may interchange summation and integration. Then, by means of the well-known integral representations of the Bessel functions, we have

$$-8\sqrt{x} * \psi(\sqrt{x}) = \frac{8}{\pi}\sum_{n=1}^{\infty}\frac{1}{n}\sqrt{x} * \sin 2\pi n\sqrt{x} = \frac{4x}{\pi}\sum_{n=1}^{\infty}\frac{1}{n^2} J_2(2\pi n\sqrt{x})\,,$$

$$4\psi(\sqrt{x}) * \psi(\sqrt{x}) = -\frac{4}{\pi}\sum_{n=1}^{\infty}\frac{1}{n}\,\psi(\sqrt{x}) * \sin 2\pi n\sqrt{x}$$
$$= \frac{4}{\pi^2}\sum_{n=1}^{\infty}\sum_{m=1}^{\infty}\frac{1}{mn}\sin 2\pi m\sqrt{x} * \sin 2\pi n\sqrt{x}$$
$$= \frac{4x}{\pi}\sum_{n=1}^{\infty}\sum_{m=1}^{\infty}\frac{J_2(2\pi\sqrt{(n^2+m^2)\,x})}{n^2+m^2}.$$

Since $J_2(z) \ll 1/\sqrt{z}$, both the series are absolutely convergent. Hence, (3.23) shows that

$$1 * R(x) = \frac{\pi}{2}x^2 + \frac{x}{\pi}\sum_{n^2+m^2>0}\frac{J_2(2\pi\sqrt{(n^2+m^2)\,x})}{n^2+m^2},$$

which proves (3.24).

Theorem 3.12 *(The Hardy Identity). The series*

$$\sum_{n=1}^{\infty}\frac{r(n)}{\sqrt{n}}J_1(2\pi\sqrt{nx})$$

is convergent for every positive value of x, uniformly convergent throughout any closed interval which does not contain an integer, and it is

$$\frac{R(x-0)+R(x+0)}{2} = \pi x + \sqrt{x}\sum_{n=1}^{\infty}\frac{r(n)}{\sqrt{n}}J_1(2\pi\sqrt{nx}). \tag{3.25}$$

Proof. Let $J_1(2\pi\sqrt{vx})/\sqrt{v}$ for $v = 0$ denote its limit $\pi\sqrt{x}$ in the sense $v \to 0$. We consider the partial sums

$$S_n(x) = \sqrt{x}\sum_{v=0}^{n}\frac{r(v)}{\sqrt{v}}J_1(2\pi\sqrt{vx})$$

and obtain by partial summation

$$S_n(x) = R(n)\sqrt{\frac{x}{n}}\,J_1(2\pi\sqrt{nx}) + \pi x\int_0^n R(t)\,J_2(2\pi\sqrt{tx})\frac{dt}{t}.$$

Integrating by parts, then we find

$$\pi^2 x\int_0^n J_2(2\pi\sqrt{tx})\,dt = -\pi\sqrt{nx}\,J_1(2\pi\sqrt{nx}) + \pi\int_0^n\sqrt{\frac{x}{t}}\,J_1(2\pi\sqrt{tx})\,dt$$
$$= -\pi\sqrt{nx}\,J_1(2\pi\sqrt{nx}) - J_0(2\pi\sqrt{nx}) + 1.$$

Hence

$$S_n(x) = 1 - J_0(2\pi\sqrt{nx}) + (R(n) - \pi n)\sqrt{\frac{x}{n}}\, J_1(2\pi\sqrt{nx})$$
$$+ \pi x \int_0^n (R(t) - \pi t)\, J_2(2\pi\sqrt{tx})\frac{dt}{t}.$$

By means of the estimates

$$R(z) - \pi z = O(\sqrt{z}), \qquad J_p(z) = O\left(\frac{1}{\sqrt{z}}\right) \qquad (z \to \infty)$$

we obtain for fixed x and $n \to \infty$

$$S_n(x) = 1 + \pi x \int_0^n (R(t) - \pi t)\, J_2(2\pi\sqrt{tx})\frac{dt}{t} + O(n^{-1/4})$$
$$= 1 + \frac{\pi x}{n} J_2(2\pi\sqrt{nx}) \int_0^n (R(t) - \pi t)\, dt$$
$$+ \pi^2 \int_0^n \left(\frac{x}{t}\right)^{3/2} J_3(2\pi\sqrt{tx})\, dt \int_0^t (R(\tau) - \pi\tau)\, d\tau + O(n^{-1/4}).$$

Since $J_2(z) \ll 1/\sqrt{z}$, the representation (3.24) shows that

$$\int_0^n (R(t) - \pi t)\, dt \ll n^{3/4}.$$

We use this estimation as well as representation (3.24). Then

$$S_n(x) = 1 + \pi x^{3/2} \sum_{\nu=1}^{\infty} \frac{r(\nu)}{\nu} \int_0^n J_2(2\pi\sqrt{\nu t})\, J_3(2\pi\sqrt{tx})\frac{dt}{\sqrt{t}} + O(n^{-1/4})$$
$$= 1 + x \sum_{\nu=1}^{\infty} \frac{r(\nu)}{\nu} \int_0^{2\pi\sqrt{nx}} J_2\left(t\sqrt{\frac{\nu}{x}}\right) J_3(t)\, dt + O(n^{-1/4}).$$

The well-known property

$$\int_0^\infty J_2(yt)\, J_3(t)\, dt = \begin{cases} y^2 & \text{for } 0 < y < 1, \\ \dfrac{1}{2} & \text{for } y = 1, \\ 0 & \text{for } y > 1 \end{cases}$$

gives

$$S_n(x) - \frac{R(x-0)+R(x+0)}{2} = -x \sum_{\nu=1}^{\infty} \frac{r(\nu)}{\nu} \int_{2\pi\sqrt{nx}}^{\infty} J_2\left(t\sqrt{\frac{\nu}{x}}\right) J_3(t)\,\mathrm{d}t + O(n^{-1/4}).$$

For the Bessel functions the asymptotic representation

$$J_p(z) = \sqrt{\frac{2}{\pi z}} \cos\left(z - \frac{\pi p}{2} - \frac{\pi}{4}\right) + O(z^{-3/2}) \tag{3.26}$$

holds. Therefore, for fixed x, we have

$$\begin{aligned} J_2\left(t\sqrt{\frac{\nu}{x}}\right) J_3(t) &= -\frac{2}{\pi t}\left(\frac{x}{\nu}\right)^{1/4} \cos\left(t\sqrt{\frac{\nu}{x}} - \frac{\pi}{4}\right) \cos\left(t + \frac{\pi}{4}\right) + O(t^{-2}\nu^{-1/4}) \\ &= -\frac{1}{\pi t}\left(\frac{x}{\nu}\right)^{1/4} \cos\left(\left(\sqrt{\frac{\nu}{x}} + 1\right) t\right) \\ &\quad - \frac{1}{\pi t}\left(\frac{x}{\nu}\right)^{1/4} \sin\left(\left(\sqrt{\frac{\nu}{x}} - 1\right) t\right) + O(t^{-2}\nu^{-1/4}). \end{aligned}$$

If x is taken from a closed interval not containing an integer, then a number α exists with $|\sqrt{\nu/x} - 1| > \alpha > 0$. For $x = \nu$ the second term falls out. Hence, whether x be integral or not,

$$\int_{2\pi\sqrt{nx}}^{\infty} J_2\left(t\sqrt{\frac{\nu}{x}}\right) J_3(t)\,\mathrm{d}t \ll n^{-1/2}\nu^{-1/4}.$$

Thus

$$S_n(x) = \frac{R(x-0)+R(x+0)}{2} + O(n^{-1/4}).$$

If $n \to \infty$, we obtain equation (3.25), and the theorem is proved.

3.2.3. *Landau's proofs of the basic estimates*

E. Landau [15], [18] developed very simple methods in order to prove the basic estimates. He used the fact that the mean-value $1 * R(x)$ is represented by an absolutely convergent series (3.24). It will be seen later that the method can be also applied to some other lattice point problems.

Proof of Theorem 3.9. We write

$$R(x) = \pi x + \Delta(x)\,,$$

and with regard to the representation (3.24) we consider the integral $\int_x^{x\pm y} \Delta(t)\,dt$, where $x > y > 0$. From the inequalities

$$\int_x^{x+y} \Delta(t)\,dt = \int_x^{x+y} (R(t) - \pi t)\,dt > y(R(x) - \pi(x+y))\,,$$

$$\int_{x-y}^{x} \Delta(t)\,dt = \int_{x-y}^{x} (R(t) - \pi t)\,dt < y(R(x) - \pi(x-y))$$

we obtain

$$\frac{1}{y}\int_{x-y}^{x} \Delta(t)\,dt - \pi y < \Delta(x) < \frac{1}{y}\int_{x}^{x+y} \Delta(t)\,dt + \pi y\,. \tag{3.27}$$

Let $z > 1$, then we get from (3.24)

$$\int_x^{x+y} \Delta(t)\,dt = \sum_{n\leqq z} \frac{r(n)}{\sqrt{n}} \int_x^{x+y} \sqrt{t}\,J_1(2\pi\sqrt{nt})\,dt$$

$$+ \sum_{n>z} \frac{r(n)}{\pi n} \{(x+y)\,J_2(2\pi\sqrt{n(x+y)}) - xJ_2(2\pi\sqrt{nx})\}\,. \tag{3.28}$$

Because of $J_1(z), J_2(z) \ll 1/\sqrt{z}$ we get

$$\int_x^{x+y} \Delta(t)\,dt \ll yx^{1/4} \sum_{n\leqq z} r(n)\,n^{-3/4} + x^{3/4} \sum_{n>z} r(n)\,n^{-5/4} \ll y(xz)^{1/4} + x^{3/4}z^{-1/4}\,.$$

Both the terms are of the same order if we choose $z = x/y^2$, provided that $y^2 < x$. Then

$$\int_x^{x+y} \Delta(t)\,dt \ll \sqrt{xy}\,.$$

Analogously, we prove

$$\int_{x-y}^{x} \Delta(t)\,dt \ll \sqrt{xy}\,.$$

Hence, (3.27) shows that $\Delta(x) \ll \sqrt{x/y} + y$. If we choose $y = x^{1/3}$, we obtain $\Delta(x) \ll x^{1/3}$, and a new proof of Theorem 3.9 has been established.

We prepare the Ω-estimate with a simple lemma, which will be also useful later on.

Lemma 3.7. *Suppose that the function $f(x)$ is integrable for $x \geqq 0$. Let*

$$f_1(x) = 1 * f(x)\,, \quad f_2(x) = 1 * f_1(x) = x * f(x)\,,$$

and let α, β *be positive numbers with* $\beta - 2 < \alpha < \beta$. *If* $f(x) = o(x^{\alpha})$, $f_2(x) = O(x^{\beta})$, then

$$f_1(x) = o\left(x^{\frac{\alpha+\beta}{2}}\right).$$

Proof. Without loss of generality suppose that

$$|f_2(x)| \leqq x^{\beta} \quad \text{for} \quad x \geqq 0\,.$$

Let $\delta > 0$. We can choose $\xi = \xi(\delta)$ such that $\xi^{2+\alpha-\beta} > \delta^{\alpha-\beta}$ and

$$|f(x)| < \delta x^{\alpha} \quad \text{for} \quad x \geqq \xi\,.$$

Assuming that $x \geqq \xi$, $0 < x < x + (x/\delta)^{(\beta-\alpha)/2} < 2x$, we obtain

$$\left| f_2\left(x + \left(\frac{x}{\delta}\right)^{\frac{\beta-\alpha}{2}}\right) - f_2(x) \right| < (2x)^{\beta} + x^{\beta} < c_1 x^{\beta}\,.$$

If z satisfies the inequality $x \leqq z \leqq x + (x/\delta)^{(\beta-\alpha)/2}$, then $|f(z)| < \delta(2x)^{\alpha} < 2c_2\delta x^{\alpha}$. Applying the representation

$$\left(\frac{x}{\delta}\right)^{\frac{\beta-\alpha}{2}} f_1(x) = -\int_x^{x+(x/\delta)^{(\beta-\alpha)/2}} (f_1(y) - f_1(x))\,\mathrm{d}y + f_2\left(x + \left(\frac{x}{\delta}\right)^{\frac{\beta-\alpha}{2}}\right) - f_2(x$$

$$= -\int_x^{x+(x/\delta)^{(\beta-\alpha)/2}} \mathrm{d}y \int_x^y f(z)\,\mathrm{d}z + f_2\left(x + \left(\frac{x}{\delta}\right)^{\frac{\beta-\alpha}{2}}\right) - f_2(x)\,,$$

we get

$$\left(\frac{x}{\delta}\right)^{\frac{\beta-\alpha}{2}} |f_1(x)| < 2c_2\delta x^{\alpha} \int_x^{x+(x/\delta)^{(\beta-\alpha)/2}} (y - x)\,\mathrm{d}y + c_1 x^{\beta}\,,$$

$$|f_1(x)| < \left(c_2 \delta^{1-\frac{\beta-\alpha}{2}} + c_1 \delta^{\frac{\beta-\alpha}{2}}\right) x^{\frac{\alpha+\beta}{2}}\,,$$

which proves the lemma.

Proof of Theorem 3.10. According to Lemma 3.7 we write

$$R(x) = \pi x + \Delta(x)\,,$$

$$\Delta_1(x) = 1 * \Delta(x)\,, \quad \Delta_2(x) = 1 * \Delta_1(x)\,.$$

(3.24) shows that

$$\Delta_1(x) = \frac{x}{\pi} \sum_{n=1}^{\infty} \frac{r(n)}{n} J_2(2\pi\sqrt{nx}) \ll x^{3/4}\,,$$

$$\Delta_2(x) = \frac{1}{\pi^2} \sum_{n=1}^{\infty} \left(\frac{x}{n}\right)^{3/2} r(n)\, J_3(2\pi\sqrt{nx}) \ll x^{5/4}\,.$$

Moreover, from (3.26) we obtain the asymptotic representation

$$\Delta_1(x) = -\frac{x^{3/4}}{\pi^2}\sum_{n=1}^{\infty} n^{-5/4} r(n)\cos\left(2\pi\sqrt{nx}-\frac{\pi}{4}\right) + O(x^{1/4}).$$

Thus the assumption $\Delta_1(x) = o(x^{3/4})$ leads to

$$\sum_{n=1}^{\infty} n^{-5/4} r(n)\cos\left(x\sqrt{n}-\frac{\pi}{4}\right) = o(1).$$

However, the equation

$$\int_0^{\omega}\cos\left(x-\frac{\pi}{4}\right)\sum_{n=1}^{\infty} n^{-5/4} r(n)\cos\left(x\sqrt{n}-\frac{\pi}{4}\right)\mathrm{d}x$$

$$= 4\int_0^{\omega}\cos^2\left(x-\frac{\pi}{4}\right)\mathrm{d}x + \sum_{n=2}^{\infty} n^{-5/4} r(n)\int_0^{\omega}\cos\left(x-\frac{\pi}{4}\right)\cos\left(x\sqrt{n}-\frac{\pi}{4}\right)\mathrm{d}x$$

$$= 2\omega + O(1),$$

where $\omega > 0$, forms a contradiction. Hence

$$\Delta_1(x) = \Omega(x^{3/4}).$$

If we assume $\Delta(x) = o(x^{1/4})$, we apply Lemma 3.7 with $f(x) = \Delta(x)$, $f_1(x) = \Delta_1(x)$, $f_2(x) = \Delta_2(x)$, $\alpha = 1/4$, $\beta = 5/4$. Then we obtain $\Delta_1(x) = o(x^{3/4})$, which contradicts the above relation. This proves Theorem 3.10.

3.2.4. *Improvements of the O-estimates*

Definition 3.1. *If $R(x) = \pi x + \Delta(x)$, we denote by ϑ the value*

$$\vartheta = \inf\{\alpha\colon \Delta(x) \ll x^{\alpha}\}.$$

It can be seen from the results of the previous sections that the inequality $1/4 \leqq \vartheta \leqq 1/3$ holds such that we have a gap of length 1/12. Whereas it has been impossible until now to find a greater lower bound for ϑ, the better and better methods of estimating exponential sums lead to smaller and smaller values of ϑ. The fundamental problem is to find an optimal estimate of a sum of type

$$\sum_{n \leqq z} r(n)\,\mathrm{e}^{2\pi i\sqrt{xn}} = \sum_{n^2+m^2 \leqq z} \mathrm{e}^{2\pi i\sqrt{x(n^2+m^2)}}.$$

If we estimate this sum only trivially and if we use representation (3.28), we obtain the result $\vartheta \leqq 1/3$ of W. Sierpiński. But in these sums one can use the theory of one-dimensional exponential sums as being the simplest possibility, or the theory of two-dimensional exponential sums which gives the better results. The improvements depend on the repeated use of Weyl's steps. It is useful to apply Weyl's steps three times. Then we can obtain the following results.

1. *One-dimensional exponential sums*. In this case we obtain Nieland's result $\vartheta \leqq 27/82$ which is contained in Theorem 3.18.

2. *Two-dimensional exponential sums*. As we have to consider two variables in this case, we have the possibility to take Weyl's steps with respect to one variable only or with respect to both. The more Weyl's steps with respect to both the variables the better the results. But great difficulties will also arise.

2a. The simplest case in applying the method of double exponential sum is by making use of three Weyl's steps with respect to one variable only. Then we obtain $\vartheta \leqq 19/58$, which will be proved in Theorem 3.13. Though we cannot find this value in the history of the circle problem we shall give a proof, because in this case one can study the two-dimensional method in the best way. In principle the proofs of the following results are quite similar, only the details are very complicated.

2b. E. C. Titchmarsh applied Weyl's steps with respect to one variable twice and with respect to both the variables once and obtained $\vartheta \leqq 15/46$. The proof of this estimate essentially depends on the fact that a certain quadratic form coming from the Hessian is positively definite.

2c. Hua Loo-Keng applied Weyl's steps with respect to one variable once and with respect to both the variables twice and obtained $\vartheta \leqq 13/40$. But one gets into trouble, because the corresponding quadratic form is not a positively definite one. But by examining it Hua Loo-Keng found that the variables of the quadratic form are not perfectly general. Furthermore, he found that for these variables the values of the form are always positive.

2d. The best result can be obtained by applying Weyl's steps to both the variables each time. But here the corresponding quadratic form vanishes along certain curves and special arguments are required in order to estimate the exponential sums in the neighbourhood of these curves. After extensive calculations Chen Jing-run obtained $\vartheta \leqq 12/37$.

Preparation for the proof of $\vartheta \leqq 19/58$. We obtain from (3.24)

$$\int_x^{x+y} \Delta(t)\, \mathrm{d}t = \left[\frac{4t}{\pi} \sum_{m=1}^{\infty} \sum_{n=0}^{\infty} \frac{J_2(2\pi \sqrt{(m^2+n^2)\,t})}{m^2+n^2}\right]_{t=x}^{t=x\pm y}$$

and assume from the first that $x^{3/10} < y < x^{1/3}$. We divide the region of summation into two parts (see Fig. 6). Let $1 < y_1 < y_2 < y_3$, where the concrete values for y_1, y_2, y_3 will be given later. Let C_1 denote the region

$$C_1 = \{(m, n)\colon 1 \leqq m \leqq y_3 \text{ and } 0 \leqq n \leqq y_1,\ 1 \leqq m \leqq y_2 \text{ and } y_1 < n \leqq y_2,\ 1 \leqq m \leqq y_1 \text{ and } y_2 < n \leqq y_3\}$$

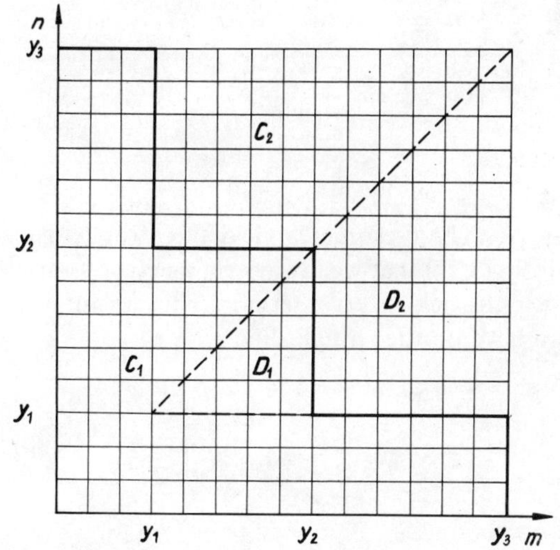

Fig. 6

and C_2 the remaining part in the first quadrant. Then we deduce as in Section 3.2.3

$$\int_x^{x\pm y} \Delta(t)\,\mathrm{d}t = 4\int_x^{x\pm y} \sum_{(m,n)\in C_1} \sqrt{\frac{t}{m^2+n^2}}\, J_1(2\pi\sqrt{(m^2+n^2)\,t})\,\mathrm{d}t$$

$$+\frac{4}{\pi}\left[\sum_{(m,n)\in C_2}\frac{t}{m^2+n^2}\,J_2(2\pi\sqrt{(m^2+n^2)\,t})\right]_{t=x}^{t=x\pm y}.$$

Therefore, applying the asymptotic representation (3.26) for the Bessel functions, we obtain

$$\int_x^{x\pm y}\Delta(t)\,\mathrm{d}t \ll \left|\int_x^{x\pm y}|\Phi_1(t)|\,t^{1/4}\,\mathrm{d}t\right| + |\Phi_2(x\pm y)|\,(x\pm y)^{3/4} + |\Phi_2(x)|\,x^{3/4}$$
$$+\, yx^{-1/4} + x^{1/4}, \tag{3.29}$$

where

$$\Phi_1(t) = \sum_{(m,n)\in C_1}(m^2+n^2)^{-3/4}\,\mathrm{e}^{2\pi\mathrm{i}\sqrt{(m^2+n^2)t}},$$

$$\Phi_2(t) = \sum_{(m,n)\in C_2}(m^2+n^2)^{-5/4}\,\mathrm{e}^{2\pi\mathrm{i}\sqrt{(m^2+n^2)t}}.$$

Certain parts of $\Phi_1(t)$ and $\Phi_2(t)$ can be dealt with directly. If $m \leqq y_1$, then

$$\sum_{\substack{(m,n)\in C_1\\ m\leqq y_1}}(m^2+n^2)^{-3/4}\,\mathrm{e}^{2\pi\mathrm{i}\sqrt{(m^2+n^2)t}} \ll \sum_{m\leqq y_1}\sum_{n=0}^{\infty}(m^2+n^2)^{-3/4}$$

$$\ll \sum_{m\leqq y_1} m^{-1/2} \ll \sqrt{y_1}\,.$$

The same result holds for $n \leqq y_1$. If $m \geqq y_3$, then

$$\sum_{\substack{(m,\,n)\in C_2\\ m\geqq y_3}} (m^2+n^2)^{-5/4}\, \mathrm{e}^{2\pi\mathrm{i}\sqrt{(n^2+m^2)t}} \ll \sum_{m\geqq y_3}\sum_{n=0}^{\infty}(m^2+n^2)^{-5/4}$$

$$\ll \sum_{m\geqq y_3} m^{-3/2} \ll y_3^{-1/2}\,.$$

The same result holds for $n \geqq y_3$. These estimates yield the error terms $yx^{1/4}y_1^{1/2}$, $x^{3/4}y_3^{-1/2}$ in (3.29). We know from (3.27) that we always get an error term y^2. Hence, we choose y_1, y_3 such that these terms yield an error term y^2. Thus we put $y_1 = y^2t^{-1/2}$, $y_3 = y^{-4}t^{3/2}$. Furthermore, we have on the straight line $m = n$

$$\sum_{m\geqq y_1} m^{-3/2}\,\mathrm{e}^{2\pi\,\mathrm{i}m\sqrt{2t}} \ll y_1^{-1/2} = y^{-1}t^{1/4}\,,$$

$$\sum_{m\geqq y_2} m^{-5/2}\,\mathrm{e}^{2\pi\,\mathrm{i}m\sqrt{2t}} \ll y_2^{-3/2} < y_1^{-3/2} = y^{-3}t^{3/4} \ll y^2t^{-3/4}\,.$$

Let D_1 and D_2 denote the regions

$$D_1 = \{(m, n)\colon m \geqq n,\ y_1 \leqq m,\ n \leqq y_2\}\,,$$

$$D_2 = \{(m, n)\colon m \geqq n,\ y_2 \leqq m \leqq y_3 \quad\text{and}\quad y_1 < n \leqq y_3\}\,.$$

Then, with the above simple estimates we get from (3.29)

$$\int_x^{x\pm y} \Delta(t)\,\mathrm{d}t \ll \left|\int_x^{x\pm y} |F_1(t)|\,t^{1/4}\,\mathrm{d}t\right|$$

$$+\,|F_2(x\pm y)|\,(x\pm y)^{3/4} + |F_2(x)|\,x^{3/4} + y^2 + x^{1/2}\,, \tag{3.30}$$

where

$$F_1(t) = \sum_{(m,\,n)\in D_1} (m^2+n^2)^{-3/4}\,\mathrm{e}^{2\pi\mathrm{i}\sqrt{(m^2+n^2)t}}\,, \tag{3.31}$$

$$F_2(t) = \sum_{(m,\,n)\in D_2} (m^2+n^2)^{-5/4}\,\mathrm{e}^{2\pi\mathrm{i}\sqrt{(m^2+n^2)t}}\,. \tag{3.32}$$

We now turn to the problem of the behaviour of sums of the form

$$S = \sum_{(m,\,n)\in D} \mathrm{e}^{2\pi\mathrm{i}\sqrt{(m^2+n^2)t}}\,, \tag{3.33}$$

where

$$D = \{(m, n)\colon m \geqq n,\ M \leqq m \leqq M' \leqq 2M,\ N \leqq n \leqq N' \leqq 2N,\ M \geqq N\}$$

and $x - y \leqq t \leqq x + y$, $y_1 \leqq M \leqq y_3$. If $M' - M \leqq y^5t^{-3/2}$, we have trivially

$$S \ll My^5t^{-3/2}\,.$$

Applying (1.16), we then obtain

$$\sum_{(m,n)\in D} (m^2+n^2)^{-3/4} \, e^{2\pi i \sqrt{(m^2+n^2)t}} \ll M^{-1/2} y^5 t^{-3/2} \ll y t^{-1/4},$$

$$\sum_{(m,n)\in D} (m^2+n^2)^{-5/4} \, e^{2\pi i \sqrt{(m^2+n^2)t}} \ll M^{-3/2} y^5 t^{-3/2} \ll y^2 t^{-3/4},$$

where we used the inequalities $y^3 \ll t$, $y_1 \leqq M$. Both the estimates yield in (3.30) an error term y^2. A similar result holds if $N' - N \leqq y^5 t^{-3/2}$. Therefore we may suppose that

$$M' - M > y^5 t^{-3/2}, \quad N' - N > y^5 t^{-3/2}. \tag{3.34}$$

In the remaining case (3.34) we estimate the sum (3.33) by applying three Weyl's steps to the variable m. We use (2.71) of Theorem 2.21 and obtain, since $|D'| \ll M^2$,

$$S \ll \frac{M^2}{\sqrt{H_1}} + \frac{M}{\sqrt{H_1}} \left(\sum_{m_1=1}^{H_1-1} |S_1| \right)^{1/2}, \tag{3.35}$$

where

$$S_1 = \Sigma_1 \, e^{2\pi i f_1(m,n)},$$

$$f_1(m,n) = \sqrt{((m+m_1)^2+n^2)\,t} - \sqrt{(m^2+n^2)\,t}$$

$$= m_1 \sqrt{t} \int_0^1 \frac{m + m_1\tau_1}{\sqrt{(m+m_1\tau_1)^2 + n^2}} \, d\tau_1,$$

$$SC(\Sigma_1): \quad (m+m_1, n), \qquad (m,n) \in D.$$

The number H_1 satisfies the condition of Theorem 2.21 if we assume by (3.34) that $1 \leqq H_1 \leqq y^5 t^{-3/2}$. Applying (2.71) again, we obtain

$$S_1 \ll \frac{M^2}{\sqrt{H_2}} + \frac{M}{\sqrt{H_2}} \left(\sum_{m_2=1}^{H_2-1} |S_2| \right)^{1/2},$$

where

$$S_2 = \Sigma_2 \, e^{2\pi i f_2(m,n)},$$

$$f_2(m,n) = m_1 m_2 n^2 \sqrt{t} \int_0^1 \int_0^1 ((m + m_1\tau_1 + m_2\tau_2)^2 + n^2)^{-3/2} \, d\tau_1 \, d\tau_2,$$

$$SC(\Sigma_2): \quad (m+m_1+m_2, n), \quad (m+m_1, n), \quad (m+m_2, n), \quad (m,n) \in D.$$

We suppose that $1 \leqq H_2 \leqq y^5 t^{-3/2}$. Then (3.35) becomes

$$S \ll \frac{M^2}{\sqrt{H_1}} + \frac{M^2}{H_2^{1/4}} + \frac{M^{3/2}}{H_1^{1/2} H_2^{1/4}} \left[\sum_{m_1=1}^{H_1-1} \left(\sum_{m_2=1}^{H_2-1} |S_2| \right)^{1/2} \right]^{1/2}.$$

The first two terms are of the same order if $H_2 = H_1^2$, Then

$$S \ll \frac{M^2}{\sqrt{H_1}} + \frac{M^{3/2}}{H_1} \left[\sum_{m_1=1}^{H_1-1} \left(\sum_{m_2=1}^{H_1^2-1} |S_2| \right)^{1/2} \right]^{1/2} . \tag{3.36}$$

For H_1 we now have the condition $1 \leqq H_1 \leqq y^{5/2} t^{-3/4}$. Applying (2.71) again, we get

$$S_2 \ll \frac{M^2}{\sqrt{H_3}} + \frac{M}{\sqrt{H_3}} \left(\sum_{m_3=1}^{H_3-1} |S_3| \right)^{1/2}, \qquad S_3 = \Sigma_3\, e^{2\pi i f_3(m,n)},$$

$$f_3(m,n) = -3 m_1 m_2 m_3 n^2 \sqrt{t} \int_0^1 \int_0^1 \int_0^1 \frac{m'}{(m'^2 + n^2)^{5/2}}\, d\tau_1\, d\tau_2\, d\tau_3 ,$$

$$m' = m + m_1\tau_1 + m_2\tau_2 + m_3\tau_3 ,$$

$SC(\Sigma_3)$: $(m + m_1 + m_2 + m_3, n), (m + m_1 + m_2, n), (m + m_2 + m_3, n),$
$(m + m_3 + m_1, n), (m + n_1, n), (m + m_2, n), (m + m_3, n),$
$(m, n) \in D$.

Again, we suppose that $1 \leqq H_3 \leqq y^5 t^{-3/2}$. Substituting the term $M^2/\sqrt{H_3}$ into (3.36), we find as before that we must put $H_3 = H_1^4$. Hence, we obtain from (3.36)

$$S \ll \frac{M^2}{\sqrt{H_1}} + \frac{M^{7/4}}{H_1^{3/2}} \left\{ \sum_{m_1=1}^{H_1-1} \left[\sum_{m_2=1}^{H_1^2-1} \left(\sum_{m_3=1}^{H_1^4-1} |S_3| \right)^{1/2} \right]^{1/2} \right\}^{1/2} \tag{3.37}$$

with the final restriction

$$1 \leqq H_1 \leqq y^{5/4} t^{-3/8} . \tag{3.38}$$

After these preparations we are in a position to estimate the sum S, where we apply Theorem 2.18 to the sum S_3.

Lemma 3.8. *Let S be defined by* (3.33). *If $y \geqq t^{11/34}$, then*

$$S \ll (M^{80} t)^{1/44} . \tag{3.39}$$

Proof. We write

$$f_3(u,v) = -3 m_1 m_2 m_3 v^2 \sqrt{t} \int_0^1 \int_0^1 \int_0^1 \frac{u'}{(u'^2 + v^2)^{5/2}}\, d\tau_1\, d\tau_2\, d\tau_3 ,$$

$$u' = u + m_1\tau_1 + m_2\tau_2 + m_3\tau_3 ,$$

so that in Theorem 2.18 (t_1, t_2) stands for (u, v). We assume that $H_1^4 = o(M)$. Then

$$\frac{\partial}{\partial u} f_3(u, v) = 3m_1m_2m_3\sqrt{t}\int_0^1\int_0^1\int_0^1 \frac{v^2(4u'^2 - v^2)}{(u'^2 + v^2)^{7/2}}\, d\tau_1\, d\tau_2\, d\tau_3$$

$$= 3m_1m_2m_3\sqrt{t}\,\frac{v^2(4u^2 - v^2)}{(u^2 + v^2)^{7/2}} + o\left(\frac{m_1m_2m_3\sqrt{t}\,N^2}{M^5}\right).$$

Therefore, we have in Theorem 2.18

$$\gamma_1 \ll \frac{m_1m_2m_3\sqrt{t}\,N^2}{M^5} \ll \frac{m_1m_2m_3\sqrt{t}\,N}{M^4}.$$

Analogously, we obtain

$$\gamma_2 \ll \frac{m_1m_2m_3\sqrt{t}\,N}{M^4}.$$

From

$$\frac{\partial^2}{\partial u^2} f_3(u, v) = 15m_1m_2m_3\sqrt{t}\int_0^1\int_0^1\int_0^1 \frac{u'v^2(3v^2 - 4u'^2)}{(u'^2 + v^2)^{9/2}}\, d\tau_1\, d\tau_2\, d\tau_3$$

we see that $\lambda_1 = \dfrac{m_1m_2m_3\sqrt{t}\,N^2}{M^6}$. Furthermore, we have

$$\frac{\partial^2}{\partial u\,\partial v} f_3(u, v) = 3m_1m_2m_3\sqrt{t}\int_0^1\int_0^1\int_0^1 \frac{v(8u'^4 - 24u'^2v^2 + 3v^4)}{(u'^2 + v^2)^{9/2}}\, d\tau_1\, d\tau_2\, d\tau_3,$$

$$\frac{\partial^2}{\partial v^2} f_3(u, v) = -3m_1m_2m_3\sqrt{t}\int_0^1\int_0^1\int_0^1 \frac{u'(2u'^4 - 21u'^2v^2 + 12v^4)}{(u'^2 + v^2)^{9/2}}\, d\tau_1\, d\tau_2\, d\tau_3.$$

From this we get

$$H(f_3(u, v)) = -27m_1^2m_2^2m_3^2t\,\frac{v^2(8u^4 + 6u^2v^2 + 3v^4)}{(u^2 + v^2)^7} + o\left(\frac{m_1^2m_2^2m_3^2tN^2}{M^{10}}\right),$$

such that $\Lambda = \dfrac{m_1^2m_2^2m_3^2tN^2}{M^{10}}$. Now

$$\frac{\partial^3}{\partial u^3} f_3(u, v) = 45m_1m_2m_3\sqrt{t}\int_0^1\int_0^1\int_0^1 \frac{v^2(8u'^4 - 12u'^2v^2 + v^4)}{(u'^2 + v^2)^{11/2}}\, d\tau_1\, d\tau_2\, d\tau_3$$

$$\ll \lambda' = \frac{m_1m_2m_3\sqrt{t}\,N^2}{M^7}.$$

It is seen that the condition $M\lambda' \ll \lambda_1$ of Theorem 2.18 is satisfied. The sum over the T-values in (2.62) is estimated trivially by $O(N/\sqrt{\lambda_1})$. Hence, we obtain from (2.62)

$$S_3 \ll \frac{m_1 m_2 m_3 \sqrt{t}\, N^2}{M^4} + \frac{M^3}{(m_1 m_2 m_3)^{1/2} t^{1/4}} + M + \frac{M^5}{m_1 m_2 m_3 \sqrt{t}\, N}$$

$$+ \frac{(m_1 m_2 m_3)^{1/2}\, t^{1/4}}{M} + N \log t\,.$$

Because of $N \leqq M \ll t^{1/4}$ the first and the fifth terms are less than $\frac{m_1 m_2 m_3 \sqrt{t}}{M^2}$ Thus, we get

$$S_3 \ll \frac{m_1 m_2 m_3 t^{1/2}}{M^2} + \frac{M^3}{(m_1 m_2 m_3)^{1/2}\, t^{1/4}} + \frac{M^5}{m_1 m_2 m_3 t^{1/2} N} + M \log t\,.$$

Using this estimate in (3.37), then we find

$$S \ll \frac{M^2}{\sqrt{H_1}} + M^{3/2} H_1^{7/8} t^{1/16} + M^{17/8} H_1^{-7/16} t^{-1/32}$$

$$+ M^{19/8} N^{-1/8} H_1^{-7/8} t^{-1/16} (\log H_1)^{1/8} + M^{15/8} (\log t)^{1/8}\,.$$

If we put $H_1 = [(M^8 t^{-1})^{1/22}]$, the first two terms are of the same order. Thus

$$S \ll (M^{80} t)^{1/44} + (M^{173} t^{-1})^{1/88} + (M^{181} N^{-11} t^{-2})^{1/88} (\log t)^{1/8}$$
$$+ M^{15/8} (\log t)^{1/8}\,.$$

Because of our assumption $y \geqq t^{11/34}$ we have

$$t^{5/34} \leqq y_1 \leqq N \leqq M \leqq y_3 \leqq t^{7/34}\,.$$

Hence, the last three terms are less than the first term. It is easily seen that the required conditions (3.38) and $H_1^4 = o(M)$ are satisfied. This proves (3.39).

Theorem 3.13.

$$R(x) = \pi x + O(x^{19/58} \log^2 x)\,. \tag{3.40}$$

Proof. In order to estimate the sums (3.31) and (3.32), we divide the domains D_1 and D_2 into subdomains by means of the straight lines $n = 2^\mu$, $m = 2^\nu$ ($\nu, \mu = 0, 1, \ldots; \nu \geqq \mu$). In each subdomain we apply Lemma 3.8. Using (1.16), we obtain from (3.39)

$$y t^{1/4} F_1(t) \ll y \sum_{2^\nu \leqq y_2} \sum_{\mu \leqq \nu} (2^{7\nu} t^6)^{1/22} \ll y (y_2^7 t^6)^{1/22} \log^2 t$$

and similarly

$$t^{3/4} F_2(t) \ll (y_2^{-15} t^{17})^{1/22} \log^2 t\,.$$

Both the estimates are of the same order if we put $y_2 = \sqrt{t}/y$. Then

$$yt^{1/4}F_1(t)\,,\ t^{3/4}F_2(t) \ll (y^{30}t^{19})^{1/44}\log^2 t$$

and, by (3.27) and (3.30),

$$\Delta(x) \ll (y^{-14}x^{19})^{1/44}\log^2 x + y + x^{1/2}y^{-1}\,.$$

If we put $y = x^{19/58}$, the condition of Lemma 3.8 is satisfied, and estimate (3.40) is proved.

3.2.5. *Hardy's method of Ω-estimates*

In this section we make some explanations to Hardy's [1] method to deal with the Ω-problem. It is essentially based on analytical properties of the function

$$\Phi(s) = \sum_{n=0}^{\infty} r(n)\,\mathrm{e}^{-s\sqrt{n}}\,.$$

This is a holomorphic function for $\mathrm{Re}(s) > 0$, and it is possible to give an analytical continuation over the whole plane with the exception of an infinity of isolated singular points, all of which lie on the imaginary axis. The behaviour of the function near the singularities then lead to the Ω-estimate. We begin with two lemmas. In Lemma 3.9 we prove a well-known functional equation for the Jacobian theta-function by means of the Poisson sum formula. On this basis we prove a representation of $\Phi(s)$ in Lemma 3.10, which gives us the desired analytical properties.

Lemma 3.9. *If* $\mathrm{Re}\,(t) > 0$, *the Jacobian theta-function* $\Theta(t)$ *is defined by*

$$\Theta(t) = \sum_{n=-\infty}^{+\infty} \mathrm{e}^{-\pi n^2 t}\,.$$

Then the functional equation

$$\Theta(t) = \frac{1}{\sqrt{t}}\,\Theta\left(\frac{1}{t}\right) \tag{3.41}$$

holds.

Proof. Applying the Poisson sum formula, we obtain

$$\Theta(t) = \lim_{N\to\infty}\sum_{n=-N}^{+N}\int_{-\infty}^{+\infty} \mathrm{e}^{-\pi\tau^2 t + 2\pi i n\tau}\,\mathrm{d}\tau = \frac{1}{\sqrt{t}}\sum_{n=-\infty}^{+\infty}\mathrm{e}^{-\pi n^2/t} = \frac{1}{\sqrt{t}}\,\Theta\left(\frac{1}{t}\right).$$

Lemma 3.10. *If* $\mathrm{Re}\,(s) > 0$, *then*

$$\sum_{n=0}^{\infty} r(n)\,\mathrm{e}^{-s\sqrt{n}} = 2\pi s\sum_{n=0}^{\infty}\frac{r(n)}{(s^2 + 4\pi^2 n)^{3/2}}\,. \tag{3.42}$$

Proof. Suppose that $s > 0$. Then by (3.41) we have

$$\Phi(s) = \sum_{n=0}^{\infty} r(n) \int_0^{\infty} \frac{1}{\sqrt{\pi t}} \mathrm{e}^{-t-ns^2/4t}\, \mathrm{d}t$$

$$= \int_0^{\infty} \frac{1}{\sqrt{\pi t}} \mathrm{e}^{-t}\, \Theta^2\left(\frac{s^2}{4\pi t}\right) \mathrm{d}t = \frac{4}{s^2} \int_0^{\infty} \sqrt{\pi t}\, \mathrm{e}^{-t}\, \Theta^2\left(\frac{4\pi t}{s^2}\right) \mathrm{d}t$$

$$= 4s \sum_{n=0}^{\infty} r(n) \int_0^{\infty} \sqrt{\pi t}\, \mathrm{e}^{-(s^2+4\pi^2 n)t}\, \mathrm{d}t = 2\pi s \sum_{n=0}^{\infty} \frac{r(n)}{(s^2 + 4\pi^2 n)^{3/2}}\,.$$

This proves (3.42) for $s > 0$. However, the analytical continuation shows that (3.42) also holds for $\mathrm{Re}\,(s) > 0$.

Moreover, it follows from (3.42) that the function $\Phi(s)$ is capable of analytical continuation into the left half plane. It is a many-valued function with isolated algebraic infinities of order 3/2 only on the imaginary axis and a pole of second order at the origin. From these properties we deduce the following theorem.

Theorem 3.14.

$$R(x) = \pi x + \Omega_{\pm}(x^{1/4})\,. \tag{3.43}$$

Proof. Let $\Delta(t) = R(t) - \pi t$. Suppose, for example, that

$$\Delta(t) \leqq K t^{1/4}\,, \quad K > 0\,, \tag{3.44}$$

for $t \geqq t_0 > 0$. It is easy to verify the equation

$$\int_0^{\infty} \mathrm{e}^{-st}\, \Delta(t^2)\, \mathrm{d}t = \frac{1}{s}\, \Phi(s) - \frac{2\pi}{s^3}\,.$$

If the functions $g(t)$ and $G(s)$ are defined by

$$g(t) = \Delta(t^2) - K\sqrt{t}\,,$$

$$G(s) = \frac{1}{s}\, \Phi(s) - \frac{2\pi}{s^3} - K_1 s^{-3/2}\,, \qquad K_1 = \frac{K}{2}\sqrt{\pi}\,,$$

we get

$$G(s) = \int_0^{\infty} \mathrm{e}^{-st}\, g(t)\, \mathrm{d}t = 2\pi \sum_{n=1}^{\infty} \frac{r(n)}{(s^2 + 4\pi^2 n)^{3/2}} - K_1 s^{-3/2}\,.$$

Now let k be a positive integer with $r(k) > 0$. If $\sigma > 0$, this equation shows that

$$G(\sigma + 2\pi i \sqrt{k}) \sim \frac{r(k)}{4\sqrt{\pi}}\, k^{-3/4}\, \mathrm{e}^{-3\pi i/4}\, \sigma^{-3/2} \tag{3.45}$$

für $\sigma \to 0$. Otherwise, we have

$$|G(\sigma + 2\pi\,\mathrm{i}\sqrt{k})| = \left|\left(\int_0^{t_0} + \int_{t_0}^{\infty}\right) \mathrm{e}^{-(\sigma+2\pi\mathrm{i}\sqrt{k})t}\, g(t)\,\mathrm{d}t\right| \leqq \int_{t_0}^{\infty} |g(t)|\,\mathrm{e}^{-\sigma t}\,\mathrm{d}t + O(1)\,.$$

If we assume (3.44) to be true, then there is

$$|G(\sigma + 2\pi\,\mathrm{i}\sqrt{k})| \leqq -G(\sigma) + O(1) = K_1\sigma^{-3/2} + O(1)\,.$$

Thus we obtain the inequality

$$K_1 \geqq \frac{r(k)}{4\sqrt{\pi}}\,k^{-3/4}$$

from (3.45). Therefore we need for K the inequality

$$K \geqq \frac{r(k)}{2\pi}\,k^{-3/4}\,,$$

and for smaller values of K (3.44) cannot hold. This proves $\Delta(t) = \Omega_+(t^{1/4})$. Similarly we get $\Delta(t) = \Omega_-(t^{1/4})$, which completes the proof of (3.43).

By a refinement of the arguments G. H. Hardy and others have proved some better Ω-estimates. They stated results of the type

$$\Delta(x) = \Omega_\pm(x^{1/4}\varrho(x))\,,$$

where $\varrho(x)$ denotes a function tending to infinity with x. But one has only found such functions $\varrho(x)$ which are increasing more weakly than any power of x.

3.2.6. *A historical outline of the development of the circle problem*

In the history of O-estimates we find the following values, where ϑ is given by Definition 3.1,

$\vartheta \leqq \frac{1}{2} = 0{,}5$ (C. F. Gauss),

$\vartheta \leqq \frac{1}{3} = 0{,}\bar{3}$ (W. Sierpiński [1], 1906),

$\vartheta \leqq \frac{1}{3} - \varepsilon = 0{,}\bar{3} - \varepsilon$ (J. G. van der Corput [6], 1923),

$\vartheta \leqq \frac{37}{112} = 0{,}3303\ldots$ (E. Landau [14], 1924, J. E. Littlewood–A. Walfisz [1], 1924),

$\vartheta \leqq \frac{163}{494} = 0{,}3299\ldots$ (A. Walfisz [1], 1927),

$\vartheta \leqq \frac{27}{82} = 0{,}3292\ldots$ (L. W. Nieland [1], 1928),

$\vartheta \leqq \frac{15}{46} = 0{,}3260\ \ldots$ (E. C. Titchmarsh [3], 1934),

$\vartheta \leqq \frac{13}{40} = 0{,}325$ (Hua Loo-Keng [1], 1942),

$\vartheta \leqq \frac{12}{37} = 0{,}3243\ \ldots$ (Chen Jing-run [3], [4], 1963),

$\vartheta \leqq \frac{35}{108} = 0{,}3240\ \ldots$ (W.-G. Nowak [12], 1984).

$\vartheta \leqq \frac{139}{429} = 0{,}32400\ \ldots$ (G. Kolesnik [7], 1985),

$\vartheta \leqq \frac{7}{22} = 0{,}3\overline{18}$ (H. Iwaniec, C. J. Mozzochi [1], 1987).

In the history of Ω-estimates we first have the trivial estimate $\vartheta \geqq 0$ and then

$\vartheta \geqq \frac{1}{4} = 0{,}25$ (E. Landau [3], 1915),
(G. H. Hardy [1], 1915).

G. H. Hardy stated a little more than E. Landau, namely $R(x) = \pi x + \Omega_{\pm}(x^{1/4})$, instead of $R(x) = \pi x + \Omega(x^{1/4-\varepsilon})$ for every positive ε.

Improvements of these results.

$R(x) = \pi x + \Omega(x^{1/4}\delta(x))$

$\delta(x) = (\log x)^{1/4}$ (G. H] Hardy [3], 1916),

$\delta(x) = (\log x)^{1/4} (\log\log x)^{(\log 2)/4} \exp(-c(\log\log\log x)^{1/2})\,, \qquad c > 0\,,$
(J. L. Hafner [1], 1981),

$R(x) = \pi x + \Omega_{+}(x^{1/4}\varrho(x))$

$\varrho(x) \to \infty$ unspecified (A. E. Ingham [1], 1940),

$\varrho(x) = (\log\log x)^{1/4} (\log\log\log x)^{1/4}\,,$
(K. S. Gangadharan [1], 1961),

$\varrho(x) = \exp(c(\log\log x)^{1/4} (\log\log\log x)^{-3/4}\,, \qquad c > 0\,,$
(K. Corrádi — I. Kátai [1], 1967).

3.3. DOMAINS WITH LAMÉ'S CURVES OF BOUNDARY

3.3.1. The basic estimates

In Section 3.3 we consider some particularities of the number of lattice points in domains which are bounded by Lamé's curves

$$|\xi|^k + |\eta|^k = x\,.$$

We always assume that $k > 2$, though the special case $k = 2$ is mostly allowed. Let $R_k(x)$ denote the number of lattice points

$$R_k(x) = \#\{(m, n): m, n \in \mathbf{Z},\ |m|^k + |n|^k \leqq x\}.$$

Theorem 3.1 shows at once that

$$R_k(x) = H_k(x) + O(x^{1/k})$$

with

$$H_k(x) = 4 \int_0^{x^{1/k}} (x - t^k)^{1/k}\, dt = c_k x^{2/k}, \qquad c_k = \frac{2\Gamma^2(1/k)}{k\Gamma(2/k)}.$$

From Theorem 3.3 it is seen that for $k > 2$

$$R_k(x) = c_k x^{2/k} + O(1)$$

cannot hold, which is also correct for $k = 2$ as we know.

Definition 3.2. *If $R_k(x) = c_k x^{2/k} + \delta_k(x)$, we denote by ω_k the value*

$$\omega_k = \inf\{\alpha_k: \delta_k(x) = O(x^{\alpha_k})\}.$$

The trivial estimates give the inequality $0 \leqq \omega_k \leqq 1/k$. The first progress concerning the upper bound was made by D. Cauer [1] in 1914, who proved that

$$\omega_k \leqq \frac{1}{k} - \frac{1}{k(2k-1)},$$

which was also confirmed by J. G. van der Corput [1] in 1919. If we apply Theorem 3.6, we obtain a much better result.

Theorem 3.15.

$$\omega_k \leqq \begin{cases} \dfrac{2}{3k} & \text{for } 2 \leqq k \leqq 3, \\[2ex] \dfrac{1}{k} - \dfrac{1}{k^2} & \text{for } k \geqq 3. \end{cases}$$

Proof. We use Theorem 3.6 with $f(t) = t^k$ and

$$\eta(\xi) = (x - \xi^k)^{1/k}.$$

Clearly, the function

$$-\eta''(\xi) = (k-1)\, x\xi^{k-2}(x - \xi^k)^{1/k-2}$$

is monotonically increasing. With $f^{-1}(t) = t^{1/k}$ Theorem 3.6 states

$$R_k(x) = c_k x^{\frac{2}{k}} + O\left(x^{\left(1-\frac{1}{k}\right)\frac{1}{k}}\right) + O\left(x^{\frac{2}{3k}}\right),$$

which proves the theorem.

For better lower bounds of ω_k we can use Theorems 3.4 and 3.8.

Theorem 3.16.

$$\omega_k \geqq \begin{cases} \dfrac{1}{2k} & for \quad 2 \leqq k \leqq 3 , \\ \dfrac{1}{k} - \dfrac{1}{k(k-1)} - \varepsilon & for \quad k > 3 \qquad (\varepsilon > 0) . \end{cases}$$

Proof. First consider Theorem 3.4 and suppose that $k > 2$. The conditions of this theorem are satisfied with $f(t) = t^k$, $\alpha = 2/k$, $H(x) = c_k x^{2/k}$. Consequently, we obtain

$$\omega_k \geqq \frac{1}{k} - \frac{1}{k(k-1)} - \varepsilon \quad \text{for} \quad k > 2 .$$

In order to apply Theorem 3.8, we consider the function

$$R_k^*(x) = \#\{(m, n)\colon m, n \in \mathbf{Z};\, m, n \geqq 0;\, [m^k] + [n^k] \leqq x\} .$$

Then

$$\frac{1}{4} R_k(x) + x^{1/k} + O(1) \leqq R_k^*(x) \leqq \frac{1}{4} R_k(x + 2) + x^{1/k} + O(1) ,$$

$$\frac{1}{4} \delta_k(x) + O(1) \leqq R_k^*(x) - \frac{c_k}{4} x^{2/k} - x^{1/k} \leqq \frac{1}{4} \delta_k(x + 2) + O(1) .$$

By putting $f(n) = [n^k]$ Theorem 3.8 now shows that

$$R_k^*(x) = \frac{c_k}{4} x^{2/k} + x^{1/k} + \Omega(x^{1/2k}) .$$

Hence, $\delta_k(x) = o(x^{1/2k})$ cannot hold. This proves the assertion.

3.3.2. *The second main term*

From Theorem 3.5 we obtain an explicit representation of $R_k(x)$. Putting there $f(t) = t^k$, $\eta(\xi) = (x - \xi^k)^{1/k}$, we get

$$R_k(x) = c_k x^{2/k} + \psi(t^k; x) + \Delta(t^k; x) \tag{3.46}$$

with

$$\psi(t^k; x) = -8 \int_0^{x^{1/k}} \xi^{k-1} (x - \xi^k)^{1/k - 1} \psi(\xi)\, d\xi , \tag{3.47}$$

$$\Delta(t^k; x) = -8 \sum_{x/2 < n^k \leqq x} \psi((x - n^k)^{1/k}) + 8 \int_0^{(x/2)^{1/k}} \xi^{k-1} (x - \xi^k)^{1/k - 1} \psi(\xi)\, d\xi + 4\psi^2 \left(\left(\frac{x}{2} \right)^{1/k} \right). \tag{3.48}$$

Lemmas 3.1 and 3.2 show that

$$\psi(t^k; x) \ll x^{1/k - 1/k^2}, \quad \Delta(t^k; x) \ll x^{2/3k}.$$

B. Randol [1] remarked in 1966 that for even integers $k \geqq 4$ the estimate

$$\delta_k(x) \ll x^{1/k - 1/k^2}$$

cannot be improved. E. Krätzel [2], [4] extended this result to odd integers $k \geqq 3$. Here we shall show that it is not necessary to consider only integers k. For this purpose we introduce the so-called generalized Bessel functions and investigate some series connected with these functions.

Definition 3.3. *Let k denote real numbers with $k \geqq 1$ and v complex numbers with* $\mathrm{Re}(v) > 1/k - 1$. *Then the generalized Bessel functions $J_v^{(k)}(x)$ are defined by the integral representation*

$$J_v^{(k)}(x) = \frac{2}{\sqrt{\pi}\,\Gamma(v + 1 - 1/k)} \left(\frac{x}{2}\right)^{kv/2} \int_0^1 (1 - t^k)^{v - 1/k} \cos xt \, dt. \tag{3.49}$$

It is well known that for $k = 2$ (3.49) gives an integral representation for the ordinary Bessel functions $J_v(x)$. In what follows an asymptotic expansion of the generalized Bessel function is required.

Lemma 3.11. *If* $k \geqq 1$, $v > 1/k - 1$, *then*

$$J_v^{(k)}(x) = \frac{1}{\sqrt{\pi}} \left(\frac{x}{2}\right)^{kv/2 - 1} \left\{ \left(\frac{x}{k}\right)^{1/k - v} \cos\left(x - \frac{\pi}{2}\left(v + 1 - \frac{1}{k}\right)\right) \right.$$

$$\left. - \sum \frac{(-1)^n \Gamma(kn + 1) \sin \dfrac{\pi k n}{2}}{n!\, \Gamma(v + 1 - 1/k - n)} x^{-kn} + O(x^{-[v - 1/k] - 1}) \right\}, \tag{3.50}$$

$$SC(\Sigma)\colon\ 1 \leqq n \leqq (v - 1/k)/k. \tag{3.51}$$

Proof. We apply a famous result on the asymptotic expansion of Fourier integrals (see E. T. Copson [1]): Let $\Phi(t)$ be $N + 1$ times continuously differentiable in $0 \leqq t \leqq 1$. Let $0 < \lambda \leqq 1$, $0 < \mu \leqq 1$. Then, as $x \to \infty$,

$$\int_0^1 t^{\lambda - 1}(1 - t)^{\mu - 1} \Phi(t)\, e^{ixt}\, dt$$

$$= \sum_{m=0}^{N} \frac{\Gamma(m + \lambda)}{m!\, x^{m+\lambda}} e^{\pi i(m+\lambda)/2} \frac{d^m}{dt^m} \{(1 - t)^{\mu - 1} \Phi(t)\}\Big|_{t=0}$$

$$+ \sum_{m=0}^{N} \frac{\Gamma(m + \mu)}{m!\, x^{m+\mu}} e^{ix + \pi i(m-\mu)/2} \frac{d^m}{dt^m} \{t^{\lambda - 1} \Phi(t)\}\Big|_{t=1} + O(x^{-N-1}). \tag{3.52}$$

In view of (3.49) we consider the integral

$$I(x) = \int_0^1 (1-t^k)^{\nu-1/k}\, e^{ixt}\, dt\,.$$

We put $N = [\nu - 1/k] + 1$, and $\mu = 1$ if $\nu - 1/k$ is an integer, and otherwise $\mu = \nu - 1/k - [\nu - 1/k]$. Then we rewrite $I(x)$ in the form

$$I(x) = \int_0^1 (1-t)^{\mu-1}\, \Phi(t)\, e^{ixt}\, dt + \int_0^1 (1-t)^{\mu-1}\, F(t)\, e^{ixt}\, dt\,,$$

where the functions $\Phi(t)$ and $F(t)$ are defined by

$$\Phi(t) = (1-t)^{1-\mu}(1-t^k)^{\nu-1/k} - F(t)\,,$$

$$F(t) = \sum_{r=0}^{N} \sum_{0 \leqq n \leqq N/k+1} (-1)^{r+n} \binom{1-\mu}{r} \binom{\nu - 1/k}{n} t^{r+kn}\,.$$

Hence, the function $\Phi(t)$ is $N+1$ times continuously differentiable in $0 \leqq t \leqq 1$ and $\Phi(t) = O(t^{N+1})$ for $t \to 0$. Therefore, we get from (3.52)

$$\int_0^1 (1-t)^{\mu-1}\, \Phi(t)\, e^{ixt}\, dt$$

$$= \sum_{m=0}^{N} \frac{\Gamma(m+\mu)}{m!\, x^{m+\mu}}\, e^{ix+\pi i(m-\mu)/2}\, \Phi^{(m)}(1) + O(x^{-N-1})$$

$$= \frac{\Gamma(\nu+1-1/k)}{x} \left(\frac{x}{k}\right)^{1/k-\nu} e^{ix-\pi i(\nu+1-1/k)/2} - \sum_{m=0}^{N} \frac{\Gamma(m+\mu)}{m!\, x^{m+\mu}}\, e^{ix+\pi i(m-\mu)/2}\, F^{(m)}(1)$$

$$+ O(x^{-[\nu-1/k]-2})\,. \tag{3.53}$$

Furthermore, we put $\lambda_n = 1$ if kn is an integer, and otherwise $\lambda_n = kn - [kn]$. Then, we obtain from (3.52)

$$\int_0^1 (1-t)^{\mu-1}\, F(t)\, e^{ixt}\, dt$$

$$= \sum_{r=0}^{N} \sum_{0 \leqq n \leqq N/k+1} (-1)^{r+n} \binom{1-\mu}{r} \binom{\nu-1/k}{n} \int_0^1 t^{\lambda_n - 1}(1-t)^{\mu-1}\, t^{r+kn+1-\lambda_n}\, e^{ixt}\, dt$$

$$= \sum_{m=0}^{N} \frac{\Gamma(m+\mu)}{m!\, x^{m+\mu}}\, e^{ix+\pi i(m-\mu)/2}\, F^{(m)}(1)$$

$$+ \sum_{0 \leqq n \leqq N/k+1} (-1)^n \binom{\nu-1/k}{n} \sum_{m=0}^{N} \frac{\Gamma(m+\lambda_n)}{m!\, x^{m+\lambda_n}}\, e^{\pi i(m+\lambda_n)/2}$$

$$\times \frac{\mathrm{d}^m}{\mathrm{d}t^m}\left\{(1-t)^{\mu-1}\left((1-t)^{1-\mu} - \sum_{r=N+1}^{\infty}(-1)^r\binom{1-\mu}{r}t^r\right)t^{kn+1-\lambda_n}\right\}\Bigg|_{t=0} + O(x^{-N-1})$$

$$= \sum_{m=0}^{N} \frac{\Gamma(m+\mu)}{m!\, x^{m+\mu}} \mathrm{e}^{\mathrm{i}x+\pi\mathrm{i}(m-\mu)/2} F^{(m)}(1)$$

$$+ \sum_{0 \leqq n \leqq N/k+1} (-1)^n \binom{\nu - 1/k}{n} \frac{\Gamma(kn+1)}{x^{kn+1}} \mathrm{e}^{\pi\mathrm{i}(kn+1)/2} + O(x^{-[\nu-1/k]-2}). \quad (3.54)$$

Adding (3.53) and (3.54), we obtain

$$I(x) = \frac{\Gamma(\nu+1-1/k)}{x}\left(\frac{x}{k}\right)^{1/k-\nu} \mathrm{e}^{\mathrm{i}x-\pi\mathrm{i}(\nu+1-1/k)/2} + \frac{\mathrm{i}}{x}$$

$$+ \sum (-1)^n \binom{\nu-1/k}{n} \frac{\Gamma(kn+1)}{x^{kn+1}} \mathrm{e}^{\pi\mathrm{i}(kn+1)/2} + O(x^{-[\nu-1/k]-2}),$$

where the summation condition is given by (3.51). This leads to the asymptotic expansion (3.50) at once.

We now define a function represented by an infinite series of the generalized Bessel functions which is related to our lattice point problem.

Definition 3.4. *Let k, ν denote real numbers with $k \geqq 1$, $\nu > 1/k - 1$. Let the function $\psi_\nu^{(k)}(x)$ bC defined by*

$$\Psi_\nu^{(k)}(x) = 2\sqrt{\pi}\,\Gamma(\nu+1-1/k) \sum_{n=1}^{\infty} \left(\frac{x}{\pi n}\right)^{k\nu/2} J_\nu^{(k)}(2\pi n x), \quad (3.55)$$

where x is not an integer, if $\nu \leqq 1/k$.

Because of the asymptotic relation (3.50) it is seen that the infinite series in (3.55) is convergent. Clearly, the function $\psi_\nu^{(k)}(x)$ is a generalization of the ψ-function. Namely for $\nu = 1/k$ we have

$$J_{1/k}^{(k)}(x) = \sqrt{\frac{2}{\pi x}} \sin x, \qquad \Psi_{1/k}^{(k)}(x) = \frac{2}{\pi} \sum_{n=1}^{\infty} \frac{\sin 2\pi n x}{n}$$

such that

$$\psi_{1/k}^{(k)}(x) = -2\psi(x)$$

for non-integers x. In the same way the following lemma generalizes the Fourier expansion of $\psi(x)$.

Lemma 3.12. *Let $k \geqq 1$, $\nu > 1/k - 1$. Then*

$$2 \sum_{0 \leqq n \leqq x}' (x^k - n^k)^{\nu - 1/k} = \frac{2\Gamma(\nu+1-1/k)\,\Gamma(1/k)}{k\Gamma(\nu+1)} x^{k\nu} + \Psi_\nu^{(k)}(x), \quad (3.56)$$

where x is not an integer, if $\nu \leqq 1/k$.

Proof. The representation (3.56) is an immediate consequence of Poisson's formula (1.11)

$$2 \sum_{0 \leqq n \leqq x}' (x^k - n^k)^{\nu - 1/k} = 2 \lim_{N \to \infty} \sum_{n=-N}^{+N} \int_0^x (x^k - t^k)^{\nu - 1/k} \, e^{2\pi \, int} \, dt$$

$$= \frac{2\Gamma(\nu + 1 - 1/k)\,\Gamma(1/k)}{k\Gamma(\nu + 1)} x^k + 4 \sum_{n=1}^{\infty} \int_0^x (x^k - t^k)^{\nu - 1/k} \cos 2\pi nt \, dt \, .$$

Applying the notations (3.49) and (3.55), the representation (3.56) follows at once.

We now obtain a representation of the function (3.47) by an *infinite series of generalized Bessel functions*:

$$\Psi(t^k; x) = \frac{8}{\pi} \sum_{n=1}^{\infty} \frac{1}{n} \int_0^{x^{1/k}} \xi^{k-1} (x - \xi^k)^{1/k - 1} \sin 2\pi n\xi \, d\xi$$

$$= 16 \sum_{n=1}^{\infty} \int_0^{x^{1/k}} (x - \xi^k)^{1/k} \cos 2\pi n\xi \, d\xi$$

$$= \frac{8}{\sqrt{\pi}} \Gamma(1 + 1/k) \, x^{1/k} \sum_{n=1}^{\infty} \frac{1}{n} J_{2/k}^{(k)}(2\pi n x^{1/k}) = 4\Psi_{2/k}^{(k)}(x^{1/k}) .$$

In (3.48) the second and third terms can be estimated by $O(1)$. Hence, (3.46) may be rewritten as

$$R_k(x) = c_k x^{2/k} + 4\psi_{2/k}^{(k)}(x^{1/k}) + \Delta_k(x) \tag{3.57}$$

with

$$\Delta_k(x) = -8 \sum_{x/2 < n^k \leqq x} \Psi((x - n^k)^{1/k}) + O(1) \, . \tag{3.58}$$

The meaning of the second term on the right-hand side of (3.57) is clear by (3.56). Applying (3.50), we obtain the asymptotic representation

$$\Psi_{2/k}^{(k)}(x^{1/k}) = \frac{2\Gamma(1/k)}{\pi k} x^{1/k - 1/k^2}$$

$$\times \sum_{n=1}^{\infty} \frac{1}{n} \left(\frac{k}{2\pi n}\right)^{1/k} \cos 2\pi \left(n x^{1/k} - \frac{1}{4}\left(1 + \frac{1}{k}\right)\right) + O(1) \, .$$

This expansion shows not only that $\psi_{2/k}^{(k)}(x^{1/k}) \ll x^{1/k - 1/k^2}$ but also that this estimation cannot be improved. Since $\Delta_k(x) \ll x^{2/3k}$, we obtain the following results.

Theorem 3.17A. *If* $\Delta_k(x) \ll x^{\alpha_k}$, *then*

$$R_k(x) = c_k x^{2/k} + \begin{cases} O(x^{1/k-1/k^2}), \\ \Omega(x^{1/k-1/k^2}) \end{cases}$$

holds for $1/k - 1/k^2 > \alpha_k$.

This result shows that the lattice point problem for $R_k(x)$ is completely solved under the conditions mentioned above. As has been known as far as now this is the case for $k > 3$. However, we can also understand the result in another sense. We interpret $4\psi^{(k)}_{2/k}(x^{1/k})$ as a second main term. Then the problem lies in a best possible estimation of the remainder $\Delta_k(x)$, which is unsolved. So far we have obtained:

Theorem 3.17B. *For* $k \geqq 2$ *the estimation*

$$R_k(x) = c_k x^{2/k} + 4\psi^{(k)}_{2/k}(x^{1/k}) + O(x^{2/3k})$$

holds.

3.3.3. Improvement of the O-estimate

We give an estimation of the remainder $\Delta_k(x)$ which contains Nieland's result for $k = 2$. We proceed in the following manner. It is not possible to apply the method of exponential sums to the whole summation interval in the sum of (3.58), because the second derivative of the function $f(t) = (x - t^k)^{1/k}$ tends to infinity for $t^k \to x$. Therefore, we carefully approach this limit point in short intervals. It will be seen that the following dissection of the sum is useful:

$$\Delta_k(x) = -8\,\Sigma_1\,\Sigma_2\,\psi((x - n^k)^{1/k}) + O(x^{1/2k}), \tag{3.59}$$

$$SC(\Sigma_1):\ 1 \leqq \nu \leqq \frac{\log x}{2k \log 2},$$

$$SC(\Sigma_2):\quad x(1 - 2^{-\nu}) < n^k \leqq x(1 - 2^{-\nu-1}).$$

We now apply the van der Corput transform to the corresponding exponential sum, and after that we estimate the new sum by using three Weyl's steps.

Lemma 3.13. *Let m be a natural number and let f(t) be defined by*

$$f(t) = -m(x - t^k)^{1/k}.$$

If the summation condition for Σ_2 *is defined as before, then*

$$\sum\nolimits_2 e^{2\pi i f(n)} = e^{\pi i/4} \sum\nolimits_3 g(n)\, e^{2\pi i F(n)} + O\left(\frac{1}{\sqrt{m}}\, x^{1/2k}\, 2^{-\nu(1-1/2k)}\right)$$
$$+ O((xm^k\, 2^{-\nu})^{2/5k}) \tag{3.60}$$

with

$$F(t) = -x^{\frac{1}{k}}\left(m^{\frac{k}{k-1}} + t^{\frac{k}{k-1}}\right)^{1-\frac{1}{k}},$$

$$g(t) = \frac{m}{\sqrt{k-1}}(mt)^{-\frac{k-2}{2(k-1)}} x^{\frac{1}{2k}}\left(m^{\frac{k}{k-1}} + t^{\frac{k}{k-1}}\right)^{-\frac{k+1}{2k}},$$

$$SC(\textstyle\sum_3)\colon\ m(2^\nu - 1)^{1-1/k} < n \leqq m(2^{\nu+1} - 1)^{1-1/k}.$$

Proof. We apply Theorem 2.9. Using the notations of that theorem, we obtain

$$f'(t) = mt^{k-1}(x - t^k)^{1/k-1},$$

$$f''(t) = m(k-1)\,x\,t^{k-2}(x - t^k)^{1/k-2},$$

$$f'''(t) = m(k-1)\,x\,t^{k-3}((k-2)\,x + (k+1)\,t^k)\,(x - t^k)^{1/k-3},$$

$$\varphi(t) = \left(\frac{xt^{\frac{k}{k-1}}}{m^{\frac{k}{k-1}} + t^{\frac{k}{k-1}}}\right)^{\frac{1}{k}},$$

$$F(t) = f(\varphi(t)) - t\varphi(t), \qquad g(t) = \frac{1}{\sqrt{f''(\varphi(t))}},$$

$$\alpha = m(2^\nu - 1)^{1-1/k}, \qquad \beta = m(2^{\nu+1} - 1)^{1-1/k},$$

$$\lambda_2 = mx^{-1/k}2^{\nu(2-1/k)}, \qquad \lambda_3 = mx^{-2/k}2^{\nu(3-1/k)}.$$

Now (3.60) follows from (2.27) immediately with the error term (2.28).

Lemma 3.14. *Let* $1 \leqq N < N' \leqq rN$ *and let* $F(t)$ *be defined as in Lemma* 3.13. *Then we have*

$$\sum_{mN < n \leqq mN'} e^{2\pi i F(n)} \ll \left(x^{\frac{1}{k}} m^{26} N^{25-\frac{1}{k-1}}\right)^{\frac{1}{30}} + \left(x^{-\frac{4}{k}} m^{121} N^{125+\frac{4}{k-1}}\right)^{\frac{1}{120}}. \tag{3.61}$$

Proof. We apply Theorem 2.6 with $k = 5$. From

$$F^{(5)}(t) = (k-1)^{-4}\, x^{\frac{1}{k}} m^{\frac{k}{k-1}} t^{\frac{k}{k-1}-5}\left(m^{\frac{k}{k-1}} + t^{\frac{k}{k-1}}\right)^{\frac{k-1}{k}-5}$$

$$\times \Big\{(k-2)(2k-3)(3k-4)\, m^{\frac{3k}{k-1}} + (k-2)(2k-1)(12k-23)$$

$$\times (m^2 t)^{\frac{k}{k-1}} + (2k-1)(18k^2 - 58k + 29)(mt^2)^{\frac{k}{k-1}}$$

$$+ (2k-1)(3k-2)(4k-3)\, t^{\frac{3k}{k-1}}\Big\}$$

we obtain

$$\lambda_5 = cx^{\frac{1}{k}} m^{-4} N^{-5-\frac{1}{k-1}}$$

with a certain constant c depending only on k. (3.61) now follows from (2.13).

Theorem 3.18.

$$\Delta_k(x) \ll x^{27/41k}. \tag{3.62}$$

Proof. We use the estimation (3.61) with $N \asymp 2^{\nu(1-1/k)}$ in (3.60). Because of

$$g(n) \asymp \left(x^{\frac{1}{k}} m^{-1} 2^{-\nu\left(2-\frac{1}{k}\right)}\right)^{\frac{1}{2}}$$

we obtain by partial summation

$$\sum\nolimits_2 e^{2\pi i f(n)} \ll \left(x^{\frac{16}{k}} m^{11} 2^{-\nu\left(5+\frac{11}{k}\right)}\right)^{\frac{1}{30}} + \left(x^{\frac{56}{k}} m^{61} 2^{\nu\left(5-\frac{61}{k}\right)}\right)^{\frac{1}{120}}$$

$$+ \left(x^{\frac{1}{k}} m^{-1} 2^{-\nu\left(2-\frac{1}{k}\right)}\right)^{\frac{1}{2}} + \left(x^{\frac{1}{k}} m\, 2^{-\frac{\nu}{k}}\right)^{\frac{2}{5}}.$$

Applying (1.18) with $s = 1$, where v is replaced by m, then

$$\sum\nolimits_2 \psi((x - n^k)^{1/k}) \ll x^{1/k} z^{-1} 2^{-\nu}$$

$$+ \left(x^{\frac{16}{k}} z^{11} 2^{-\nu\left(5+\frac{11}{k}\right)}\right)^{\frac{1}{30}} + \left(x^{\frac{56}{k}} z^{61} 2^{\nu\left(5-\frac{61}{k}\right)}\right)^{\frac{1}{120}}$$

$$+ \left(x^{\frac{1}{k}} 2^{-\nu\left(2-\frac{1}{k}\right)}\right)^{\frac{1}{2}} + \left(x^{\frac{1}{k}} z\, 2^{-\frac{\nu}{k}}\right)^{\frac{2}{5}}.$$

The first two terms are of the same order by putting $z = \left(x^{\frac{14}{k}} 2^{-\nu\left(25-\frac{11}{k}\right)}\right)^{\frac{1}{41}}$. Then

$$\sum\nolimits_2 \psi((x - n^k)^{1/k}) \ll \left(x^{\frac{27}{k}} 2^{-\nu\left(16+\frac{11}{k}\right)}\right)^{\frac{1}{41}} + \left(x^{\frac{105}{k}} 2^{-\nu\left(44+\frac{61}{k}\right)}\right)^{\frac{1}{164}}$$

$$+ \left(x^{\frac{1}{k}} 2^{-\nu\left(2-\frac{1}{k}\right)}\right)^{\frac{1}{2}} + \left(x^{\frac{11}{k}} 2^{-\nu\left(5+\frac{6}{k}\right)}\right)^{\frac{2}{41}}.$$

If we use this estimate in (3.59), the estimate (3.62) now follows.

3.3.4. *The Ω-estimate*

The aim of this section is to give a lower bound for the number α_k in

$$\Delta_k(x) \ll x^{\alpha_k} .$$

Though the result corresponds to that of the circle problem, the preliminary arrangements to proofs are much more complicated than there. Namely, the analytical background of Landau's and Hardy's methods mentioned above lies in the functional equation

$$\theta(t) = \frac{1}{\sqrt{t}}\, \theta\left(\frac{1}{t}\right)$$

of the Jacobian theta-function. This could be directly seen in Hardy's proof in Section 3.2.5. But also Landau's proof which uses the representation (3.24) is based on this functional equation in the long run. For, we can obtain equation (3.24) from this property. Let $c > 0$, then

$$\int_0^x R(t)\,\mathrm{d}t = \sum_{n \leq x} (x-n)\, r(n) = \frac{1}{2\pi\,\mathrm{i}} \int_{c-\mathrm{i}\infty}^{c+\mathrm{i}\infty} \sum_{n=0}^{\infty} r(n)\, \mathrm{e}^{(x-n)\,s} \frac{\mathrm{d}s}{s^2}$$

$$= \frac{1}{2\pi\,\mathrm{i}} \int_{c-\mathrm{i}\infty}^{c+\mathrm{i}\infty} \theta^2\left(\frac{s}{\pi}\right) \mathrm{e}^{xs} \frac{\mathrm{d}s}{s^2} = \frac{1}{2\pi\,\mathrm{i}} \int_{c-\mathrm{i}\infty}^{c+\mathrm{i}\infty} \frac{\pi}{s^3}\, \theta^2\left(\frac{\pi}{s}\right) \mathrm{e}^{xs}\,\mathrm{d}s$$

$$= \frac{1}{2\pi\,\mathrm{i}} \int_{c-\mathrm{i}\infty}^{c+\mathrm{i}\infty} \frac{\pi}{s^3} \sum_{n=0}^{\infty} r(n)\, \mathrm{e}^{xs-\pi^2 n/s}\,\mathrm{d}s .$$

Term-by-term integration, which can be justified, and the known integral representation of the Bessel functions show that

$$\int_0^x R(t)\,\mathrm{d}t = \frac{\pi}{2} x^2 + \frac{x}{\pi} \sum_{n=1}^{\infty} \frac{r(n)}{n} J_2(2\pi\sqrt{nx})$$

is an immediate *consequence of the above functional equation of the Jacobian theta-function.*

The lattice point problem $R_k(x)$ is associated with the analytic function

$$\sum_{n=0}^{\infty} \mathrm{e}^{-n^k t}, \qquad \mathrm{Re}\,(t) > 0 .$$

But it is clear that this function cannot have a similar functional equation for $k > 2$. This makes difficulties.

In 1981 L. Schnabel [1] found a very precise asymptotic representation of $1 * R_k(t)$ and in compliance with Landau's method he obtained

$$\Delta_k(x) = \Omega(x^{1/3k}) .$$

Later E. Krätzel [16] found the improvement

$$\Delta_k(x) = \Omega_{\pm}(x^{1/2k})$$

for integers $k \geqq 3$, where the proof is in conformity with Hardy's method. Here we would like to represent the latter result for all $k > 2$. The actual proof is quite the same as the proof of Theorem 3.14, but for preparing it we need two lemmas.

Lemma 3.15. *Let k, m, n be real numbers with $k \geqq 2$, $m, n \neq 0$. Then the function $I(s)$, defined if* Re $(s) > 0$ *by*

$$I(s) = \int_0^\infty \int_0^\infty e^{2\pi i(n\tau + mt) - s(\tau^k + t^k)^{1/k}} d\tau \, dt ,$$

is capable of analytic continuation into the left half-plane. It is a many-valued function with a finite number of isolated singular points, all of which lie on the imaginary axis. There is at most one algebraic infinity of order 3/2. The other possible inifinities are of smaller order. For $s \to 0$ we have $I(s) = O(1)$.

Proof. Substituting $t \to \tau t$ and integrating with respect to τ, we obtain

$$I(s) = \int_0^\infty (s(1 + t^k)^{1/k} - 2\pi \, i(n + mt))^{-2} \, dt \, .$$

This shows that all the singularities lie on the imaginary axis. It is clear that $I(s) = O(1)$ for $s \to 0$. Now we write

$$I(s) = \int_0^\infty (1 + t^k)^{-2/k} (s - 2\pi \, ih(t))^{-2} \, dt$$

with $h(t) = (n + mt)(1 + t^k)^{-1/k}$. The analytic continuation of the function $I(s)$ is possible into all points $s = 2\pi \, ih(t)$ with $h'(t) \neq 0$, provided that t is not an endpoint of the integration interval.

Now let $h'(t_0) = 0$. Then we find from

$$h'(t_0) = (1 + t_0^k)^{-1/k-1} (m - nt_0^{k-1}) = 0$$

the value $t_0 = \left(\frac{m}{n}\right)^{1/(k-1)}$, provided that the numbers m, n are of the same sign. Then

$$h(t_0) = \pm \left(|m|^{\frac{k}{k-1}} + |n|^{\frac{k}{k-1}}\right)^{1-\frac{1}{k}} .$$

where there is the upper sign if m, n are both positive and otherwise there is the lower sign. At such a point $s = 2\pi \, ih(t_0)$ the function $I(s)$ has a singularity. We put

$s = \sigma + 2\pi\,\mathrm{i}h(t_0)$, $\sigma > 0$, and investigate the behaviour of $I(s)$ for $\sigma \to 0$. Using the first terms of the Taylor series of $h(t)$ at $t = t_0$, we get for $\sigma \to 0$

$$I(s) = \int_{t_0/2}^{2t_0} (1 + t^k)^{-2/k} (\sigma + 2\pi\,\mathrm{i}h(t_0) - 2\pi\,\mathrm{i}h(t))^{-2}\,\mathrm{d}t + O(1)$$

$$\ll \left| \int_{t_0/2}^{2t_0} (\sigma - \pi\mathrm{i}h''(t_0)\,(t - t_0)^2)^{-2}\,\mathrm{d}t \right| \ll \sigma^{-3/2}.$$

Since $h'(0) = m \neq 0$, the associated infinity is of order 1. In case $t \to \infty$, we find the same property. This proves the lemma.

We now consider the function

$$\Phi_k(s) = \sum_{m=-\infty}^{+\infty} \sum_{n=-\infty}^{+\infty} \mathrm{e}^{-s(|m|^k + |n|^k)^{1/k}} = 4 \sum_{m=0}^{\infty}{}' \sum_{n=0}^{\infty}{}' \mathrm{e}^{-s(m^k + n^k)^{1/k}}.$$

Twice applying the Poisson sum formula and writing

$$\sum_{n=-\infty}^{+\infty}{}^* = \lim_{N \to \infty} \sum_{n=-N'}^{+N},$$

then

$$\Phi_k(s) = 4 \sum_{m=-\infty}^{+\infty}{}^* \sum_{n=-\infty}^{+\infty}{}^* \int_0^\infty \int_0^\infty \mathrm{e}^{2\pi\,\mathrm{i}(mt + n\tau) - s(t^k + \tau^k)^{1/k}}\,\mathrm{d}t\,\mathrm{d}\tau.$$

We now put

$$\Phi_k(s) = \Phi_{k,0}(s) + \Phi_{k,1}(s) + \Phi_{k,2}(s)$$

with

$$\Phi_{k,0}(s) = 4 \int_0^\infty \int_0^\infty \mathrm{e}^{-s(t^k + \tau^k)^{1/k}}\,\mathrm{d}t\,\mathrm{d}\tau = \frac{4\Gamma^2(1/k)}{k\Gamma(2/k)} \cdot \frac{1}{s^2},$$

$$\Phi_{k,1}(s) = 16 \sum_{n=1}^{\infty} \int_0^\infty \int_0^\infty \mathrm{e}^{-s(t^k + \tau^k)^{1/k}} \cos 2\pi n\tau\,\mathrm{d}\tau\,\mathrm{d}t$$

$$\Phi_{k,2}(s) = 4 \sum_{mn \neq 0}{}^* \sum{}^* \int_0^\infty \int_0^\infty \mathrm{e}^{2\pi\,\mathrm{i}(mt + n\tau) - s(t^k + \tau^k)^{1/k}}\,\mathrm{d}t\,\mathrm{d}\tau.$$

It is easily seen that

$$\int_0^\infty R_k(t^k)\,\mathrm{e}^{-st}\,\mathrm{d}t = \frac{1}{s}\,\Phi_k(s),$$

$$\int_0^\infty \frac{2\Gamma^2(1/k)}{k\Gamma(2/k)}\,t^2\,\mathrm{e}^{-st}\,\mathrm{d}t = \frac{1}{s}\,\Phi_{k,0}(s),$$

$$\int_0^\infty 4\psi_{2/k}^{(k)}(t)\,\mathrm{e}^{-st}\,\mathrm{d}t = \frac{1}{s}\,\Phi_{k,1}(s)\,,$$

and (3.57) now shows that

$$\int_0^\infty \Delta_k(t^k)\,\mathrm{e}^{-st}\,\mathrm{d}t = \frac{1}{s}\,\Phi_{k,2}(s)\,.$$

Note that $\Phi_{k,2}(0) = 0$. T' ... following lemma is now an immediate consequence of Lemma 3.15.

Lemma 3.16. *The function $\Phi_{k,2}(s)$ is capable of analytic continuation into the left half-plane. It is a many-valued function with isolated singularities only on the imaginary axis. There are algebraic infinities of order 3/2. The other infinities are of smaller order. For $s > 0$ and $s \to 0$ we have $\Phi_{k,2}(s) = O(s)$.*

Theorem 3.19.

$$R_k(x) = c_k x^{2/k} + 4\psi_{2/k}^{(k)}(x^{1/k}) + \Phi_\pm(x^{1/2k})\,. \tag{3.63}$$

Proof. We use notation (3.57). Corresponding to the proof of Theorem 3.14 suppose that there exists a positive constant K such that for $t \geqq t_0$

$$\Delta_k(t) \leqq K t^{1/2k}\,. \tag{3.64}$$

If we put

$$g(t) = \Delta_k(t^k) - K\sqrt{t}\,,$$

$$G(s) = \frac{1}{s}\,\Phi_{k,2}(s) - K_1 s^{-3/2}\,, \qquad K_1 = \frac{K}{2}\sqrt{\pi}\,,$$

we obtain

$$G(s) = \int_0^\infty g(t)\,\mathrm{e}^{-st}\,\mathrm{d}t\,.$$

It is seen from Lemma 3.16 that $G(s)$ has isolated algebraic infinities of maximal order 3/2 on the imaginary axis and that

$$G(s) = -K_1 s^{-3/2} + O(1) \qquad (s > 0, s \to 0)\,.$$

Moreover, there exists a number $\tau \neq 0$ such that

$$G(\sigma + \mathrm{i}\tau) \sim c\sigma^{-3/2}$$

with a certain constant $c \neq 0$ and $\sigma > 0$, $\sigma \to 0$. On the other hand we have

$$|G(\sigma + \mathrm{i}\tau)| \leqq -G(\sigma) + O(1) = K_1\sigma^{-3/2} + O(1)\,.$$

This shows that we must have $|c| \leqq K_1$. Hence, for smaller values of K_1 and K, respectively, inequality (3.64) cannot hold for all sufficiently large t. This proves

$$\Delta_k(t) = \Omega_+(t^{1/2k})\,.$$

Analogously we obtain

$$\Delta_k(t) = \Omega_-(t^{1/2k}) .$$

This completes the proof of the theorem.

NOTES ON CHAPTER 3

Section 3.1

3.1.1. Theorems 3.2—3.4 were proved by E. K. Hayashi [1], [2].

3.1.2. Theorem 3.6 is an unpublished result of E. Krätzel. It is possible to extend it to more general domains. Let $D(x)$ be a closed domain bounded by a smooth Jordan curve which is defined by $\Phi(\xi/x, \eta/x) = 0$. Suppose that $\Phi(u, v)$ is an analytic function with grad $\Phi \neq (0, 0)$. Y. Colin de Verdière [1], W.-G. Nowak [11] and K. Haberland [1] proved similar results for (3.10) under the assumption that the curve of boundary has a rational slope at each point with curvature 0.

3.1.3. For the proof of Theorem 3.7 see D. J. Newman [1]. Theorem 3.8 sharpens a result of R. C. Vaughan [1] by means of the method of W. Jurkat, which is described in E. K. Hayashi [1].

A. Sárközy [1] also considered the number of solutions of the inequality $f(n) + g(n) \leqq x$, where the difference $f(n) - g(n)$ is not too large.

It may be added that H.-E. Richert [6] obtained a multiplicative analogue of the Erdös-Fuchs Theorem.

Section 3.2

The proof of Theorem 3.13, Section 3.2.4, follows the lines of Hua Loo-Keng's [1] proof of $\vartheta \leqq 13/40$. The reader will now have no difficulty in studying Hua-Loo-Keng's proof.

Note that the interesting asymptotic relation

$$\int_0^x \Delta^2(t)\, dt \sim cx^{3/2}$$

holds. For results in this direction see E. Landau [12] and K. Chandrasekharan — R. Narasimhan [3].

Section 3.3

In 1966 B. Randol [1] proved Theorem 3.17 A for even integers $k \geqq 4$ and in 1967 E. Krätzel [2] for integers $k > 3$. Theorem 3.17 B is due to E. Krätzel [4] 1969. Note that the remainder $\Delta_k(x)$ in (3.57) can be expressed by

$$\Delta_k(x) = \frac{1}{\pi}\, \psi^{(k)}_{1/k-1/2}(x^{1/k}) * \psi^{(k)}_{1/k-1/2}(x^{1/k})$$

for non-integers x. Theorem 3.18 was obtained by E. Krätzel [9] in 1981 and Theorem 3.19 by the same authors in 1981.

See also the papers of S. B. Abljalimov, G. Hamm and W.-G. Nowak.

Chapter 4

Many-dimensional additive problems

In this chapter we investigate the number of lattice points in generalized spheres. More precisely, we would like to estimate the function $R_{k,n}(x)$.

Definition 4.1. *Let $R_{k,n}(x)$ denote the number of lattice points for $k \geqq 2$ inside and on the surface*

$$|\xi_1|^k + |\xi_2|^k + \dots + |\xi_n|^k = x .$$

This means

$$R_{k,n}(x) = \#\{(a_1, a_2, \dots, a_n): a_1, a_2, \dots, a_n \in \mathbf{Z},\ |a_1|^k + |a_2|^k + \dots + |a_n|^k \leqq x\} .$$

The problem of estimating the number of lattice points in spheres ($k = 2$) is a classical one. For $n = 4$ and $n \geqq 5$ there is a particular situation. Let $V_n(x)$ denote the volume of the n-dimensional sphere. Then we have

$$R_{2,n}(x) = V_n(x) + \begin{cases} O(x^{n/2-1}), \\ \Omega(x^{n/2-1}), \end{cases} \quad \text{for } n \geqq 5$$

and

$$R_{2,4}(x) = V_4(x) + \begin{cases} O(x(\log x)^{2/3}), \\ \Omega(x), \end{cases}$$

such that the problem being completely solved for $n \geqq 5$ and almost completely for $n = 4$. But the required methods do not available here. Those readers who are interested in these problems are referred to F. Fricker [1] and A. Walfisz [2]. Here we shall only investigate the case of the three-dimensional sphere. The main aim of this chapter is the estimation of $R_{k,n}(x)$ for $k > 2$. In all cases the possibility $k = 2$ is allowed, but the results are very bad.

4.1. LATTICE POINTS IN SPHERES

We begin with trivial estimates

$$R_{2,3}(x) = \sum_{a_1^2+a_2^2+a_3^2 \leqq x} 1 = \sum_{a^2 \leqq x} R_{2,2}(x - a^2)$$

$$= \sum_{a^2 \leqq x} \{\pi(x - a^2) + O(\sqrt{x - a^2})\} = \frac{4\pi}{3} x^{3/2} + O(x) .$$

Here we have used the trivial estimate of the number of lattice points in a circle. It is easily seen that we can improve this result slightly if we take non-trivial estimates of the circle problem. So Sierpiński's result $R_{2,2}(x) = \pi x + O(x^{1/3})$ gives

$$R_{2,3}(x) = \frac{4\pi}{3} x^{3/2} + O(x^{5/6}) .$$

In the other direction it is easy to prove that

$$R_{2,3}(x) = \frac{4\pi}{3} x^{3/2} + \Omega(x^{1/2}) .$$

For, assuming that

$$R_{2,3}(x) = \frac{4\pi}{3} x^{3/2} + o(x^{1/2}) ,$$

we obtain for integers x

$$0 = R_{2,3}\left(x + \frac{1}{2}\right) - R_{2,3}(x) = \frac{4\pi}{3}\left\{\left(x + \frac{1}{2}\right)^{3/2} - x^{3/2}\right\} + o(x^{1/2})$$

$$= \pi x^{1/2} + o(x^{1/2}),$$

which forms a contradiction. At this point it may be remarked that we are not able to improve the Ω-result substantially. That means, we cannot replace the value 1/2 in the exponent by a larger one.

We now prove by means of Titchmarsh's method the first non-trivial result. Let us begin with a simple lemma.

Lemma 4.1. *Let D denote the domain*

$$D = \{(t_1, t_2): 0 \leqq t_2 \leqq t_1,\ 2t_t^2 + t_2^2 \leqq x\} . \tag{4.1}$$

Then

$$R_{2,3}(x) = \frac{4\pi}{3} x^{3/2} - 48 \sum_{(a_1, a_2) \in D} \psi(\sqrt{x - a_1^2 - a_2^2}) + O(\sqrt{x}) . \tag{4.2}$$

Proof. The sphere may be divided into 48 parts such that

$$R_{2,3}(x) = 48\, \Sigma''\, 1 + O(\sqrt{x}) ,$$

$$SC(\Sigma''):\ a_1^2 + a_2^2 + a_3^2 \leqq x,\ 0 \leqq a_2 \leqq a_1 \leqq a_3; \quad \text{for} \quad 0 = a_2,\ a_2 = a_1 ,$$
$a_1 = a_3$ the corresponding terms get factors 1/2.

Since in the main term the lattice points with $a_1 = a_2 = a_3$ and $0 = a_1 = a_2 < a_3$ are not counted correctly, the error term $O(\sqrt{x})$ occurs. Now

$$R_{2,3}(x) = 48 \sum_{(a_1, a_2) \in D}'' \left([\sqrt{x - a_1^2 - a_2^2}] + \frac{1}{2} - a_1\right) + O(\sqrt{x})$$

$$= S - 48 \sum_{(a_1, a_2) \in D} \psi(\sqrt{x - a_1^2 - a_2^2}) + O(\sqrt{x}) , \tag{4.3}$$

where

$$S = 48 \sum_{(a_1, a_2) \in D}'' (\sqrt{x - a_1^2 - a_2^2} - a_1) = 48 \int \Sigma_1'' \, dt_3 ,$$

$SC(\int \Sigma_1'')$: $a_1^2 + a_2^2 + t_3^2 \leqq x$, $0 \leqq a_2 \leqq a_1 \leqq t_3$; for $0 = a_2$ and $a_1 = a_2$ the corresponding terms get factors 1/2.

Summing over a_1, we get

$$S = 48 \int \Sigma_2' \left\{[\sqrt{x - a_2^2 - t_3^2}] + \frac{1}{2} - a_2\right\} dt_3$$

$$+ 48 \int \Sigma_3' \left\{[t_3] + \frac{1}{2} - a_2\right\} dt_3 ,$$

$$SC(\int \Sigma_2'): \quad \frac{x - a_2^2}{2} < t_3^2 \leqq x - 2a_2^2 , \qquad 0 \leqq a_2 ,$$

$$SC(\int \Sigma_3'): \quad a_2^2 \leqq t_3^2 \leqq \frac{x - a_2^2}{2} , \qquad 0 \leqq a_2 .$$

Further

$$S = 48 \int \Sigma_4' \, dt_1 \, dt_3 - 48 \int \Sigma_2' \, \psi(\sqrt{x - a_2^2 - t_3^2}) \, dt_3 - 48 \int \Sigma_3' \, \psi(t_3) \, dt_3 ,$$

$$SC(\int \Sigma_4'): \quad a_2^2 + t_1^2 + t_3^2 \leqq x , \qquad 0 \leqq a_2 \leqq t_1 \leqq t_3 .$$

Since the integral $\int_y^z \psi(t) \, dt$ is uniformly bounded, we see at once that the last term is of order $\sqrt{x}$. To the last but one term we apply integration by parts and we obtain the same result. Hence

$$S = 48 \int \Sigma_4' \, dt_1 \, dt_3 + O(\sqrt{x})$$

$$= \iiint_{t_1^2 + t_2^2 + t_3^2 \leqq x} dt_1 \, dt_2 \, dt_3 + O(\sqrt{x}) = \frac{4\pi}{3} x^{3/2} + O(\sqrt{x}) .$$

The result (4.2) now follows from (4.3) and from this representation of S.

Theorem 4.1.

$$R_{2,3}(x) = \frac{4\pi}{3} x^{3/2} + O(x^{3/4} \log x) . \tag{4.4}$$

Proof. We apply the representation (4.2). The domain D defined by (4.1) is part of the rectangle

$$D' = \left\{(t_1, t_2)\colon 0 \leqq t_1 \leqq \sqrt{\frac{x}{2}}, \, 0 \leqq t_2 \leqq \sqrt{\frac{x}{3}}\right\}.$$

We may use the simple Theorem 2.16 with the function

$$f(t_1, t_2) = \sqrt{x - t_1^2 - t_2^2}\,.$$

The derivatives and the Hessian are given by

$$f_{t_1 t_1} = -\frac{x - t_2^2}{(x - t_1^2 - t_2^2)^{3/2}}\,, \quad f_{t_2 t_2} = -\frac{x - t_1^2}{(x - t_1^2 - t_2^2)^{3/2}}\,,$$

$$f_{t_1 t_2} = -\frac{t_1 t_2}{(x - t_1^2 - t_2^2)^{3/2}}\,, \quad H(f) = \frac{x}{(x - t_1^2 - t_2^2)^2}\,.$$

It is easily seen that

$$c_1 \asymp c_2 \asymp \sqrt{x}\,, \quad |D'| \asymp x\,, \quad \lambda_1 = \lambda_2 = 1/\sqrt{x}\,, \quad c_1 \sqrt{\lambda_1} \asymp c_2 \sqrt{\lambda_2}\,.$$

The bound of D is given by straight lines or the curve

$$t_1 = \varrho(t_2) = \sqrt{\frac{x - t_2^2}{2}}\,.$$

Then we have

$$-\varrho''(t_2) = \frac{x}{\sqrt{2}\,(x - t_2^2)^{3/2}} \ll \frac{1}{\sqrt{x}} = r\,.$$

Finally, we obtain $R \asymp \log x$. The result (4.4) is now an immediate consequence of (2.57).

In order to improve the estimate (4.4), we investigate the exponential sum

$$\sum_{(a_1, a_2) \in D} \mathrm{e}^{-2\pi i m \sqrt{x - a_1^2 - a_2^2}}\,.$$

The transformation of this sum is formulated in the following lemma. A careful estimate of the transformed sum leads to the result of Theorem 4.2.

Lemma 4.2. *Let m denote a positive integer and D the domain* (4.1). *Then*

$$\sum_{(a_1, a_2) \in D} \mathrm{e}^{-2\pi i m \sqrt{x - a_1^2 - a_2^2}} = \frac{i}{2} \sum_{m_1 = 0}^{m} \sum_{m_2 = 0}^{m} \frac{m \sqrt{x}}{m^2 + m_1^2 + m_2^2}\, \mathrm{e}^{-2\pi i \sqrt{x(m^2 + m_1^2 + m_2^2)}}$$

$$+ O(\sqrt{m}\, x^{1/4} \log(mx)) + O(\sqrt{x} \log(mx))\,. \quad (4.5)$$

Proof. We apply Theorem 2.24, where the domain D is given by (4.1) and the function $f(t_1, t_2)$ by

$$f(t_1, t_2) = -m \sqrt{x - t_1^2 - t_2^2}\,.$$

It is easily seen that

$$c_1 \asymp c_2 \asymp \sqrt{x}\,, \quad \lambda_1 = \frac{m}{\sqrt{x}}\,, \quad \Lambda = \frac{m^2}{x}\,.$$

Since

$$f_{t_1t_1t_1} = \frac{3mt_1(x - t_2^2)}{(x - t_1^2 - t_2^2)^{5/2}} \leqq \frac{12m}{x - t_2^2} \ll \frac{m}{x},$$

we can put $\lambda' = m/x$, and the condition $c_1\lambda' \ll \lambda_1$ is satisfied. Of course, we have $G = 1, G_1 = 0$.

Let D_1 be the image of D under the mapping

$$y_1 = f_{t_1}(t_1, t_2) = \frac{mt_1}{\sqrt{x - t_1^2 - t_2^2}}, \qquad y_2 = t_2 .$$

Then

$$\gamma_1 = m, \qquad |D_1| \asymp m\sqrt{x} .$$

The function $f_1(y_1, y_2)$ is defined by

$$f_1(y_1, y_2) = f(\varphi_1, y_2) - y_1\varphi_1 ,$$

where $\varphi_1 = \varphi_1(y_1, y_2)$ is the solution of $f_{t_1}(\varphi_1, y_2) = y_1$. We obtain

$$\varphi_1(y_1, y_2) = y_1 \sqrt{\frac{x - y_2^2}{m^2 + y_1^2}},$$

$$f_1(y_1, y_2) = -\sqrt{(x - y_2^2)(m^2 + y_1^2)},$$

$$\frac{\partial^3}{\partial y_2^3} f_1(y_1, y_2) = \frac{3xy_2\sqrt{m^2 + y_1^2}}{(x - y_2^2)^{5/2}} \ll \frac{m}{x}.$$

Thus we can put $\lambda'' = m/x$, and the condition $c_2\lambda_1\lambda'' \ll \Lambda$ is satisfied. The function

$$\frac{\partial}{\partial y_2} \frac{1}{\sqrt{f_{t_1t_2}(\varphi_1, y_2)}} = \frac{\partial}{\partial y_2} \frac{m(x - y_2^2)^{1/4}}{(m^2 + y_1^2)^{3/4}} = -\frac{my_2}{2}(m^2 + y_1^2)^{-3/4}(x - y_2^2)^{-3/4}$$

is monotonic with respect to y_2, and we may put $G_2 = \frac{1}{\sqrt{m}} x^{-1/4}$. The image D_2 of D under the mapping

$$y_1 = f_{t_1}(t_1, t_2) = \frac{mt_1}{\sqrt{x - t_1^2 - t_2^2}}, \qquad y_2 = f_{t_2}(t_1, t_2) = \frac{mt_2}{\sqrt{x - t_1^2 - t_2^2}}$$

is given by

$$D_2 = \{(y_1, y_2) : 0 \leqq y_2 \leqq y_1 \leqq m\} .$$

In particular we get $\gamma_2 = m$.

Hence, we obtain for the error terms in Theorem 2.24

$$\Delta \ll \sqrt{m}\, x^{1/4} \log(mx) + \sqrt{x} \log(mx), \qquad G_2\Delta_2 \ll \sqrt{m}\, x^{1/4} .$$

Finally we need the Hessian $H(f(\varphi_1, \varphi_2))$ and the function

$$f_2(y_1, y_2) = f(\varphi_1, \varphi_2) - y_1\varphi_1 - y_2\varphi_2 ,$$

where $\varphi_1 = \varphi_1(y_1, y_2)$, $\varphi_2 = \varphi_2(y_1, y_2)$ are solutions of

$$f_{t_1}(\varphi_1, \varphi_2) = y_1 , \qquad f_{t_2}(\varphi_1, \varphi_2) = y_2 .$$

We obtain

$$\varphi_1^2 = \frac{y_1^2 x}{m^2 + y_1^2 + y_2^2}, \qquad \varphi_2^2 = \frac{y_2^2 x}{m^2 + y_1^2 + y_2^2},$$

$$f_2(y_1, y_2) = -\sqrt{x(m^2 + y_1^2 + y_2^2)} ,$$

$$H(f(\varphi_1, \varphi_2)) = \frac{1}{m^2 x} (m^2 + y_1^2 + y_2^2)^2 .$$

Now (2.78) shows that

$$\sum_{(a_1, a_2) \in D} e^{-2\pi i m \sqrt{x - a_1^2 - a_2^2}} = i \sum_{(m_1, m_2) \in D_2} \frac{mx}{m^2 + m_1^2 + m_2^2} e^{-2\pi i \sqrt{x(m^2 + m_1^2 + m_2^2)}}$$
$$+ O(\textstyle\sum_1 T(\varrho(n_2), n_2)) + O(\sqrt{m}\, x^{1/4} \log (mx))$$
$$+ O(\sqrt{x} \log (mx)) , \tag{4.6}$$

$SC(\Sigma_1)$: n_2 runs over integers with $0 \leqq n_2 \leqq \sqrt{x/3}$ such that the points (t_1, n_2) lie on the bound of D with $t_1 = \varrho(n_2)$, $\varrho(t_2) = t_2$ or

$$\varrho(t_2) = \sqrt{\frac{x - t_2^2}{2}} .$$

If $\varrho(t) = t$, we obtain

$$\frac{\mathrm{d}}{\mathrm{d}t} f_{t_1}(\varrho(t), t) = \frac{\mathrm{d}}{\mathrm{d}t} \frac{mt}{\sqrt{x - 2t^2}} = \frac{mx}{(x - 2t^2)^{3/2}} \gg \frac{m}{\sqrt{x}} = \sqrt{\Lambda} .$$

In case of $\varrho(t) = \sqrt{\frac{x - t^2}{2}}$

$$f_{t_1}(\varrho(t), t) = m$$

is an integer. Hence, by means of (2.79) we get

$$\Sigma_1\, T(\varrho(n_2), n_2) \ll \sqrt{m}\, x^{1/4} + \sqrt{x} \log x .$$

If we substitute this estimate into (4.6), we get (4.5) immediately.

Theorem 4.2.

$$R_{2,3}(x) = \frac{4\pi}{3} x^{3/2} + O(x^{7/10} \log^4 x) . \tag{4.7}$$

Proof. Of course, we apply Theorem 1.8. In view of (1.17) and Lemma 4.1 we have to consider a sum of type

$$S = \sum_{m=1}^{\infty} c_m \sum_{(a_1, a_2) \in D} \mathrm{e}\,(-m \sqrt{x - a_1^2 - a_2^2}) .$$

Applying Theorem 1.8 with $s = 2$, then $c_m \ll K_m$, where K_m is defined by

$$K_m = \min\left(\frac{z^2}{m^3}, \frac{1}{m}\right), \qquad z > 1 .$$

(4.5) shows that

$$S = \frac{\mathrm{i}}{2} \sum_{m=1}^{\infty} \sum_{m_1=0}^{m} \sum_{m_2=0}^{m} \frac{m c_m \sqrt{x}}{m^2 + m_1^2 + m_2^2} \mathrm{e}(-\sqrt{x(m^2 + m_1^2 + m_2^2)})$$

$$+ O(\sqrt{z}\, x^{1/4} \log (zx)) + O(\sqrt{x} \log^2 (zx)) . \tag{4.8}$$

We first consider the sum

$$S_M = \sum_{m=M}^{2M} \sum_{m_1=0}^{m} \sum_{m_2=0}^{m} \frac{m c_m \sqrt{x}}{m^2 + m_1^2 + m_2^2} \mathrm{e}\,(-\sqrt{x(m^2 + m_1^2 + m_2^2)}) .$$

Since $0 \leqq m - m_1, m - m_2 \leqq 2M$, this sum can be written in the following form:

$$S_M = \frac{1}{16M^2} \sum_{r=1}^{4M} \sum_{s=1}^{4M} \sum_{k_1=0}^{2M} \sum_{k_2=0}^{2M} \sum_{m=M}^{2M} \sum_{m_1=0}^{2M} \sum_{m_2=0}^{2M} \frac{m c_m \sqrt{x}}{m^2 + m_1^2 + m_2^2}$$

$$\times \mathrm{e}\left(\frac{1}{4M} (r(m - m_1 - k_1) + s(m - m_2 - k_2)) - \sqrt{x(m^2 + m_1^2 + m_2^2)}\right) .$$

Let $\varrho(n)$ denote the number of representations of n by $n = m^2 + m_1^2$ with $M \leqq m \leqq 2M$, $0 \leqq m_1 \leqq 2M$, and let μ denote that number $s/4M$, $1 \leqq s \leqq 4M$, for which the following sum takes its maximum value. Then

$$S_M \ll \frac{K_M \sqrt{x}}{M} \sum_{r=1}^{4M} \sum_{s=1}^{4M} \left| \sum_{k_1=0}^{2M} \sum_{k_2=0}^{2M} \mathrm{e}\left(\frac{1}{4M} (rk_1 + sk_2)\right)\right|$$

$$\times \sum_{n=M^2}^{8M^2} \varrho(n) \left| \sum_{m_2=0}^{2M} \frac{1}{n + m_2^2} \mathrm{e}\,(-\mu m_2 - \sqrt{x(n + m_2^2)}) \right| .$$

Now

$$\sum_{r=1}^{4M}\left|\sum_{k=0}^{2M} \mathrm{e}\left(\frac{rk}{4M}\right)\right| \ll \sum_{r=1}^{4M} \min\left(M, \frac{1}{\left|\sin\frac{\pi r}{4M}\right|}\right) \ll M \log M .$$

Hence

$$S_M \ll \sqrt{x}\, K_M M \log^2 M \sum_{n=M^2}^{8M^2} \varrho(n) \left|\sum_{m_2=0}^{2M} \frac{\mathrm{e}\,(-\mu m_2 - \sqrt{x(n+m_2^2)})}{n+m_2^2}\right| .$$

Applying the inequality of Schwarz and using the fact that

$$\sum_{n=M^2}^{8M^2} \varrho^2(n) \ll M^2 \log M ,$$

we obtain

$$S_M^2 \ll x K_M^2 M^4 \log^5 M \sum_{n=M^2}^{8M^2} \left|\sum_{m_2=0}^{2M} \frac{\mathrm{e}\,(-\mu m_2 - \sqrt{x(n+m_2^2)})}{n+m_2^2}\right|^2$$

$$\ll x K_M^2 M^4 \log^5 M \sum_{u=0}^{2M} \sum_{v=0}^{2M} \sum_{n=M^2}^{8M^2} \frac{\mathrm{e}(f(n))}{(n+u^2)\,(n+v^2)} ,$$

where

$$f(t) = -\mu(u-v) - \sqrt{x}(\sqrt{t+u^2} - \sqrt{t+v^2}) .$$

In case of $u = v$ we get an error term of order $x K_M^2 M^3 \log^5 M$. Now suppose that $u > v$. We estimate the sum

$$T = \sum_{n=M^2}^{M'} \mathrm{e}^{2\pi i f(n)} \qquad (M^2 < M' \leqq 8M^2)$$

by means of van der Corput's Theorem 2.1. From

$$f''(t) = \frac{1}{4}\sqrt{x}\,((t+u^2)^{-3/2} - (t+v^2)^{-3/2})$$

it is seen that

$$|f''(t)| \gg \frac{u^2-v^2}{M^5}\sqrt{x} .$$

Therefore, we obtain from (2.6)

$$T \ll \sqrt{\frac{u^2-v^2}{M}}\, x^{1/4} + \frac{M^{5/2}}{\sqrt{u^2-v^2}}\, x^{-1/4} .$$

A similar result holds in case of $v > u$. Hence

$$S_M^2 \ll xK_M^2 \log^5 M \sum_{v<u\leq 2M} \left(\sqrt{\frac{u^2-v^2}{M}}\, x^{1/4} + \frac{M^{5/2}}{\sqrt{u^2-v^2}}\, x^{-1/4} \right)$$

$$+ xK_M^2 M^3 \log^5 M$$

$$\ll xK_M^2 \log^6 M (M^{5/2} x^{1/4} + M^{7/2} x^{-1/4} + M^3) .$$

Then

$$S_M \ll x^{5/8} K_M M^{5/4} \log^3 M$$

provided that $M^2 \ll x$. In the opposite case we use trivial estimation for the corresponding part of S. If z denotes a number with $z^2 < x$, we obtain from (4.8)

$$S \ll x^{5/8} z^{1/4} \log^4 x + \sum_{m \geq \sqrt{x}} \left(\frac{z}{m}\right)^2 \sqrt{x} + \sqrt{z}\, x^{1/4} \log x + \sqrt{x} \log^2 x$$

$$\ll x^{5/8} z^{1/4} \log^4 x + z^2 + \sqrt{z}\, x^{1/4} \log x + \sqrt{x} \log^2 x .$$

Now (1.17) shows that

$$\sum_{(a_1, a_2) \in D} \Psi(\sqrt{x - a_1^2 - a_2^2}) \ll \frac{x}{z} + x^{5/8} z^{1/4} \log^4 x + z^2$$

$$+ \sqrt{z}\, x^{1/4} \log x + \sqrt{x} \log^2 x .$$

The first two terms are of the same order apart from the logarithmic factor if we put $z = x^{3/10}$. This proves the estimate (4.7).

A historical outline of the development of the sphere problem

Let $\Delta_{2,3}(x)$ be defined by

$$R_{2,3}(x) = \frac{4\pi}{3} x^{3/2} + \Delta_{2,3}(x) ,$$

and let ϑ denote the value

$$\vartheta = \inf \{\alpha : \Delta_{2,3}(x) \ll x^\alpha\} .$$

Then the inequality $1/2 \leq \vartheta \leq 1$ follows from the trivial estimates. There are the following improvements of the upper bound:

$\vartheta \leq \frac{5}{6} = 0{,}8\overline{3}$ (W. Sierpiński [2], 1909),

$\vartheta \leq \frac{3}{4} = 0{,}75$ (E. Landau [1], 1912),

$$\vartheta \leqq \frac{7}{10} = 0{,}7 \qquad \text{(I. M. Vinogradov [1], 1935),}$$

$$\vartheta \leqq \frac{113}{162} = 0{,}6975\ldots \qquad \text{(I. M. Vinogradov [2], 1949),}$$

$$\vartheta \leqq \frac{11}{16} = 0{,}6875 \qquad \text{(I. M. Vinogradov [3], 1955),}$$

$$\vartheta \leqq \frac{19}{28} = 0{,}6785\ldots \qquad \text{(I. M. Vinogradov [4], 1960),}$$

$$\vartheta \leqq \frac{35}{52} = 0{,}6730\ldots \qquad \text{(Chen Jing-run [5], 1963),}$$

$$\vartheta \leqq \frac{2}{3} = 0{,}\bar{6} \qquad \text{(I. M. Vinogradov [6], 1963).}$$

G. Szegö [1] proved the result

$$R_{2,3}(x) = \frac{4\pi}{3} x^{3/2} + \Omega(\sqrt{x \log x})$$

in the other direction in 1926.

4.2. LATTICE POINTS IN GENERALIZED SPHERES

4.2.1. Preliminaries

First of all we consider the trivial estimates for the function $R_{k,n}(x)$ given by Definition 4.1. Again, we assume that $k \geqq 2$. For the sake of simplicity even the number k should be an integer in Theorem 4.4.

Theorem 4.3. *Let* $V_{k,n}$ *be defined by*

$$V_{k,n} = \left(\frac{2}{k}\right)^n \frac{\Gamma^n(1/k)}{\Gamma(1 + n/k)}\,. \tag{4.9}$$

Then

$$R_{k,n}(x) = V_{k,n} x^{n/k} + O(x^{(n-1)/k})\,. \tag{4.10}$$

Proof. From Section 3.3.1 it is seen that (4.10) holds for $n = 2$. Now suppose (4.10) to be true for $n - 1$. Then

$$R_{k,n}(x) = \sum_{|a|^k \leq x} R_{k,n-1}(x - |a|^k)$$

$$= \sum_{|a|^k \leqq x} \{V_{k,n-1}(x - |a|^k)^{(n-1)/k} + O((x - |a|^k)^{(n-2)/k})\}$$

$$= 2V_{k,n-1} \int_0^{x^{1/k}} (x - t^k)^{(n-1)/k}\, dt + O(x^{(n-1)/k})$$

$$= V_{k,n} x^{n/k} + O(x^{(n-1)/k}) .$$

Hence (4.10) is proved for n, and the induction is complete.

Theorem 4.4. *Suppose that k is an integer. If $n \geqq k$, then*

$$R_{k,n}(x) = V_{k,n} x^{n/k} + \Omega(x^{n/k-1}) , \tag{4.11}$$

and if $n < k$, we have for every $\varepsilon > 0$

$$R_{k,n}(x) = V_{k,n} x^{n/k} + \Omega\left(x^{\frac{n-1}{k}\left(1-\frac{1}{k-n+1}\right)-\varepsilon}\right). \tag{4.12}$$

Proof. In case of $n \geqq k$ suppose that on the contrary,

$$R_{k,n}(x) = V_{k,n} x^{n/k} + o(x^{n/k-1}) .$$

If x is an integer, we get

$$0 = R_{k,n}(x + 1/2) - R_{k,n}(x) = V_{k,n}((x + 1/2)^{n/k} - x^{n/k}) + o(x^{n/k-1})$$
$$= \frac{n}{2k} V_{k,n} x^{n/k-1} + o(x^{n/k-1}) .$$

This gives a contradiction, and (4.11) is proved.

We are going to prove (4.12). Let a, x, y be integers such that $a^k = x + 1$ and $2 < y < x$. Then

$$R_{k,n}(x + y) - R_{k,n}(x) = \Sigma_1\, 1 \geqq \Sigma_2\, 1 ,$$
$$SC(\Sigma_1):\quad x < |a_1|^k + \ldots + |a_{n-1}|^k + |a_n|^k \leqq x + y ,$$
$$SC(\Sigma_2):\quad |a_1|^k + \ldots + |a_{n-1}|^k \leqq y - 1 .$$

Hence

$$R_{k,n}(x + y) - R_{k,n}(x) \geqq R_{k,n-1}(y - 1) \gg y^{(n-1)/k} .$$

Assuming that

$$R_{k,n}(x) = V_{k,n} x^{\frac{n}{k}} + o\left(x^{\frac{n-1}{k}\left(1-\frac{1}{k-n+1}\right)-\varepsilon}\right),$$

we obtain

$$y^{\frac{n-1}{k}} \ll R_{k,n}(x + y) - R_{k,n}(x) \ll x^{\frac{n}{k}-1} y + o\left(x^{\frac{n-1}{k}\left(1-\frac{1}{k-n+1}\right)-\varepsilon}\right).$$

Choosing

$$y = \left[x^{1-\frac{1}{k-n+1}-\varepsilon}\right],$$

this inequality forms a contradiction. This proves (4.12).

The first progress in estimating $R_{k,n}(x)$ was obtained by B. Randol [2] in case of even k. He proved

$$R_{k,n}(x) = V_{k,n}x^{n/k} + \begin{cases} O(x^{(n-1)(k-1)/k^2}) & \text{for} \quad k > n+1\,, \\ O(x^{n(n-1)/k(n+1)} \log x) & \text{for} \quad k \leqq n+1\,. \end{cases}$$

Another approach to this problem was developed by E. Krätzel [8] who showed that B. Randol's result remains to be true if k is an odd integer. Analyzing the proof it is readily seen that the restriction of k to an integer is unnecessary. Moreover, B. Randol remarked that the estimation for even $k > n + 1$ cannot be improved. This follows from E. Krätzel's estimation

$$R_{k,n}(x) = V_{k,n}x^{n/k} + nV_{k,n-1}\psi_{n/k}^{(k)}(x^{1/k}) + O(x^{\alpha}\log^2 x)\,,$$

$$\alpha = \frac{(3n-4)\,k - 5n + 8}{3k(k-1)}\,,$$

for every $k > n + 1$, where the ψ-function is defined by (3.55). Note that we obtain a second main term analogously to the case $n = 2$ considered in Section 3.3.2. In the next section we shall explain the basic estimates. Though it is possible to improve all these estimates slightly, we shall do this only in case of $n = 3$, because here we even can find a third main term. In principle, the Ω-estimation in the last section shows that we may have up to $n - 1$ main terms. However, with our method we are not in a position to prove sufficiently good O-estimates.

4.2.2. *The basic estimate*

We first consider a special representation of $R_{k,n}(x)$. It is widely known that a sum can be approximated by an integral in most cases. Therefore, we replace in the representation

$$R_{k,n}(x) = \Sigma\, 1\,, \qquad SC(\Sigma)\colon \ |a_1|^k + |a_2|^k + \ldots + |a_n|^k \leqq x$$

each sum by an integral and correct the mistake by the difference of sum and integral. It may be convenient to suppose that $n \geqq 1$. Then the following statement is straightforward.

Lemma 4.3. *If $n \geqq 1$, then*

$$R_{k,n}(x) = V_{k,n}x^{n/k} + \sum_{r=1}^{n-1} H_{k,n,r}(x) + \Delta_{k,n}(x)\,, \tag{4.13}$$

where

$$H_{k,n,r}(x) = \binom{n}{r}\frac{n-r}{k}V_{k,n-r}\int_0^x (x-t)^{\frac{n-r}{k}-1}\Delta_{k,r}(t)\,dt \tag{4.14}$$

for $1 \leqq r \leqq n-1$,

$$\Delta_{k,n}(x) = \sum_{j=0}^{n} (-1)^{n-j} \binom{n}{j} \sum \int \mathrm{d}t_{j+1} \cdot \ldots \cdot \mathrm{d}t_n \,, \tag{4.15}$$

$$SC(\Sigma \textstyle\int)\colon \; |a_1|^k + \ldots + |a_j|^k + |t_{j+1}|^k + \ldots + |t_n|^k \leqq x \,.$$

In particular, we have

$$\Delta_{k,1}(x) = -2\psi(x^{1/k})$$

and

$$H_{k,n,1}(x) = nV_{k,n-1}\psi^{(k)}_{n/k}(x^{1/k}) \tag{4.16}$$

for $n > 1$.

Proof. We have

$$R_{k,n}(x) = \sum_{j=0}^{n} \binom{n}{j} \sum_{r=j}^{n} (-1)^{r-j} \binom{n-j}{r-j} \sum \int \mathrm{d}t_{j+1} \cdot \ldots \cdot \mathrm{d}t_n$$

$$= \sum_{r=0}^{n} \binom{n}{r} \sum_{j=0}^{r} (-1)^{r-j} \binom{r}{j} \sum \int \mathrm{d}t_{j+1} \cdot \ldots \cdot \mathrm{d}t_n \,,$$

where the summation and integration condition in $\Sigma \int$ is the same as in the lemma. It is clear that the term $r = 0$ leads to the main term $V_{k,n}x^{n/k}$ and that the term $r = n$ gives the remainder $\Delta_{k,n}(x)$. Putting

$$H_{k,n,r}(x) = \binom{n}{r} \sum_{j=0}^{r} (-1)^{r-j} \binom{r}{j} \sum \int \mathrm{d}t_{j+1} \cdot \ldots \cdot \mathrm{d}t_n$$

it is easily seen that

$$H_{k,n,r}(x) = \binom{n}{r} V_{k,n-r} \sum_{j=0}^{r} (-1)^{r-j} \binom{r}{j} \sum \int_1 T \,\mathrm{d}t_{j+1} \cdot \ldots \cdot \mathrm{d}t_r \,,$$

$$T = (x - |a_1|^k - \ldots - |a_j|^k - |t_{j+1}|^k - \ldots - |t_r|^k)^{\frac{n-r}{k}} \,,$$

$$SC(\Sigma \textstyle\int_1)\colon \; |a_1|^k + \ldots + |a_j|^k + |t_{j+1}|^k + \ldots + |t_r|^k \leqq x \,.$$

Using the notation (4.15), the representation (4.14) now follows immediately. Obviously, we have

$$R_{k,1}(x) = 2x^{1/k} - 2\psi(x^{1/k})$$

such that $\Delta_{k,1}(x)$ is given as above. Now (4.14) shows that

$$H_{k,n,1}(x) = -\frac{2n(n-1)}{k} V_{k,n-1} \int_0^x (x-t)^{\frac{n-1}{k}-1} \psi(t^{1/k}) \,\mathrm{d}t$$

for $n > 1$. By using the Fourier expansion of the ψ-function and by integrating by parts, we obtain

$$H_{k,n,1}(x) = \frac{2n(n-1)}{\pi} V_{k,n-1} \sum_{m=1}^{\infty} \frac{1}{m} \int_0^{x^{1/k}} (x - t^k)^{\frac{n-1}{k}} t^{k-1} \sin 2\pi mt \, dt$$

$$= 4nV_{k,n-1} \sum_{m=1}^{\infty} \int_0^{x^{1/k}} (x - t^k)^{\frac{n-1}{k}} \cos 2\pi mt \, dt$$

$$= nV_{k,n-1} \psi_{n/k}^{(k)}(x^{1/k}) .$$

This completes the proof of the lemma.

Lemma 4.4. *The representations* (4.13) *and* (4.14) *hold with*

$$\Delta_{k,r}(x) = -2^r r! \sum \psi((x - a_2^k - a_3^k - \ldots - a_r^k)^{1/k}) + O\left(x^{\frac{r-2}{k}}\right), \tag{4.17}$$

$$SC(\Sigma): \ 0 \leqq x - a_2^k - a_3^k - \ldots - a_r^k < a_2^k \leqq a_3^k \leqq \ldots \leqq a_r^k$$

if $r = 2, 3, \ldots, n$ *and*

$$\Delta_{k,1}(x) = -2\psi(x^{1/k}) .$$

Proof. Let $R_{k,n}(x)$ be written in the form

$$R_{k,n}(x) = 2^n n! \sum{}' 1 + O\left(x^{\frac{n-2}{k}}\right),$$

$SC(\Sigma')$: $a_1^k + a_2^k + \ldots + a_n^k \leqq x, 0 \leqq a_1 \leqq a_2 \leqq \ldots \leqq a_n$. Each term with $a_r = a_{r+1} = \ldots = a_{r+s}$, where $0 \leqq r \leqq n - 1, 1 \leqq s \leqq n, a_0 = 0$, gets a factor 2^{-s}.

Taking the sum over a_1, we obtain

$$R_{k,n}(x) = 2^n n! \, \Sigma'_{1,1} \left\{[(x - a_2^k - \ldots - a_n^k)^{1/k}] + \frac{1}{2}\right\}$$

$$+ 2^n n! \, \Sigma'_{1,2}\, a_2 + O\left(x^{\frac{n-2}{k}}\right),$$

$$SC(\Sigma'_{1,1}): \ 0 \leqq x - a_2^k - \ldots - a_n^k < a_2^k \leqq a_3^k \leqq \ldots \leqq a_n^k ,$$

$$SC(\Sigma'_{1,2}): \ a_2^k \leqq x - a_2^k - \ldots - a_n^k , \ a_2 \leqq a_3 \leqq \ldots \leqq a_n .$$

Using the representation $[y] + 1/2 = y - \psi(y)$, we can form an integral and moreover, we obtain the remainder $\Delta_{k,n}(x)$ in the notation of (4.17). Thus

$$R_{k,n}(x) = 2^n n! \, \Sigma' \textstyle\int_1 dt_1 + \Delta_{k,n}(x) + O\left(x^{\frac{n-2}{k}}\right),$$

$$SC(\Sigma' \textstyle\int_1): \ t_1^k + a_2^k + \ldots + a_n^k \leqq x , \ 0 \leqq t_1 \leqq a_2 \leqq \ldots \leqq a_n .$$

Summing now over a_2, then

$$R_{k,n}(x) = 2^n n!\, \Sigma' \int_{2,1} \{[(x - t_1^k - a_3^k - \ldots - a_n^k)^{1/k}] - [t_1]\}\, dt_1$$

$$+ 2^n n!\, \Sigma' \int_{2,2} \left\{a_3 - \frac{1}{2} - [t_1]\right\} dt_1 + \Delta_{k,n}(x) + O\left(x^{\frac{n-2}{k}}\right),$$

$$SC(\Sigma' \textstyle\int_{2,1}):\ 0 \leqq x - t_1^k - a_3^k - \ldots - a_n^k < a_3^k \leqq a_4^k \leqq \ldots \leqq a_n^k,\ t_1^k \leqq \frac{x}{n},$$

$$SC(\Sigma' \textstyle\int_{2,2}):\ a_3^k \leqq x - t_1^k - a_3^k - \ldots - a_n^k,\ 0 \leqq t_1 \leqq a_3 \leqq a_4 \leqq \ldots \leqq a_n.$$

Replacing again the greatest integers by the corresponding values and ψ-functions, we can form a second integral. Hence

$$R_{k,n}(x) = 2^n n!\, \Sigma' \int_2 dt_1\, dt_2 - 2^n n!\, \Sigma' \int_{2,1} \psi((x - t_1^k - a_3^k - \ldots - a_n^k)^{1/k})\, dt_1$$

$$+ 2^n n! \{\Sigma' \textstyle\int_{2,1} + \Sigma' \int_{2,2} \psi(t_1)\}\, dt_1 + \Delta_{k,n}(x) + O\left(x^{\frac{n-2}{k}}\right),$$

$$SC(\Sigma' \textstyle\int_2):\ t_1^k + t_2^k + a_3^k + \ldots + a_n^k \leqq x,\ 0 \leqq t_1 \leqq t_2 \leqq a_3 \leqq \ldots \leqq a_n.$$

Since an integral of the ψ-function is bounded, the third term is of order $x^{(n-2)/k}$. In the second term the restriction $t_1^k \leqq \frac{x}{n}$ is unnecessary, because it is easy to verify by partial integration that the additional integral is of order $x^{(n-2)/k}$. Consequently, we obtain for the second term

$$2n \int_0^{x^{1/k}} \Delta_{k,n-1}(x - t_1^k)\, dt_1 + O\left(x^{\frac{n-2}{k}}\right) = H_{k,n,n-1}(x) + O\left(x^{\frac{n-2}{k}}\right).$$

Therefore

$$R_{k,n}(x) = 2^n n!\, \Sigma' \int_2 dt_1\, dt_2 + H_{k,n,n-1}(x) + \Delta_{k,n}(x) + O\left(x^{\frac{n-2}{k}}\right).$$

We now proceed in the same way such that we obtain after the last but one step

$$R_{k,n}(x) = 2^n n!\, \Sigma' \int_{n-1} dt_1 \cdot \ldots \cdot dt_{n-1} + \sum_{r=2}^{n-1} H_{k,n,r}(x) + \Delta_{k,n}(x) + O\left(x^{\frac{n-2}{k}}\right),$$

$$SC(\Sigma' \textstyle\int_{n-1}):\ t_1^k + \ldots + t_{n-1}^k + a_n^k \leqq x,\ 0 \leqq t_1 \leqq \ldots \leqq t_{n-1} \leqq a_n.$$

Finally, we perform the sum over a_n:

$$R_{k,n}(x) = 2^n n! \int_{n,1} \{[(x - t_1^k - \ldots - t_{n-1}^k)^{1/k}] - [t_{n-1}]\} \, dt_1 \cdot \ldots \cdot dt_{n-1} + \sum_{r=2}^{n-1} H_{k,n,r}(x) + \Delta_{k,n}(x) + O\left(x^{\frac{n-2}{k}}\right),$$

$$IC(\textstyle\int_{n,1}): \; t_1^k + \ldots + t_{n-2}^k + 2t_{n-1}^k \leqq x, \; 0 \leqq t_1 \leqq t_2 \leqq \ldots \leqq t_{n-1},$$

$$R_{k,n}(x) = 2^n n! \int_n dt_1 \cdot \ldots \cdot dt_n - 2^n n! \int_{n,1} \{\psi((x - t_1^k - \ldots - t_{n-1}^k)^{1/k}) - \psi(t_{n-1})\} \, dt_1 \cdot \ldots \cdot dt_{n-1} + \sum_{r=2}^{n-1} H_{k,n,r}(x) + \Delta_{k,n}(x) + O\left(x^{\frac{n-2}{k}}\right),$$

$$IC(\textstyle\int_n): \; t_1^k + \ldots + t_n^k \leqq x, \; 0 \leqq t_1 \leqq t_2 \leqq \ldots \leqq t_n.$$

Again making use of the fact that an integral over the ψ-function is bounded, we see that the integral over $\psi(t_{n-1})$ is of order $x^{(n-2)/k}$ and that the restriction is unnecessary with respect to the integral of the other ψ-function. Therefore, this integral becomes

$$-2^n n! \int_{n,1} \psi((x - t_1^k - \ldots - t_{n-1}^k)^{1/k}) \, dt_1 \cdot \ldots \cdot dt_{n-1}$$

$$= -2^n n! \int_{n,2} \psi((x - t_1^k - \ldots - t_{n-1}^k)^{1/k}) \, dt_1 \cdot \ldots \cdot dt_{n-1} + O\left(x^{\frac{n-2}{k}}\right)$$

$$= 2^{n-1} n \int_{n,3} \Delta_{k,1}(x - t_1^k - \ldots - t_{n-1}^k) \, dt_1 \cdot \ldots \cdot dt_{n-1} + O\left(x^{\frac{n-2}{k}}\right)$$

$$= 2^{n-1} \frac{n}{k} \int_{n,4} (x - t_1^k - \ldots - t_{n-2}^k - t)^{1/k-1} \Delta_{k,1}(t) dt_1 \cdot \ldots \cdot dt_{n-2} dt + O\left(x^{\frac{n-2}{k}}\right)$$

$$= \frac{n(n-1)}{k} V_{k,n-1} \int_0^x (x - t)^{\frac{n-1}{k} - 1} \Delta_{k,1}(t) \, dt + O\left(x^{\frac{n-2}{k}}\right)$$

$$= H_{k,n,1}(x) + O\left(x^{\frac{n-2}{k}}\right),$$

$$IC(\textstyle\int_{n,2}): \; t_1^k + \ldots + t_{n-1}^k \leqq x, \; 0 \leqq t_1 \leqq t_2 \leqq \ldots \leqq t_{n-1},$$

$$IC(\int_{n,3}):\ t_1^k + \dots + t_{n-1}^k \leqq x\,,\quad t_1, t_2, \dots, t_{n-1} \geqq 0\,,$$

$$IC(\int_{n,4}):\ t_1^k + \dots + t_{n-2}^k + t \leqq x\,,\quad t_1, \dots, t_{n-2}, t \geqq 0\,.$$

Because of

$$2^n n! \int_n dt_1 \cdot \dots \cdot dt_n = V_{k,n} x^{n/k}$$

we conclude that

$$R_{k,n}(x) = V_{k,n} x^{n/k} + \sum_{r=1}^{n-1} H_{k,n,r}(x) + \Delta_{k,n}(x) + O\left(x^{\frac{n-2}{k}}\right).$$

This means that (4.13) and (4.14) hold with the asymptotic representations (4.17) of $\Delta_{k,r}(x)$. Now the lemma is proved.

A first non-trivial estimate of $R_{k,n}(x)$ can be obtained by J. G. van der Corput's Theorem 2.3. Even this estimate shows that we can get a second main term corresponding to Theorem 3.17B.

Theorem 4.5.

$$R_{k,n}(x) = V_{k,n} x^{n/k} + nV_{k,n-1} \psi_{n/k}^{(k)}(x^{1/k}) + O(x^{(n-4/3)/k})\,. \tag{4.18}$$

Proof. Let $f(t) = (x - t^k - a_3^k - \dots - a_r^k)^{1/k}$. Then

$$f''(t) = -(k-1)\, t^{k-2}\, (x - a_3^k - \dots - a_r^k)\, (x - t^k - a_3^k - \dots - a_r^k)^{1/k-2}\,.$$

From (2.8) and (4.17) we obtain for $r = 2$

$$\Delta_{k,r}(x) \ll \sum \int |f''(t)|^{1/3}\, dt + O(x^{(r-3/2)/k}),$$

$$SC(\Sigma \int):\ 0 \leqq x - t^k - a_3^k - \dots - a_r^k < t^k \leqq a_3^k \leqq \dots \leqq a_r^k\,.$$

Clearly, the application of Theorem 2.3 requires the inequality

$$0 < x - t^k - a_3^k - \dots - a_r^k\,,$$

and (4.17) requires $t \geqq 1$. But it is easily seen that the estimation with the mentioned summation condition is correct. Hence

$$\Delta_{k,r}(x) \ll x^{(r-4/3)/k}\,.$$

From (4.14) it is easily seen that

$$H_{k,n,r}(x) \ll x^{(n-4/3)/k}$$

for $r \geqq 2$. Now (4.18) follows from (4.13) and (4.16).

From (3.50) and (3.55) it is seen that the second main term in (4.38) has the precise order $x^{(n-1)(k-1)/k^2}$. Note that this exponent is larger than the exponent of the error term in (4.18) if $k > 3(n-1)$.

By much more work it is possible to improve the result (4.18). The remainder $\Delta_{k,n}(x)$ is given by (4.17) which may be written by

$$\Delta_{k,n}(x) = 2^n n! \sum_{\mathbf{n} \in D} \psi(f(\mathbf{n})) + O\left(x^{\frac{n-2}{k}}\right), \tag{4.19}$$

where

$$f(\mathbf{t}) = (x - t_1^k - t_2^k - \dots - t_{n-1}^k)^{1/k}, \tag{4.20}$$

$\mathbf{t} = (t_1, t_2, \dots, t_{n-1})$, and $\mathbf{n}$ runs over all lattice points of the domain

$$D = \left\{\mathbf{t}\colon \frac{1}{2}(x - t_2^k - \dots - t_{n-1}^k) \leqq t_1^k \leqq x - t_2^k - \dots - t_{n-1}^k,\right.$$
$$\left. 0 \leqq t_1 \leqq t_2 \leqq \dots \leqq t_{n-1}\right\}. \tag{4.21}$$

We divide the domain D into subdomains $D(\mathbf{v})$ described in the following lemma and apply Theorem 2.29 with $p = n - 1$ to each subdomain. By means of the resulting estimate (4.23) it is then easy to get an estimate for $\Delta_{k,n}(x)$.

Lemma 4.5. *Let* $\mathbf{v} = (v_1, v_2, \dots, v_{n-1})$, *where the* v_j *are non-negative integers. Let* $f(\mathbf{t})$ *be defined by* (4.20), *and let* $D(\mathbf{v})$ *denote the domain*

$$D(\mathbf{v}) = \{\mathbf{t}\colon q_{v_j}(x - t_{j+1}^k - \dots - t_{n-1}^k) \leqq t_j^k \leqq q_{v_j+1}(x - t_{j+1}^k - \dots - t_{n-1}^k),$$
$$j = 1, 2, \dots, n-1,\ 0 \leqq t_1 \leqq t_2 \leqq \dots \leqq t_{n-1}\}, \tag{4.22}$$

where q_{v_j} *is defined by*

$$q_{v_j} = 2^{\frac{k}{k-1}v_j}\left(j + 2^{\frac{k}{k-1}v_j}\right)^{-1}.$$

If

$$2^{\gamma_j} \ll x^{1/k}, \qquad \gamma_j = \frac{k}{k-1} v_j + \frac{1}{k-1} \sum_{r=j+1}^{n-1} v_r$$

for $j = 1, 2, \dots, n-1$, *then*

$$\sum_{\mathbf{n} \in D(\mathbf{v})} \psi(f(\mathbf{n})) \ll 2^{-\vartheta(\mathbf{v})} x^{\frac{n(n-1)}{(n+1)k}} + x^{\frac{n-2}{k}} \log^2 x, \tag{4.23}$$

where $\vartheta(\mathbf{v})$ *is given by*

$$\vartheta(\mathbf{v}) = \frac{1}{k-1}\left(1 + \frac{k-2}{n+1}\right) \sum_{r=1}^{n-1} r v_r.$$

Proof. We apply Theorem 2.29 with $p = n - 1$, $-f(t)$, where $f(t)$ is given by (4.20), and the domain $D = D(\mathbf{v})$. Since the method of exponential sums yields the same estimates for the sums $\Sigma\,\psi(f(\boldsymbol{n}))$ and $\Sigma\,\psi(-f(\boldsymbol{n}))$, we may replace the function $f(\boldsymbol{t})$ by $-f(\boldsymbol{t})$, which gives some formal simplifications.

We first compute the volume of $D(\boldsymbol{v})$. t_j varies in an interval of length

$$(q_{v_{j+1}}^{1/k} - q_{v_j}^{1/k})\,(x - t_{j+1}^k - \dots - t_{n-1}^k)^{1/k}$$

$$\asymp 2^{-\frac{k}{k-1}v_j}(x - t_{j+1}^k - \dots - t_{n-1}^k)^{1/k} \asymp 2^{-\gamma_j}x^{1/k}\,.$$

Hence

$$|D(\boldsymbol{v})| \asymp 2^{-\delta}x^{(n-1)/k}\,, \qquad \delta = \sum_{j=1}^{n-1}\gamma_j\,.$$

Because of $2^{\gamma_j} \ll x^{1/k}$ the number of lattice points in $D(\boldsymbol{v})$ is of order of the volume.

Later we need the volumes of the domains $D_j(\boldsymbol{v})$ and of the suitably chosen domains $R_{j,\mu}$ which may be computed at this point. $D_j(\boldsymbol{v})$ denotes the image of $D(\boldsymbol{v})$ under the mapping

$$y_\varrho = -f_{t_\varrho}(t) = t_\varrho^{k-1}(x - t_1^k - \dots - t_{n-1}^k)^{1/k-1} \quad \text{for} \quad \varrho = 1, 2, \dots, j\,,$$
$$y_\varrho = t_\varrho \qquad\qquad \text{for} \quad \varrho = j+1, \dots, n-1.$$

Note that $y_1 \leqq y_2 \leqq \dots \leqq y_j$ and $y_r \asymp 2^{v_1 + \dots + v_r}$ for $r = 1, 2, \dots, j$. Hence

$$|D_j(\boldsymbol{v})| \asymp 2^{\omega_j - \delta_j}x^{\frac{n-j-1}{k}}\,, \qquad \omega_j = \sum_{r=1}^{j}(j - r + 1)\,v_r\,, \qquad \delta_j = \sum_{r=j+1}^{n-1}\gamma_r\,.$$

The domain $R_{j,\mu}$ described in (H) of Section 2.3 can be chosen such that

$$|R_{j,\mu}| \asymp \begin{cases} 2^{-v_1 - \dots - v_\mu}\,|D_j(\boldsymbol{v})| & \text{for} \quad \mu \leqq j\,, \\ 2^{\gamma_\mu}x^{-1/k}\,|D_j(\boldsymbol{v})| & \text{for} \quad \mu > j\,. \end{cases}$$

Note that the conditions (G) and (H) of Section 2.3 are satisfied. In particular, the number of lattice points of $D_j(\boldsymbol{v})$ is of order of their volume.

We now compute the Hessian $H_j(-f)$. We have

$$-f_{t_i} = t_i^{k-1}(x - t_1^k - \dots - t_{n-1}^k)^{1/k-1}\,,$$
$$-f_{t_it_i} = (k-1)\,t_i^{k-2}(x + t_i^k - t_1^k - \dots - t_{n-1}^k)\,(x - t_1^k - \dots - t_{n-1}^k)^{1/k-2}\,,$$
$$-f_{t_it_j} = (k-1)\,(t_it_j)^{k-1}\,(x - t_1^k - \dots - t_{n-1}^k)^{1/k-2} \qquad (i \neq j)\,.$$

It is easily seen that the Hessian $H_j(-f)$ can be written in the form

$$H_j(-f) = (k-1)^j\,(t_1 \cdot \ldots \cdot t_j)^{k-2}\,(x - t_1^k - \dots - t_{n-1}^k)^{j/k-2j}\,K_j(x)$$

$(j = 1, 2, \dots, n-1)$, where $K_j(x)$ is defined by the determinant

$$K_j(x) = \begin{vmatrix} z + t_1^k & t_1 t_2^{k-1} & \cdots & t_1 t_j^{k-1} \\ t_2 t_1^{k-1} & z + t_2^k & & \vdots \\ \vdots & & \ddots & t_{j-1} t_j^{k-1} \\ t_j t_1^{k-1} & \cdots & t_j t_{j-1}^{k-1} & z + t_j^k \end{vmatrix}$$

and z is given by $z = x - t_1^k - \ldots - t_{n-1}^k$. It is easy to verify that the function $K_j(x)$ contains the factors z^{j-1} and $x - t_{j+1}^k - \ldots - t_{n-1}^k$. Obviously, the coefficient of x^j is 1. Hence

$$K_j(x) = (x - t_{j+1}^k - \ldots - t_{n-1}^k)\,(x - t_1^k - \ldots - t_{n-1}^k)^{j-1}\,.$$

It can now easily seen that

$$|H_j(-f)| \asymp \Lambda_j = 2^{\sigma_j} x^{-j/k}\,,$$

$$\sigma_j = \sum_{r=1}^{j} \left(j - r + 2 + \frac{r}{k-1} \right) v_r + \frac{j}{k-1} \sum_{r=j+1}^{n-1} v_r\,.$$

The function $f_{j-1}(y)$ is defined by

$$f_{j-1}(y) = f(\varphi_1, \ldots, \varphi_{j-1}, y_j, \ldots, y_{n-1}) - \sum_{\varrho=1}^{j-1} y_\varrho \varphi_\varrho(y)\,,$$

in $D_{j-1}(v)$, where the functions $\varphi_\varrho(y)$ are solutions of

$$-f_{t_\varrho}(\varphi_1, \ldots, \varphi_{j-1}, y_j, \ldots, y_{n-1}) = y_\varrho \qquad (\varrho = 1, 2, \ldots, j-1)\,.$$

We find

$$\varphi_\varrho^k(y) = y^{\frac{1}{k-1}}(x - y_j^k - \ldots - y_{n-1}^k) \left(1 + y_1^{\frac{k}{k-1}} + \ldots + y_{j-1}^{\frac{k}{k-1}}\right)^{-1},$$

$$f_{j-1}(y) = -\left(1 + y_1^{\frac{k}{k-1}} + \ldots + y_{j-1}^{\frac{k}{k-1}}\right)^{1-\frac{1}{k}} (x - y_j^k - \ldots - y_{n-1}^k)^{\frac{1}{k}}\,.$$

From the third derivative

$$\begin{aligned} \frac{\partial^3}{\partial y_j^3} f_{j-1}(y) = (k-1) &\left(1 + y_1^{\frac{k}{k-1}} + \ldots + y_{j-1}^{\frac{k}{k-1}}\right)^{1-\frac{1}{k}} \\ &\times y_j^{k-3}(x - y_{j+1}^k - \ldots - y_{n-1}^k)\,(x - y_j^k - \ldots - y_{n-1}^k)^{1/k-3} \\ &\times \{(k-2)\,(x - y_{j+1}^k - \ldots - y_{n-1}^k) + (k+1)\, y_j^k\} \end{aligned}$$

we obtain that

$$\left| \frac{\partial^3}{\partial y_j^3} f_{j-1}(y) \right| \asymp L_j$$

with

$$L_j = 2^{\tau_j}(x - y_{j+1}^k - \ldots - y_{n-1}^k)^{-2/k}, \qquad \tau_j = \sum_{r=1}^{j-1} v_r + \frac{3k-1}{k-1} v_j .$$

Then

$$L_j \ll \lambda_j = 2^{\sigma_j - \sigma_{j-1} + \gamma_j} x^{-2/k}$$

and, because of $\gamma_j \leqq \sigma_j - \sigma_{j-1}$,

$$\lambda_j \ll \Lambda_j^2 \Lambda_{j-1}^{-2} .$$

Finally, we have

$$\begin{aligned} &H_{j-1}(-f(\varphi_1, \ldots, \varphi_{j-1}, y_j, \ldots, y_{n-1})) \\ &= (k-1)^{j-1} (\varphi_1 \cdot \ldots \cdot \varphi_{j-1})^{k-2} (x - y_j^k - \ldots - y_{n-1}^k) \\ &\quad \times (x - \varphi_1^k - \ldots - \varphi_{j-1}^k - y_j^k - \ldots - y_{n-1}^k)^{\frac{j-1}{k} - j} \\ &= (k-1)^{j-1} \left(1 + y_1^{\frac{k}{k-1}} + \ldots + y_{j-1}^{\frac{k}{k-1}}\right)^{1 + \frac{j-1}{k}} . \\ &\quad \times (y_1 \cdot \ldots \cdot y_{j-1})^{\frac{k-2}{k-1}} (x - y_j^k - \ldots - y_{n-1}^k)^{-\frac{j-1}{k}} \end{aligned}$$

and so

$$g_j(\mathbf{y}) = \frac{\partial}{\partial y_j} \frac{1}{\sqrt{H_{j-1}}} = -\frac{j-1}{2} \frac{y_j^{k-1}}{x - y_j^k - \ldots - y_{n-1}^k} \frac{1}{\sqrt{H_{j-1}}} .$$

This shows that $g_j(\mathbf{y})$ is monotonic with respect to y_j and that

$$|g_j(y)| \asymp G_j = 2^{\gamma_j} x^{-1/k} \frac{1}{\sqrt{\Lambda_{j-1}}} .$$

At the bound of $D(\mathbf{v})$ we have to consider the parts with $t_1 = \varrho(t_2, \ldots, t_{n-1})$, where ϱ is defined by

$$\varrho = (x - t^k - \ldots - t_{n-1}^k)^{\frac{1}{k}} 2^{\frac{1}{k-1} v_1} \left(1 + 2^{\frac{k}{k-1} v_1}\right)^{-\frac{1}{k}}$$

(or the same function with $v_1 + 1$ instead of v_1) or $\varrho = t_2$. In the first case $-f_{t_1}(\varrho, t_2, \ldots, t_{n-1}) = 2^{v_1}$ is an integer. In the second case it turns out that

$$-\frac{\partial}{\partial y_2} f_{t_1}(t_2, t_2, t_3, \ldots, t_{n-1})$$

$$= -f_{t_1t_1} - f_{t_1t_2} = (k-1)(t_1t_2)^{\frac{k-2}{2}}(x - t_3^k - \ldots - t_{n-1}^k)(x - t_1^k - \ldots - t_{n-1}^k)^{1/k-2}$$

$$= \left(\frac{x - t_3^k - \ldots - t_{n-1}^k}{x - t_1^k - \ldots - t_{n-1}^k}\right)^{1/2} \sqrt{H_2(f)} \gg \sqrt{\Lambda_2}\,.$$

In order to apply (2.90), the estimates of $D(v)\,\Lambda_{n-1}^{1/(n+1)}$ and Δ_2 are required. Straightforward calculations show that

$$|D(v)|\,\Lambda_{n-1}^{\frac{1}{n+1}}\,2^{-\delta+\sigma_{n-1}/(n+1)}\,x^{\frac{n(n-1)}{(n+1)k}} = 2^{-\vartheta(v)}\,x^{\frac{n(n-1)}{(n+1)k}}.$$

This gives the main term on the right-hand side of (4.23). Several results are required for the estimation of Δ_2:

$$\frac{|R_{2,1}|}{\sqrt{\Lambda_2}} \ll 2^{-\nu_1}\frac{|D_2(v)|}{\sqrt{\Lambda_2}} \ll 2^{-\nu_1+\omega_2-\delta_2-\sigma_2/2}\,x^{\frac{n-2}{k}} \ll x^{\frac{n-2}{k}},$$

$$\frac{|R_{1,2}|}{\sqrt{\Lambda_2}} \ll 2^{\gamma_2}x^{-1/k}\frac{|D_1(v)|}{\sqrt{\Lambda_2}} \ll 2^{\gamma_2+\omega_1-\delta_1-\sigma_2/2}\,x^{\frac{n-2}{k}} \ll x^{\frac{n-2}{k}},$$

$$\Lambda_{n-1}^{\frac{2-j}{2n+2}}\frac{|R_{j-1,j}|}{\sqrt{\Lambda_j}} \ll 2^{\frac{2-j}{2n+2}\sigma_{n-1}+\gamma_j}\,x^{-\frac{(2-j)(n-1)}{(2n+2)k}-\frac{1}{k}}\frac{|D_{j-1}(v)|}{\sqrt{\Lambda_j}}$$

$$\ll 2^{\frac{2-j}{2n+2}\sigma_{n-1}+\gamma_j+\omega_{j-1}-\delta_{j-1}-\sigma_j/2}\,x^{\left(n-2-\frac{j-2}{n+1}\right)\frac{1}{k}} \ll x^{\frac{n-2}{k}} \quad (j \geqq 3),$$

$$\Lambda_{n-1}^{\frac{1-j}{2n+2}}\frac{|R_{j-1,j}|}{\sqrt{\Lambda_{j-1}}} \ll 2^{\frac{1-j}{2n+2}\sigma_{n-1}+\gamma_j}\,x^{-\frac{(1-j)(n-1)}{(2n+2)k}-\frac{1}{k}}\frac{|D_{j-1}(v)|}{\sqrt{\Lambda_{j-1}}}$$

$$\ll 2^{\frac{1-j}{2n+2}\sigma_{n-1}+\gamma_j+\omega_{j-1}-\delta_{j-1}-\sigma_{j-1}/2}\,x^{\left(n-2-\frac{j-1}{n+1}\right)\frac{1}{k}} \ll x^{\frac{n-2}{k}}$$

$$(j \geqq 1).$$

Of course, each logarithm is of order $\log x$. Hence

$$\Delta_2 \ll x^{\frac{n-2}{k}}\log^2 x\,.$$

This gives the second term of the right-hand side of (4.23). Thus, (4.23) is completely proved.

Theorem 4.6.

$$\Delta_{k,n}(x) \ll x^{\frac{n(n-1)}{(n+1)k}}. \tag{4.24}$$

Proof. $\Delta_{k,n}(x)$ is given by (4.19). We divide the domain D defined by (4.21) into subdomaines $D(\mathbf{v})$ described by (4.22) with the restriction $2^{\gamma_j} \ll x^{1/k}$ for $j = 1, 2, \dots, n-1$ and in a remaining part. In this part the inequality $2^{\gamma_j} \gg x^{1/k}$ holds at least for one j. Thus it is seen that the volume of this remainder is at most of order $x^{(n-2)/k}$. Hence

$$\Delta_{k,n}(x) = 2^n n!\, \Sigma_1 \sum_{\mathbf{n} \in D(\mathbf{v})} \psi(f(\mathbf{n})) - O\left(x^{\frac{n-2}{k}}\right),$$

$$SC(\Sigma_1):\quad \text{Sum over all } v = (v_1, v_2, \dots, v_{n-1}) \text{ with } 2^{\gamma_j} \ll x^{1/k},$$

$$\gamma_j = \frac{k}{k-1} v_j + \frac{1}{k-1} \sum_{r=j+1}^{n-1} v_r, \qquad j = 1, 2, \dots, n-1.$$

The number of terms in Σ_1 is at most of order $(\log x)^{n-1}$. Hence, (4.23) shows that

$$\Delta_{k,n}(x) \ll \Sigma_1\, 2^{-\vartheta(\mathbf{v})} x^{\frac{n(n-1)}{(n+1)k}} + x^{\frac{n-2}{k}} (\log x)^{n+1}.$$

Since $\vartheta(\mathbf{v})$ has positive coefficients for each v_r, we may take the sum over all $\mathbf{v}$ with non-negative $v_1, v_2, \dots, v_{n-1}$. Now (4.24) follows at once.

In order to estimate the terms $H_{k,n,r}(x)$ in (4.13) an estimation of $\int_0^x \Delta_{k,r}(t)\, dt$ is required. Therefore, we prove the following lemma, which gives the simplest estimate one can imagine.

Lemma 4.6. *If $r \geqq 2$, then*

$$\int_0^x \Delta_{k,r}(t)\, dt \ll x^{1+\frac{r-2}{k}}. \tag{4.25}$$

Proof. From (4.17) we obtain for the remainder

$$\Delta_{k,r}(x) = -2^r r!\, \Sigma_1\, \psi((x - n_1^k - \dots - n_{r-1}^k)^{1/k}) + O\left(x^{\frac{r-2}{k}}\right),$$

$$SC(\Sigma_1):\quad \frac{1}{2}(x - n_2^k - \dots - n_{r-1}^k) \leqq n_1^k \leqq x - n_2^k - \dots - n_{r-1}^k,$$

$$0 \leqq n_1 \leqq n_2 \leqq \dots \leqq n_{r-1},$$

Then

$$\int_0^x \Delta_{k,r}(t)\, dt = -2^r r! \int_0^x \Sigma_1\, \psi((t - n_1^k - \dots - n_{r-1}^k)^{1/k})\, dt + O\left(x^{1+\frac{r-2}{k}}\right)$$

$$= -2^r r!\, \Sigma_2 \int_1 \psi((t - n_1^k - \dots - n_{r-1}^k)^{1/k})\, dt + O\left(x^{1+\frac{r-2}{k}}\right),$$

where in $SC(\Sigma_1)$ x is to replace by t and

$$SC(\Sigma_2):\ n_1^k + \dots + n_{r-1}^k \leqq x\,, \qquad 0 \leqq n_1 \leqq n_2 \leqq \dots \leqq n_{r-1}\,,$$
$$IC(\textstyle\int_1):\quad n_1^k + \dots + n_{r-1}^k \leqq t \leqq 2n_1^k + n_2^k + \dots + n_{r-1}^k\,, \qquad t \leqq x\,.$$

Let $\psi_1(x)$ denote the function

$$\psi_1(x) = \int_0^x \psi(t)\,dt\,.$$

Integrating by parts, then we find

$$\int_0^x \Delta_{k,r}(t)\,dt$$
$$= -2^r r!\,k\,\Sigma_1 (x - n_1^k - \dots - n_{r-1}^k)^{1-1/k}\,\psi_1((x - n_1^k - \dots - n_{r-1}^k)^{1/k})$$
$$+ 2^r r!(k-1) \int_0^x \Sigma_1\,(t - n_1^k - \dots - n_{r-1}^k)^{-1/k}$$
$$\times\,\psi_1((t - n_1^k - \dots - n_{r-1}^k)^{1/k})\,dt + O\left(x^{1+\frac{r-2}{k}}\right).$$

Since $\psi_1(x)$ is bounded, trivial estimates give (4.25) immediately.

Theorem 4.7. *If* $k \leqq n + 1$, *then*

$$R_{k,n}(x) = V_{k,n}x^{\frac{n}{k}} + O\left(x^{\frac{n(n-1)}{(n+1)k}}\right) \tag{4.26}$$

and if $k > n + 1$, *then*

$$R_{k,n}(x) = V_{k,n}x^{\frac{n}{k}} + nV_{k,n-1}\psi^{(k)}_{n/k}\left(x^{\frac{1}{k}}\right) + O\left(x^{\frac{3n-4}{3k} - \frac{2(n-2)}{3k^2}}\right). \tag{4.27}$$

Proof. We use the representation (4.13) of $R_{k,n}(x)$. The estimate of $\Delta_{k,n}(x)$ is given by (4.24). From (4.16), (3.50) and (3.55) it is known that

$$H_{k,n,1}(x) \ll x^{(n-1)(k-1)/k^2} + x^{(n-1)/k-1}\,. \tag{4.28}$$

It remains to estimate the terms $H_{k,n,r}(x)$ for $r = 2, 3, \dots, n-1$. If $n - r > k$ in (4.14), then

$$H_{k,n,r}(x) = c_2 \int_0^x (x-t)^{\frac{n-r}{k}-2}\,dt \int_0^t \Delta_{k,r}(\tau)\,d\tau\,,$$

$$c_1 = \binom{n}{r}\frac{n-r}{k}V_{k,n-r}\,, \qquad c_2 = \left(\frac{n-r}{k} - 1\right)c_1$$

and by means of (4.25)

$$H_{k,n,r}(x) \ll x^{\frac{n-2}{k}}\,. \tag{4.29}$$

We now consider the case $n - r \leqq k$. If $n - r = k$, then

$$H_{k,n,r}(x) = c_1 \int_0^x \Delta_{k,r}(t)\, dt \ll x^{1+\frac{r-2}{k}}.$$

If $n - r < k$, we use partial integration. Let $0 < z < x$. Then

$$H_{k,n,r}(x) = c_1 \left\{ \int_0^{x-z} + \int_{x-z}^{x} \right\} (x-t)^{\frac{n-r}{k}-1} \Delta_{k,r}(t)\, dt$$

$$= c_1 z^{\frac{n-r}{k}-1} \int_0^{x-z} \Delta_{k,r}(t)\, dt + c_2 \int_0^{x-z} (x-t)^{\frac{n-r}{k}-2} dt \int_0^t \Delta_{k,r}(\tau)\, d\tau$$

$$+ c_1 \int_{x-z}^{x} (x-t)^{\frac{n-r}{k}-1} \Delta_{k,r}(t)\, dt$$

$$\ll z^{\frac{n-r}{k}-1} x^{1+\frac{r-2}{k}} + z^{\frac{n-r}{k}} x^{\frac{r(r-1)}{(r+1)k}}.$$

The two terms are of the same order by putting $z = x^{1-\frac{2}{(r+1)k}}$. Then

$$H_{k,n,r}(x) \ll x^{\frac{n-r}{k}\left(1-\frac{2}{(r+1)k}\right)+\frac{r(r-1)}{(r+1)k}}, \tag{4.30}$$

which also holds for $n - r = k$. It is easily seen that $\frac{n-r}{k}\left(1 - \frac{2}{(r+1)k}\right) + \frac{r(r-1)}{(r+1)k}$ is monotonically increasing if $k = n + 1$ and decreasing otherwise. Therefore, (4.26) and (4.27) follow from (4.13) and the estimates (4.24), (4.28), (4.29) and (4.30).

Remark. It is known that the second main term of (4.27) has the precise order $x^{\frac{n-1}{k}\left(1-\frac{1}{k}\right)}$. And this exponent is larger than the exponent of the error term. Consequently, we have the sharp estimate

$$R_{k,n}(x) = V_{k,n} x^{\frac{n}{k}} + \begin{cases} O\left(x^{\frac{n-1}{k}\left(1-\frac{1}{k}\right)}\right), \\ \Omega\left(x^{\frac{n-1}{k}\left(1-\frac{1}{k}\right)}\right), \end{cases}$$

provided that $k > n + 1$. However, the result (4.27) shows that besides the main term $V_{k,n}x^{n/k}$ a second main term is possible. Moreover, it is seen that the quality of the estimate (4.27) essentially depends on the estimate (4.25). Clearly, this result is a trivial one, because we have used the asymptotic representation (4.17) of $\Delta_{k,r}(x)$ with the bad error term $x^{(r-2)/k}$. Only in case of $r = 2$ the exact representation (3.48) is known, so that it is worthwhile investigating the three-dimensional case separately.

4.2.3. *The three-dimensional case*

If $n = 3$, Theorem 4.7 shows that

$$R_{k,3}(x) = V_{k\,3}x^{3/k} + O(x^{3/2k})$$

for $k \leqq 4$ and

$$R_{k,3}(x) = V_{k,3}x^{\frac{3}{k}} + 3V_{k,2}\psi^{(k)}_{3/k}\left(x^{\frac{1}{k}}\right) + O\left(x^{\frac{1}{3k}\left(5-\frac{2}{k}\right)}\right)$$

for $k > 4$, where the second main term has the precise order $x^{\frac{2}{k}\left(1-\frac{1}{k}\right)}$. In order to improve the estimates of the error terms, we first replace Lemma 4.6 by a precise estimate of the integral $\int\limits_0^x \Delta_{k,2}(t)\,dt$.

Lemma 4.7.

$$\int\limits_0^x \Delta_{k,2}(t)\,dt \ll x^{1-1/2k}\,. \tag{4.31}$$

Proof. (3.48) shows that

$$\Delta_{k,2}(x) = -8 \sum_{x/2<n^k\leqq x} \psi((x-n^k)^{1/k})$$

$$+8\int\limits_0^{(x/2)^{1/k}} t^{k-1}(x-t^k)^{1/k-1}\,\psi(t)\,dt + 4\psi^2\left(\left(\frac{x}{2}\right)^{1/k}\right).$$

Let $\psi_1(x)$ denote the function

$$\psi_1(x) = \int\limits_0^x \psi(t)\,dt = \frac{1}{2}(x-[x])^2 - \frac{1}{2}(x-[x]) = \frac{1}{2}\psi^2(x) - \frac{1}{8}\,.$$

The Fourier expansion of $\psi_1(x)$ is given by

$$\psi_1(x) = \frac{1}{2\pi^2}\sum_{n=1}^{\infty}\frac{\cos 2\pi nx}{n^2} - \frac{1}{12},$$

so that $\psi_1(x) + 1/12$ is bounded and $\int\limits_0^x (\psi_1(t) + 1/12)\,dt$ as well. Thus, integrating by parts we obtain

$$\Delta_{k,2}(x) = -8\sum_{x/2<n^k\leqq x}\psi((x-n^k)^{1/k}) + \frac{1}{3} + 16\left\{\psi_1\left(\left(\frac{x}{2}\right)^{1/k}\right) + \frac{1}{12}\right\} + O(x^{-1/k})$$

and

$$\int_0^x \Delta_{k,2}(t)\,dt = -8\int_0^x \sum_{t/2<n^k\leq t} \psi((t-n^k)^{1/k})\,dt + \frac{x}{3} + O(x^{1-1/k})$$

$$= -\frac{2k}{3}\sum_{n^k\leq x/2} n^{k-1} - 8k\sum_{x/2<n^k\leq x}(x-n^k)^{1-1/k}$$

$$\times\left\{\psi_1((x-n^k)^{1/k}) + \frac{1}{12}\right\}$$

$$+8(k-1)\int_0^x \sum_{t/2<n^k\leq t}(t-n^k)^{-1/k}\left\{\psi_1((t-n^k)^{1/k}) + \frac{1}{12}\right\}dt$$

$$+\frac{x}{3} + O(x^{1-1/k})$$

$$= -8k\sum_{x/2<n^k\leq x}(x-n^k)^{1-1/k}\left\{\psi_1((x-n^k)^{1/k}) + \frac{1}{12}\right\} + O(x^{1-1/k}).$$

Partial summation now gives

$$\int_0^x \Delta_{k,2}(t)\,dt$$

$$= -8k(k-1)\int_{(x/2)^{1/k}}^{x^{1/k}} t^{k-1}(x-t^k)^{-1/k}\sum_{x/2<n^k\leq t^k}\left\{\psi_1((x-n^k)^{1/k}) + \frac{1}{12}\right\}dt$$

$$+ O(x^{1-1/k}). \tag{4.32}$$

It remains to estimate the sum

$$S = \sum_{x/2<n^k\leq t^k}\left\{\psi_1((x-n^k)^{1/k}) + \frac{1}{12}\right\}$$

$$= \frac{1}{2\pi^2}\sum_{m=1}^{\infty}\frac{1}{m^2}\sum_{x/2<n^k\leq t^k}\cos 2\pi m(x-n^k)^{1/k}.$$

We split up the sum over n into sums with the restrictions

$$x(1-2^{-\nu}) < n^k \leq x(1-2^{-\nu-1}), \qquad n\leq t, \tag{4.33}$$

where $\nu = 1, 2, \ldots$ We obtain from Theorem 2.1

$$\sum e^{2\pi i m(x-n^k)^{1/k}} \ll (m^k\, 2^{-\nu}x)^{1/2k},$$

where the summation condition is given by (4.33). Taking the sum over v and m it is seen that $S \ll x^{1/2k}$. Result (4.31) now follows from (4.32) immediately.

Lemma 4.8. *Let* $\Delta_{k,2}(t) \ll \Delta^*_{k,2}(x)$ *for* $1 \leqq t \leqq x$. *Then*

$$H_{k,3,2}(x) \ll x^{\frac{1}{k}\left(1-\frac{1}{2k}\right)} (\Delta^*_{k,2}(x))^{1-\frac{1}{k}} . \tag{4.34}$$

Proof. We have, by (4.14) and (4.9),

$$H_{k,3,2}(x) = \frac{6}{k} \int\limits_0^x (x-t)^{1/k-1} \Delta_{k,2}(t)\, dt$$

$$= \frac{6}{k} \left\{ \int\limits_0^{x-y} + \int\limits_{x-y}^{x} \right\} (x-t)^{1/k-1} \Delta^*_{k,2}(t)\, dt , \tag{4.35}$$

where y is a suitably chosen value with $0 < y < x$. In the first integral partial integration may be applied such that (4.31) can be used. In the second integral we replace $|\Delta_{k,2}(t)|$ by its maximum. Then

$$H_{k,3,2}(x) \ll y^{1/k-1} x^{1-1/2k} + y^{1/k} \Delta^*_{k,2}(x) .$$

Choosing $y = \dfrac{x^{1-1/2k}}{\Delta^*_{k,2}(x)}$, we obtain (4.34) at once.

Theorem 3.17B shows that $\Delta_{k,2}(x) \ll x^{2/3k}$ which is the simplest non-trivial estimate. However, Theorem 3.18 gives the better estimate $\Delta_{k,2}(x) \ll x^{27/41k}$. Hence, we obtain the following theorem from (4.13), (4.24) and (4.34) with $\Delta^*_{k,2}(x) = x^{27/41k}$ at once.

Theorem 4.8. *If* $4 < k \leqq 95/13$, *then*

$$R_{k,3}(x) = V_{k,3} x^{3/k} + 3V_{k,2} \psi^{(k)}_{3/k}(x^{1/k}) + O(x^{3/2k}) ,$$

and if $k > 95/13$, *then*

$$R_{k,3}(x) = V_{k,3} x^{\frac{3}{k}} + 3V_{k,2} \psi^{(k)}_{3/k}\left(x^{\frac{1}{k}}\right) + O\left(x^{\frac{1}{82k}\left(136-\frac{95}{k}\right)}\right) .$$

In this place it may be remarked that Theorem 4.10 which is proved in the next section shows that

$$H_{k,3,2}(x) = \Omega(x^{3/2k-1/k^2}) .$$

We now see that this exponent is only a little smaller than $3/2k$. Therefore, if we give a slightly better estimate of the remainder $\Delta_{k,3}(x)$, *we can consider* $H_{k,3,2}(x)$

represented by (4.35) *as a third main term.* For this purpose we transform the exponential sum

$$\sum_{\boldsymbol{n}\in D(\boldsymbol{v})} e^{2\pi i f(\boldsymbol{n})},$$

where $f(\boldsymbol{t})$ and $D(\boldsymbol{v})$ are defined by (4.20) and (4.22), respectively, for $n = 3$. The transformed exponential sum then will be estimated by applying two Weyl's steps.

Lemma 4.9. *Let m denote a positive integer. Let*

$$f(t_1, t_2) = -m(x - t_1^k - t_2^k)^{1/k},$$

$$D = D(v_1, v_2) = \{(t_1, t_2)\colon 0 \leqq t_1 \leqq t_2,\ q_{v_1}(x - t_2^k) \leqq t_1^k \leqq q_{v_1+1}(x - t_2^k),\ q_{v_2}x \leqq t_2^k \leqq q_{v_2+1}x\},$$

where v_1, v_2 are non-negative integers and

$$q_{v_j} = 2^{\frac{k}{k-1}v_j}\left(j + 2^{\frac{k}{k-1}v_j}\right)^{-1} \quad (j = 1, 2).$$

If $2^{\gamma_1}, 2^{\gamma_2} \ll x^{1/k}$, $\gamma_1 = \dfrac{1}{k-1}(kv_1 + v_2)$, $\gamma_2 = \dfrac{k}{k-1}v_2$, then

$$\sum_{(n_1, n_2)\in D} e^{2\pi i f(n_1, n_2)} = \sum_{(m_1, m_2)\in D_2} \frac{i}{\sqrt{H_2(f(\varphi_1, \varphi_2))}} e^{2\pi i f_2(m_1, m_2)} + O(x^{1/k}\log(mx)) + O((m^k x)^{1/2k}\log(mx)), \tag{4.36}$$

where

$$D_2 = D_2(v_1, v_2) = \Big\{(y_1, y_2)\colon y_1 \leqq y_2,\ m\,2^{v_1} \leqq y_1 \leqq m\,2^{v_1+1},\ \left(m^{\frac{k}{k-1}} + y_1^{\frac{k}{k-1}}\right)2^{v_2\frac{k}{k-1}-1} \leqq y_2^{\frac{k}{k-1}} \leqq \left(m^{\frac{k}{k-1}} + y_1^{\frac{k}{k-1}}\right)\cdot 2^{(v_2+1)\frac{k}{k-1}-1}\Big\},$$

$$f_2(y_1, y_2) = -x^{\frac{1}{k}}\left(m^{-\frac{k}{k-1}} + y_1^{\frac{k}{k-1}} + y_2^{\frac{k}{k-1}}\right)^{1-\frac{1}{k}},$$

$$H_2(f(\varphi_1, \varphi_2)) = (k-1)^2 m^{\frac{k}{k-1}}(y_1y_2)^{\frac{k-2}{k-1}} x^{-\frac{2}{k}}\left(m^{\frac{k}{k-1}} + y_1^{\frac{k}{k-1}} + y_2^{\frac{k}{k-1}}\right)^{\frac{2}{k}+1}.$$

Proof. We apply Theorem 2.28 with $p = 2$. It is unnecessary to check all the conditions of this theorem, because we did this in the proof of Lemma 4.5. Therefore we can use all the calculations of the proof of Lemma 4.5 in the special case $n = 3$, where we only have to take into consideration that the function $f(t_1, t_2)$ now contains a factor m. Note that $D_2(v_1, v_2)$ is the image of $D(v_1, v_2)$ under the mapping

$$y_j = mt_j^{k-1}(x - t_1^k - t_2^k)^{1-1/k} \qquad (j = 1, 2).$$

The function $f_2(y_1, y_2)$ is defined by

$$f_2(y_1, y_2) = f(\varphi_1, \varphi_2) - y_1\varphi_1(y_1, y_2) - y_2\varphi_2(y_1, y_2),$$

where the functions $\varphi_j(y_1, y_2)$ are given by

$$\varphi_j^k(y_1, y_2) = xy_j^{\frac{k}{k-1}}\left(m^{\frac{k}{k-1}} + y_1^{\frac{k}{k-1}} + y_2^{\frac{k}{k-1}}\right)^{\frac{1}{k}-1} \qquad (j = 1, 2).$$

From the proof of Lemma 4.5 it is seen that the Hessian $H_2(f(\varphi_1, \varphi_2))$ is given by

$$H_2(f(\varphi_1, \varphi_2)) = m^2(k-1)^2\, x(\varphi_1\varphi_2)^{k-2}\,(x - \varphi_1^k - \varphi_2^k)^{2/k-3}.$$

Hence, we obtain the exponential sum on the right-hand side of (4.36) from (2.85) at once. The two error terms in (2.85) may be replaced, because of the assumptions for the bound of $D(v_1, v_2)$, by the error term Δ_1 defined in Theorem 2.29. Then, corresponding to the estimations in the proof of Lemma 4.5,

$$\begin{aligned}\Delta_1 &= \frac{|R_{2,1}| + |R_{1,2}|\log(c_2+1)}{\sqrt{\Lambda_2}}\\ &\quad + |R_{0,1}|\log(c_1\Lambda_1 + 2) + \frac{|R_{1,2}|}{\sqrt{\Lambda_1}}\log\left(\frac{c_2\Lambda_2}{\Lambda_1} + 2\right)\\ &\ll x^{1/k}\log(mx) + (m^k x)^{1/2}\log(mx).\end{aligned}$$

Thus, (4.36) follows from (2.85).

Lemma 4.10. *Let* $f(t_1, t_2)$ *and* $D = D(v_1, v_2)$ *be defined as in Lemma* 4.9. *Then*

$$\begin{aligned}\sum_{(n_1, n_2)\in D} e^{2\pi i f(n_1, n_2)} &\ll \left(m^{11} x^{\frac{15}{k}} 2^{\frac{4k-11}{k-1}v_1 - \frac{15}{k-1}v_2}\right)^{\frac{1}{14}}\\ &\quad + \left(m^{27} x^{\frac{26}{k}} 2^{\frac{13k-27}{k-1}v_1 - \frac{26}{k-1}v_2}\right)^{\frac{1}{28}}\\ &\quad + x^{1/k}\log(mx) + (m^k x)^{1/2k}\log(mx). \qquad (4.37)\end{aligned}$$

Proof. We consider the exponential sum $\sum_{a<m_1\leqq b} e^{2\pi i f_2(m_1, y_2)}$, where $[a, b]$ is any subinterval of $[m2^{\nu_1}, m2^{\nu_1+1}]$, and $f_2(y_1, y_2)$ defined in Lemma 4.9 is considered as a function of y_1. We apply Theorem 2.6 with $k = 4$. Because of

$$\frac{\partial^4}{\partial y_1^4} f_2(y_1, y_2) = -\frac{x^{1/k}}{(k-1)^3} y_1^{\frac{k}{k-1}-4}\left(m^{\frac{k}{k-1}} + y_2^{\frac{k}{k-1}}\right)\left(m^{\frac{k}{k-1}} + y_1^{\frac{k}{k-1}} + y_2^{\frac{k}{k-1}}\right)^{\frac{k-1}{k}-4}$$
$$\times\left\{(k-2)(2k-3)\left(m^{\frac{k}{k-1}} + y_2^{\frac{k}{k-1}}\right)^2 + (2k-1)(3k-7)\right.$$
$$\left.\times\left(m^{\frac{k}{k-1}} + y_2^{\frac{k}{k-1}}\right) y_1^{\frac{k}{k-1}} + (2k-1)(3k-2)\, y_1^{\frac{2k}{k-1}}\right\}$$

we put

$$\lambda_4 = m^{-3} x^{1/k} 2^{-3\nu_1 - \frac{k}{k-1}\nu_2}$$

in Theorem 2.6. Then (2.13) shows that

$$\sum_{a<m_1\leqq b} e^{2\pi i f_2(m_1, y_2)} \ll (b-a)\,\lambda_4^{1/14} + (b-a)^{3/4}\,\lambda_4^{-1/14}$$
$$\ll \left(m^{11} x^{\frac{1}{k}} 2^{11\nu_1 - \frac{1}{k-1}\nu_2}\right)^{\frac{1}{14}} + \left(m^{27} x^{-\frac{2}{k}} 2^{27\nu_1 + \frac{2}{k-1}\nu_2}\right)^{\frac{1}{28}}.$$

Since

$$H_2(f(\varphi_1, \varphi_2)) \asymp m^2 x^{-\frac{2}{k}} 2^{\frac{3k-2}{k-1}\nu_1 + \frac{2k}{k-1}\nu_2},$$

summation by parts gives

$$\sum_{(m_1, m_2)\in D_2} \frac{i}{\sqrt{H_2(f(\varphi_1, \varphi_2))}}\, e^{2\pi i f_2(m_1, m_2)}$$
$$\ll \left(m^{11} x^{\frac{15}{k}} 2^{\frac{4k-11}{k-1}\nu_1 - \frac{15}{k-1}\nu_2}\right)^{\frac{1}{14}} + \left(m^{27} x^{\frac{26}{k}} 2^{\frac{13k-27}{k-1}\nu_1 - \frac{26}{k-1}\nu_2}\right)^{\frac{1}{28}}.$$

Now (4.37) follows from (4.36).

Lemma 4.11.

$$\Delta_{k,3}(x) \ll x^{37/25k}. \tag{4.38}$$

Proof. We use the representation (4.19) for $\Delta_{k,3}(x)$. We divide the domain D into $O(\log^2 x)$ subdomains $D(\nu_1, \nu_2)$ with $0 \leqq \nu_1, \nu_2 \ll \log x$ such that

$$\Delta_{k,3}(x) = 48 \sum_{\nu_1, \nu_2 \ll \log x}\ \sum_{(n_1, n_2)\in D(\nu_1, \nu_2)} \psi(f(n_1, n_2)) + O(x^{1/k}\log^2 x).$$

Now we obtain from (1.18) and (4.37)

$$\sum_{(n_1, n_2) \in D(v_1, v_2)} \psi(f(n_1, n_2)) \ll \frac{|D(v_1, v_2)|}{z}$$

$$+ \sum_{m=1}^{\infty} \min\left(\frac{z}{m^2}, \frac{1}{m}\right) \left| \sum_{(n_1, n_2) \in D(v_1, v_2)} e^{2\pi i m f(n_1, n_2)} \right|$$

$$\ll \frac{1}{z} x^{\frac{2}{k}} 2^{-\frac{k}{k-1} v_1 - \frac{k+1}{k-1} v_2}$$

$$+ \left(z^{11} x^{\frac{15}{k}} 2^{\frac{4k-11}{k-1} v_1 - \frac{15}{k-1} v_2} \right)^{\frac{1}{14}}$$

$$+ \left(z^{27} x^{\frac{26}{k}} 2^{\frac{13k-27}{k-1} v_1 - \frac{26}{k-1} v_2} \right)^{\frac{1}{28}}$$

$$+ x^{\frac{1}{k}} \log^2 x + (z^k x)^{\frac{1}{2k}} \log^2 x \,,$$

provided that $z < x$. The first two terms are of the same order if

$$z = \left(x^{\frac{13}{k}} 2^{-\frac{18k-11}{k-1} v_1 - \frac{14k-1}{k-1} v_2} \right)^{\frac{1}{25}} .$$

Hence

$$\sum_{(n_1, n_2) \in D(v_1, v_2)} \psi(f(n_1, n_2)) \ll \left(x^{\frac{37}{k}} 2^{-\frac{7k+11}{k-1} v_1 - \frac{11k+26}{k-1} v_2} \right)^{\frac{1}{25}}$$

$$+ x^{\frac{143}{100k}} + x^{\frac{1}{k}} \log^2 x + x^{\frac{38}{50k}} \log^2 x \,.$$

Summing over v_1, v_2, estimate (4.38) now follows.

We now collect the results in the following theorem where the Ω-results are obtained from Theorem 4.10. There k must be an integer.

Theorem 4.9. *$R_{k,3}(x)$ is represented by*

$$R_{k,3}(x) = V_{k,3} x^{3/k} + H_{k,3,1}(x) + H_{k,3,2}(x) + \Delta_{k,3}(x) \,,$$

where the following representations and estimates hold:

$$H_{k,3,1}(x) = 3 V_{k,2} \psi_{3/k}^{(k)}(x^{1/k}) = \begin{cases} O\left(x^{\frac{2}{k}\left(1 - \frac{1}{k}\right)} \right), \\ \Omega\left(x^{\frac{2}{k}\left(1 - \frac{1}{k}\right)} \right), \end{cases}$$

$$H_{k,3,2}(x) = \frac{6}{k}\int_0^x (x-t)^{1/k-1}\,\Delta_{k,2}(t)\,dt = \begin{cases} O\left(x^{\frac{1}{82k}\left(136-\frac{95}{k}\right)}\right), \\ \Omega\left(x^{\frac{1}{k}\left(\frac{3}{2}-\frac{1}{k}\right)}\right), \end{cases}$$

$$\Delta_{k,3}(x) = \begin{cases} O(x^{37/25k}) \\ \Omega(x^{1/k}) \end{cases}$$

4.2.4. *The Ω-estimate*

In this section the investigations of Section 3.3.4 are generalized to the n-dimensional case. We proceed in a similar way.

Lemma 4.12. *Let $k \geqq 2$ be a natural number. Let m, n be integers with $1 \leqq m \leqq n$, $n \geqq 2$ and $v_1, v_2, \ldots, v_m$ non-zero real numbers. Then the function $I_m(s)$, defined if* $\mathrm{Re}(s) > 0$ *by*

$$I_m(s) = \int_0^\infty \ldots \int_0^\infty e^{2\pi i(v_1t_1+\ldots+v_mt_m)-s(t_1^k+\ldots+t_n^k)^{1/k}}\,dt_1\cdot\ldots\cdot dt_n\,,$$

is capable of analytic continuation into the left halfplane. It is a many-valued function with a finite number of isolated singular points, all of which lie on the imaginary axis. There is at most one algebraic infinity of order $(n-m)(1-1/k) + (m+1)/2$. The other possible infinities for $s \neq 0$ are of smaller order. For $s \to 0$ $(s > 0)$ we have $I_m(s) = O(s^{m-n})$.

Proof. *Substituting $t_r \to t_1 t_r$ $(r = 2, 3, \ldots, n)$ and integrating with respect to t_1,* we obtain

$$I_m(s) = \int_0^\infty \ldots \int_0^\infty \frac{(n-1)!\,dt_2\cdot\ldots\cdot dt_n}{(s(1+t_2^k+\ldots+t_n^k)^{1/k} - 2\pi\,i(v_1+v_2t_2+\ldots+v_mt_m))^n}\,.$$

This shows that all the singularities lie on the imaginary axis. It is clear by substituting $t_\varrho \to \frac{1}{s}t_\varrho$ $(\varrho = m+1, m+2, \ldots, n)$ that for $s \to 0$ $(s > 0)$

$$I_m(s) = \int_0^\infty \ldots \int_0^\infty \frac{s^{m-n}(n-1)!\,dt_2\cdot\ldots\cdot dt_n}{((t_{m+1}^k+\ldots+t_n^k)^{1/k} - 2\pi\,i(v_1+v_2t_2+\ldots+v_mt_m))^n}$$
$$+\,O(s^{m-n+1})\,.$$

Now we write

$$I_m(s) = \int_0^\infty \ldots \int_0^\infty \frac{(n-1)!\,(1+t_2^k+\ldots+t_n^k)^{-n/k}\,dt_2\cdot\ldots\cdot dt_n}{(s-2\pi\,ih(t_2,\ldots,t_n))^n}$$

with

$$h = h(t_2, \dots, t_n) = (v_1 + v_2 t_2 + \dots + v_m t_m)(1 + t_2^k + \dots + t_n{}^k)^{-1/k}.$$

The analytic continuation of the function $I_m(s)$ is possible into all points $s = 2\pi\, ih$, where at least one of the first partial derivatives of h is different from zero, provided that the t_j are not endpoints of the integration intervals. Hence, we find singularities at points with

$$\frac{1}{h}\frac{\partial h}{\partial t_r} = \frac{v_r}{v_1 + v_2 t_2 + \dots + v_m t_m} - \frac{t_r^{k-1}}{1 + t_2^k + \dots + t_n^k} = 0$$

for $r = 2, 3, \dots, m$ and

$$\frac{1}{h}\frac{\partial h}{\partial t_\varrho} = -\frac{t_\varrho^{k-1}}{1 + t_2^k + \dots + t_n^k} = 0$$

for $\varrho = m + 1, m + 2, \dots, n$. This system has the solution $(t_2^0, t_3{}^0, \dots, t_n^0)$ with

$$t_r^0 = \left(\frac{v_r}{v_1}\right)^{1/(k-1)}, \qquad t_\varrho^0 = 0,$$

provided that the numbers $v_1, v_2, \dots, v_m$ are of the same sign. Then

$$h(t_2^0, \dots, t_n^0) = \pm\left(|v_1|^{\frac{k}{k-1}} + \dots + |v_m|^{\frac{k}{k-1}}\right)^{1-\frac{1}{k}}$$

is under the upper sign if $v_1, v_2, \dots, v_m$ are positive and under the lower sign otherwise. At such a point $s = 2\pi\, ih(t_2^0, \dots, t_n^0)$ the function $I_m(s)$ has a singularity. We put

$$s = \sigma + 2\pi\, ih(t_2^0, t_3^0, \dots, t_n^0), \qquad \sigma > 0,$$

and investigate the behaviour of $I_m(s)$ for $\sigma \to 0$. Using the first terms of the Taylor series of $h(t_2, \dots, t_n)$ at the point $(t_2^0, \dots, t_n^0)$, we get for $\sigma \to 0$

$$I_m(s) = \int_U \!\!\dots\!\! \int \frac{(-1)^n (n-1)!\,(1 + t_2^k + \dots + t_n^k)^{-n/k}\, dt_2 \cdot \dots \cdot dt_n}{(\sigma + 2\pi\, ih(t_2^0, \dots, t_n^0) - 2\pi\, ih(t_2, \dots, t_n))^n} + O(1)$$

$$\ll \left| \int_U \!\!\dots\!\! \int \frac{dt_2 \cdot \dots \cdot dt_n}{(\sigma - g)^n} \right|,$$

where U is a suitable neighbourhood of the point $(t_2^0, t_3^0, \dots, t_n^0)$ and where g is given by

$$g = 2\pi\, i\left(\sum_{r_1=2}^{m} \sum_{r_2=2}^{m} c_{r_1, r_2}(t_{r_1} - t_{r_1}^0)(t_{r_2} - t_{r_2}^0) + \sum_{\varrho=m+1}^{n} d_\varrho t_\varrho^k\right),$$

and c_{r_1, r_2}, d_ϱ are well-defined constants. Hence, we have for $\sigma \to 0$

$$I_m(s) \ll \sigma^{-(n-m)(1-1/k)-(m+1)/2} .$$

At the bound of integration the first derivatives with respect to t_r $(r = 2, 3, \ldots, m)$ are different from zero such that the associated infinities are of smaller order. This proves the lemma.

We now consider the function

$$\Phi_{k,n}(s) = \sum_{\nu_1=-\infty}^{+\infty} \ldots \sum_{\nu_n=-\infty}^{+\infty} \mathrm{e}^{-s(|\nu_1|^k + \ldots + |\nu_n|^k)^{1/k}}$$

$$= 2^n \sum_{\nu_1=0}^{\infty}{}' \ldots \sum_{\nu_n=0}^{\infty}{}' \mathrm{e}^{-s(\nu_1^k + \ldots + \nu_n^k)^{1/k}} .$$

n times applying the Poisson sum formula and again writing $\sum_{n=-\infty}^{+\infty}{}^* = \lim_{N\to\infty} \sum_{n=-N}^{+N}$, then

$$2^{-n}\Phi_{k,n}(s) = \sum_{\nu_1=-\infty}^{+\infty}{}^* \ldots \sum_{\nu_n=-\infty}^{+\infty}{}^* \int_0^\infty \ldots \int_0^\infty \mathrm{e}^{2\pi i(\nu_1 t_1 + \ldots + \nu_n t_n) - s f^{1/k}} \mathrm{d}t_1 \cdot \ldots \cdot \mathrm{d}t_n$$

with

$$f = f(t_1, \ldots, t_n) = t_1^k + \ldots + t_n^k .$$

We now put

$$\Phi_{k,n}(s) = \sum_{r=0}^{n} \Phi_{k,n,r}(s)$$

with

$$\binom{n}{r}^{-1} 2^{-n}\Phi_{k,n,r}(s) = \sum_{\nu_1 \cdot \ldots \cdot \nu_r \neq 0}{}^* \ldots \sum{}^* \int_0^\infty \ldots \int_0^\infty \mathrm{e}^{2\pi i(\nu_1 t_1 + \ldots + \nu_r t_r) - s f^{1/k}} \mathrm{d}t_1 \cdot \ldots \cdot \mathrm{d}t_n$$

if $r \geqq 1$ and

$$\Phi_{k,n,0}(s) = 2^n \int_0^\infty \ldots \int_0^\infty \mathrm{e}^{-s f^{1/k}} \mathrm{d}t_1 \cdot \ldots \cdot \mathrm{d}t_n = \frac{V_{k,n} n!}{s^n} .$$

It is easily seen that

$$\int_0^\infty R_{k,n}(t^k)\, \mathrm{e}^{-st}\, \mathrm{d}t = \frac{1}{s} \Phi_{k,n}(s) , \qquad \int_0^\infty V_{k,n} t^n\, \mathrm{e}^{-st}\, \mathrm{d}t = \frac{1}{s} \Phi_{k,n,0}(s) .$$

If $r \geqq 1$, it is seen from (4.14) and (4.15) that

$$H_{k,n,r}(x) = \binom{n}{r} \sum_{j=0}^{r} (-1)^{r-j} \binom{r}{j} \sum \int \mathrm{d}t_{j+1} \cdot \ldots \cdot \mathrm{d}t_n ,$$

$$SC(\Sigma \textstyle\int): \quad |a_1|^k + \ldots + |a_j|^k + |t_{j+1}|^k + \ldots + |t_n|^k \leqq x .$$

Now j times applying the Poisson sum formula, we get

$$\binom{n}{r}^{-1} H_{k,n,r}(x) = \sum_{j=0}^{r} (-1)^{r-j} \binom{r}{j} \sum_{\nu_1}^{*} \dots \sum_{\nu_j}^{*} \int e^{2\pi i(\nu_1 t_1 + \dots + \nu_j t_j)} \, dt_1 \cdot \dots \cdot dt_n ,$$

$$IC(\textstyle\int): \ |t_1|^k + \dots + |t_n|^k \leq x .$$

Thus, we obtain for $1 \leq r < n$

$$\int_0^\infty H_{k,n,r}(t^k) \, e^{-st} \, dt = \frac{1}{s} \binom{n}{r} 2^n \sum_{j=0}^{r} (-1)^{r-j} \binom{r}{j} \sum_{\nu_1}^{*} \dots \sum_{\nu_j}^{*} I_j(s) = \frac{1}{s} \Phi_{k,n,r}(s) .$$

Furthermore, we have

$$\int_0^\infty \Delta_{k,n}(t^k) \, e^{-st} \, dt = \frac{1}{s} \Phi_{k,n,n}(s) .$$

The following lemma is now an immediate consequence of Lemma 4.12.

Lemma 4.13. *The function*

$$G(s; m) = \sum_{r=m}^{n} \Phi_{k,n,r}(s) \qquad (1 \leq m \leq n) \tag{4.40}$$

is capable of analytic continuation into the left halfplane. It is a many-valued function with isolated singularities only on the imaginary axis. There are algebraic infinities of order $(n - m)(1 - 1/k) + (m + 1)/2$. *The other infinities are of smaller order, provided that* $s \neq 0$.

The behaviour of $G(s; m)$ in the neightbourhood of the point $s = 0$ is easily obtained from (4.40) and the asymptotic representation (4.39). Supposing that $r > 1$, (4.39) shows that

$$I_r(s) = \frac{g(\nu_1, \dots, \nu_{r-1})}{\nu_r} s^{r-n} + O(s^{r-n+1})$$

with a certain function $g(\nu_1, \dots, \nu_{r-1})$ für $s > 0$ and $s \to 0$. Hence, the sum over the positive and negative values of ν_r vanishes.

This gives

$$\Phi_{k,n,r}(s) = O(s^{r-n+1}) .$$

This result also holds in case of $r = 1$. Putting $t_2^k + \dots + t_n^k = z^k$, we obtain from (4.39)

$$I_1(s) = A_1 \int_0^\infty \frac{z^{n-2} \, dz}{(z - 2\pi i \nu_1)^n} s^{1-n} + O(s^{2-n}) = A_2 \frac{s^{1-n}}{\nu_1} + O(s^{2-n})$$

with certain constants A_1, A_2. Again, the sum over the positive and negative values of ν_1 vanishes. Thus, the following lemma is clear.

Lemma 4.14. *If* $1 \leqq m \leqq n$, $s > 0$ *and* $s \to 0$, *then*

$$G(s; m) = O(s^{m-n+1}). \tag{4.41}$$

Theorem 4.10. *If* $k > 2$ *is an integer and if* $n \geqq 2$, $1 \leqq m \leqq n$; $2n - k < (k+2)m$, *then*

$$R_{k,n}(x) = V_{k,n} x^{\frac{n}{k}} + \sum_{r=1}^{m-1} H_{k,n,r}(x) + \Omega_{\pm}\left(x^{\frac{n-m}{k}\left(1-\frac{1}{k}\right)+\frac{m-1}{2k}}\right). \tag{4.42}$$

Proof. Corresponding to the proofs of Theorems 3.14 and 3.19 suppose that there exists a positive constant K such that for $t \geqq t_0$

$$g(t; m) \leqq K t^{(n-m)(1-1/k)+(m-1)/2}, \tag{4.43}$$

where $g(t; m)$ is defined by

$$g(t; m) = R_{k,n}(t^k) - V_{k,n} t^n - \sum_{r=1}^{m-1} H_{k,n,r}(t^k).$$

Then

$$\int_0^\infty g(t; m)\, e^{-st}\, dt = \frac{1}{s}\Phi_{k,n}(s) - \frac{1}{s}\sum_{r=0}^{m-1}\Phi_{k,n,r}(s) = \frac{1}{s}G(s; m).$$

Putting

$$g_1(t; m) = g(t; m) - K t^{(n-m)(1-1/k)+(m-1)/2},$$

$$G_1(s; m) = \frac{1}{s}G(s; m) - K_1 s^{-(n-m)(1-1/k)-(m+1)/2},$$

$$K_1 = K\Gamma\left((n-m)\left(1-\frac{1}{k}\right)+\frac{m+1}{2}\right),$$

we obtain

$$G_1(s; m) = \int_0^\infty g_1(t; m)\, e^{-st}\, dt.$$

It is seen from Lemmas 4.13 and 4.14 that $G_1(s; m)$ has isolated algebraic infinities of maximal order $(n-m)(1-1/k) + (m+1)/2$ on the imaginary axis and that because of $2n - k < (k+2)m$

$$G_1(s; m) \sim -K_1 s^{-(n-m)(1-1/k)-(m+1)/2}, \qquad (s > 0,\ s \to 0).$$

Moreover, there exists a number $\tau \neq 0$ such that

$$G_1(\sigma + i\tau; m) \sim c\sigma^{-(n-m)(1-1/k)-(m+1)/2}$$

with a certain constant $c \neq 0$ and $\sigma > 0$, $\sigma \to 0$. On the other hand we have

$$G_1(\sigma + i\tau; m) = \left| \left(\int_0^{t_0} + \int_0^{\infty} \right) g_1(t; m)\, \mathrm{e}^{-(\sigma + i\tau)t}\, \mathrm{d}t \right|$$

$$\leqq \int_{t_0}^{\infty} |g_1(t; m)|\, \mathrm{e}^{-\sigma t}\, \mathrm{d}t + O(1) \leqq -G_1(\sigma; m) + O(1)$$

$$\leqq K_1 \sigma^{-(n-m)(1-1/k)-(m+1)/2} + O(1)\,.$$

This shows that we must have $|c| \leqq K_1$. Hence, for smaller values of K_1 and K, respectively, inequality (4.43) cannot hold for all sufficiently large values of t. This proves (4.42) under the upper sign. A similar proof holds for the lower sign in (4.42).

NOTES ON CHAPTER 4

Section 4.1

Theorem 4.1 was first proved by E. Landau [1]. Later M. N. Bleicher and I. M. Knopp [1], [2] gave a much simpler proof. The transformation formula proved in Lemma 4.2 is due to I. M. Vinogradov [2], [7]. It will be remarked that Theorem 2.24 is based on Vinogradov's proof of Lemma 4.2. Theorem 4.2 was obtained by I. M. Vinogradov [1].

Section 4.2

The good and simple estimate (4.12) of Theorem 4.4 is contained in a more general estimate due to E. K. Hayashi [1], [2]. All the other results are based on two papers of E. Krätzel [8], [16] and on papers of H. Menzer [1] and L. Schnabel [1].

Chapter 5

Plane multiplicative problems

One of the most important and unsolved problems in analytic number theory is that of the estimation of the mean value

$$D(x) = \sum_{n \leqq x} d(n),$$

where $d(n)$ denotes the number of the divisors of n. P. G. L. Dirichlet [1] proved that

$$D(x) = x \log x + (2C - 1)\, x + O(\sqrt{x}),$$

where C is Euler's constant. The problem which is called *Dirichlet's divisor problem* is that of determining as precisely as possible the maximum order of the error term. But in reality this is a lattice point problem. For, if we write

$$D(x) = \sum_{mn \leqq x} 1 = \# \{(m, n)\colon m, n \in \mathbf{Z};\ m, n > 0,\ mn \leqq x\},$$

we see that $D(x)$ counts the number of lattice points under the hyperbola $\xi\eta = x$ with $\xi > 0, \eta > 0$. In 1903 G. Voronoi [1] proved that more than the above estimation is true. And W. Sierpiński [1] could show in 1906 that Voronoi's method also is applicable for the circle problem. Since that time the theory of both the problems has undergone extensive developments with similar results.

Here as in Chapter 3 we could consider very general curves as $f(\xi)\, g(\eta) = x$. But there arise some more difficulties so that we only refer to the special case of curves of type $\xi^a \eta^b = x$, where Dirichlet's divisor problem is included. Questions of such kind arise in many problems of number theory. There the numbers a, b are integers in general. But this property is not required in most representations and estimations of this chapter. *Therefore, we shall always suppose that a, b are real numbers with the restriction* $1 \leqq a \leqq b$. In case of $a = b$ we put $a = b = 1$ without loss of generality.

In this chapter the main object is to estimate the number of lattice points under the curve $\xi^a \eta^b = x$ with $\xi > 0$, $\eta > 0$. Consequently we are interested in the function

$$D(a, b; x) = \# \{(m, n)\colon m, n \in \mathbf{Z};\ m, n \geqq 1;\ m^a n^b \leqq x\}.$$

In particular, if the numbers a, b are integers, we define the divisor function

$$d(a, b; n) = \#\{(n_1, n_2): n_1, n_2 \in \mathbf{Z}; n_1, n_2 \geqq 1; n_1^a n_2^b = n\}$$

such that $D(a, b; x)$ is represented by

$$D(a, b; x) = \sum_{n \leqq x} d(a, b; n) .$$

5.1. THE BASIC ESTIMATES

We first consider a basic representation of $D(a, b; x)$.

Theorem 5.1. *The representation*

$$D(a, b; x) = H(a, b; x) + \Delta(a, b; x) \tag{5.1}$$

holds with the main term

$$H(a, b; x) = \begin{cases} \zeta\left(\dfrac{b}{a}\right) x^{1/a} + \zeta\left(\dfrac{a}{b}\right) x^{1/b} & for \quad 1 \leqq a < b\,, \\ x \log x + (2C - 1)\, x & for \quad 1 = a = b\,, \end{cases} \tag{5.2}$$

where $\zeta(s)$ denotes Riemann's zeta-function and C Euler's constant, and the remainder

$$\Delta(a, b; x) = -\sum_{n^{a+b} \leqq x} \left\{\psi\left(\left(\frac{x}{n^b}\right)^{1/a}\right) + \psi\left(\left(\frac{x}{n^a}\right)^{1/b}\right)\right\} + O(1)\,. \tag{5.3}$$

Proof. We split up the sum $D(a, b; x) = \sum_{m^a n^b \leqq x} 1$ into the regions $m \leqq n$ and $m \geqq n$, where in each sum the lattice points on $m = n$ are counted with a factor 1/2. Then

$$\begin{aligned} D(a, b; x) &= \sum_{\substack{m^a n^b \leqq x \\ m \leqq n}}' 1 + \sum_{\substack{m^a n^b \leqq x \\ n \leqq m}}' 1 = \sum_{n^{a+b} \leqq x} \left\{\left[\left(\frac{x}{n^b}\right)^{1/a}\right] + \left[\left(\frac{x}{n^a}\right)^{1/b}\right] - 2n + 1\right\} \\ &= \sum_{n^{a+b} \leqq x} \left\{\left(\frac{x}{n^b}\right)^{1/a} + \left(\frac{x}{n^a}\right)^{1/b}\right\} - x^{\frac{2}{a+b}} + 2x^{\frac{1}{a+b}} \psi\left(x^{\frac{1}{a+b}}\right) \\ &\quad - \sum_{n^{a+b} \leqq x} \left\{\psi\left(\left(\frac{x}{n^b}\right)^{1/a}\right) + \psi\left(\left(\frac{x}{n^a}\right)^{1/b}\right)\right\} + O(1)\,. \end{aligned}$$

If we suppose $a < b$, we can use the asymptotic formula

$$\sum_{1 \leqq n \leqq z} \frac{1}{n^s} = \zeta(s) + \frac{z^{1-s}}{1-s} - \frac{\psi(z)}{z^s} + O\left(\frac{1}{z^{s+1}}\right)$$

for $s \neq 1$. In case of $s = a = b = 1$ this formula takes the form

$$\sum_{1 \leqq n \leqq z} \frac{1}{n} = \log z + C - \frac{\psi(z)}{z} + O\left(\frac{1}{z^2}\right).$$

Now the representation (5.1) with the terms (5.2) and (5.3) follows at once.

Theorem 5.2.

$$\Delta(a, b; x) \ll x^{\frac{1}{a+b}}. \tag{5.4}$$

If a, b are integers, than

$$\Delta(a, b; x) = \begin{cases} \Omega(1) & for \quad 1 \leqq a < b\,, \\ \Omega(\log x) & for \quad 1 = a = b\,. \end{cases} \tag{5.5}$$

Proof. (5.4) is an immediate consequence of the representation (5.3) by trivial estimation of the sums.

In order to prove (5.5), we first assume that $1 < a < b$. Let y, z be positive integers and $x = y^a z^b$. Then

$$D(a, b; x + 1) - D(a, b; x - 1) \geqq 1\,.$$

If we suppose that $\Delta(a, b; x) = o(1)$, then we obtain contrarily to this result

$$D(a, b; x + 1) - D(a, b; x - 1) = o(1)\,.$$

Note that in this case it is not required that a, b are integers.

If $1 = a < b$, we have for integers x

$$D(1, b; x + 1/2) - D(1, b; x) = 0\,.$$

Supposing that $D(1, b; x) = \zeta(b)\, x + \zeta(1/b)\, x^{1/b} + o(1)$, we obtain

$$D(1, b; x + 1/2) - D(1, b; x) = \frac{1}{2}\zeta(b) + o(1)$$

which forms a contradiction.

In case of $a = b = 1$ we again have for integers x

$$D(1, 1; x + 1/2) - D(1, 1; x) = 0\,.$$

If we suppose that $D(1, 1; x) = x \log x + (2C - 1)\, x + o(\log x)$, then the deduction

$$D(1, 1; x + 1/2) - D(1, 1; x) = \frac{1}{2}\log x + o(\log x)$$

leads to a contradiction. So (5.5) is proved in all parts.

Finally, we give the first non-trivial estimate based on van der Corput's Theorem 2.3. In case of $a = b = 1$ this result was first proved by G. Voronoi in a very complicated manner. If we apply van der Corput's theorem, the proof is now very simple. It will be seen that new difficulties arise in the unsymmetrical case $a < b$.

Theorem 5.3.

$$\Delta(a, b; x) \ll \begin{cases} x^{\frac{1}{2a+b}} & for \quad 1 \leqq a < b\,, \\ x^{1/3} \log x & for \quad 1 = a = b\,. \end{cases} \tag{5.6}$$

Proof. We begin with the case $a = b = 1$. Here (5.3) shows that

$$\Delta(1, 1; x) = -2 \sum_{n \leqq \sqrt{x}} \psi\left(\frac{x}{n}\right) + O(1)\,.$$

Now we apply Theorem 2.3 with the function $f(t) = x/t$. It is easily seen that this function satisfies all the conditions of the theorem in the interval $1 \leqq t \leqq \sqrt{x}$. Hence, we obtain from (2.8)

$$\Delta(1, 1; x) \ll \int_1^{\sqrt{x}} x^{1/3} \frac{dt}{t} + x^{1/4} \ll x^{1/3} \log x\,.$$

In case of $a < b$ we first consider the sum

$$S(a, b; x) = \sum_{n^{a+b} \leqq x} \psi\left(\left(\frac{x}{n^a}\right)^{1/b}\right).$$

Here we apply Theorem 2.3 as before. We have

$$f(t) = \left(\frac{x}{t^a}\right)^{1/b}, \qquad f''(t) = \frac{a}{b}\left(\frac{a}{b} + 1\right)\left(\frac{x}{t^{a+2b}}\right)^{1/b},$$

and because of $a + 2b < 3b$ we obtain from (2.8)

$$S(a, b; x) \ll \int_1^{x^{1/(a+b)}} \left(\frac{x}{t^{a+2b}}\right)^{1/3b} dt + x^{\frac{1}{2}\cdot\frac{1}{a+b}} \ll x^{\frac{2}{3}\cdot\frac{1}{a+b}}.$$

If we proceed in the same way for the sum $S(b, a; x)$, we would obtain

$$S(b, a; x) \ll x^{1/3a}\,,$$

since $b + 2a > 3a$. This bad estimate can be improved by dividing the interval $[1, x^{1/(a+b)}]$ into two subintervals $[1, y]$, $(y, x^{1/(a+b)}]$ with a suitable value y. We again use van der Corput's method in the second interval, but we estimate the sum trivially in the first interval. Thus we obtain

$$S(b, a; x) \ll y + \int_y^{x^{1/(a+b)}} \left(\frac{x}{t^{b+2a}}\right)^{1/3a} \mathrm{d}t + x^{\frac{1}{2}\cdot\frac{1}{a+b}}$$

$$\ll y + (y^{a-b}x)^{\frac{1}{3a}} + x^{\frac{1}{2}\cdot\frac{1}{a+b}}.$$

The first two terms are of the same order if we put $y = x^{\frac{1}{2a+b}}$. This gives

$$S(b, a; x) \ll x^{\frac{1}{2a+b}}.$$

Applying the above estimates of $S(a, b; x)$, the result (5.6) follows by means of (5.3).

5.2. THE REPRESENTATION PROBLEM

This section is concerned with the representation of $D(a, b; x)$ by an infinite series. We begin with a useful integral representation for the integral

$$\int_1^x D(a, b; t)\frac{\mathrm{d}t}{t}$$

which we can expand into an infinite series. Now we try to find an expansion of $D(a, b; x)$ by differentiating this infinite series. This is possible without great difficulty in the symmetrical case $a = b = 1$. But in the unsymmetrical case $a < b$ we have additionally to consider the sum $\sum_{m^a n^b \leqq x} m^{a-1} n^{b-1}$. For this reason we first consider the much more general function

$$D_{\nu,\mu}(a, b; x) = \sum_{m^a n^b \leqq x} m^{a\nu} n^{b\mu}$$

with any real values ν, μ. If, in particular, a, b are integers, then $D_{\nu,\mu}(a, b; x)$ represents a sum over a divisor function which is defined by

$$\sigma_{\nu,\mu}(a, b; k) = \sum_{m^a n^b = k} m^{a\nu} n^{b\mu}.$$

Then

$$D_{\nu,\mu}(a, b; x) = \sum_{k \leqq x} \sigma_{\nu,\mu}(a, b; k).$$

Finally, we establish a non-trivial Ω-estimation as an application of the asymptotic expansion of the above integral.

5.2.1. *An integral representation*

The purpose of this section is to find an integral representation for the integral

$$\int_1^x D_{\nu,\mu}(a, b; t) \frac{dt}{t^{\nu+1}}.$$

Because of the equation

$$D_{\nu,\mu}(a, b; x) = x^{\mu} D_{\nu-\mu,0}(a, b; x) - \mu \int_1^x t^{\mu-1} D_{\nu-\mu,0}(a, b; t)\, dt \tag{5.7}$$

it is sufficient to consider the special case $\mu = 0$.

Theorem 5.4. *If* $-1/b < \nu < 1/a$, *then*

$$\int_1^x D_{\nu,0}(a, b; t) \frac{dt}{t^{\nu+1}} = K_{\nu,0}(a, b; x) + \int_0^\infty \psi(t^{1/b})\, \psi\left(\left(\frac{x}{t}\right)^{1/a}\right) \frac{dt}{t^{\nu+1}}, \tag{5.8}$$

where

$$K_{\nu,0}(a, b; x) = \frac{a}{1 + a\nu} \zeta\left(\frac{b}{a}(1 + a\nu)\right) x^{\frac{1}{a}} + \frac{b}{1 - b\nu} \zeta\left(\frac{a}{b}(1 - b\nu)\right) x^{\frac{1}{b} - \nu}$$
$$+ \frac{\zeta(-a\nu)}{2\nu} x^{-\nu} - \frac{\zeta(b\nu)}{2\nu} \tag{5.9}$$

for $\nu \neq 1/b - 1/a, 0, 1/b$. *The excluded cases follow by considering the limiting values. In particular, we have*

$$K_{0,0}(a, b; x) = a\zeta\left(\frac{b}{a}\right) x^{1/a} + b\zeta\left(\frac{a}{b}\right) x^{1/b} + \frac{1}{4} \log x + \frac{a+b}{4} \log 2\pi$$

for $a < b$,

$$K_{0,0}(1, 1; x) = x \log x + (2C - 1)\, x + \frac{1}{4} \log x + \frac{1}{2} \log 2\pi,$$

where C *is Euler's constant,*

$$K_{\nu,0}(a, b; x) = x^{1/a} \log x + ((a + b)\, C - 1)\, x^{1/a}$$
$$+ \frac{ab}{2(a - b)} \zeta\left(1 - \frac{a}{b}\right) x^{1/a - 1/b} + \frac{ab}{2(b - a)} \zeta\left(1 - \frac{b}{a}\right)$$

for $\nu = 1/b - 1/a$, $a < b$.

Proof. We obtain for the integral on the left-hand side of (5.8)

$$\int_1^x D_{\nu,0}(a,b;t)\frac{dt}{t^{\nu+1}} = \int_1^x [t^{1/b}]\left[\left(\frac{x}{t}\right)^{1/a}\right]\frac{dt}{t^{\nu+1}}$$

$$= \int_1^x \left\{t^{1/b} - \frac{1}{2} - \psi(t^{1/b})\right\}\left\{\left(\frac{x}{t}\right)^{1/a} - \frac{1}{2} - \psi\left(\left(\frac{x}{t}\right)^{1/a}\right)\right\}\frac{dt}{t^{\nu+1}}.$$

We write

$$\int_1^x \left\{\left(\frac{x}{t}\right)^{1/a} - \frac{1}{2}\right\}\psi(t^{1/b})\frac{dt}{t^{\nu+1}}$$

$$= \int_1^\infty \left\{\left(\frac{x}{t}\right)^{1/a} - \frac{1}{2}\right\}\psi(t^{1/b})\frac{dt}{t^{\nu+1}} - \int_x^\infty \psi(t^{1/b})\,\psi\left(\left(\frac{x}{t}\right)^{1/a}\right)\frac{dt}{t^{\nu+1}},$$

$$\int_1^x \left\{t^{1/b} - \frac{1}{2}\right\}\psi\left(\left(\frac{x}{t}\right)^{1/a}\right)\frac{dt}{t^{\nu+1}}$$

$$= x^{-\nu}\int_1^\infty \left\{\left(\frac{x}{t}\right)^{1/b} - \frac{1}{2}\right\}\psi(t^{1/a})\frac{dt}{t^{1-\nu}} - \int_0^1 \psi(t^{1/b})\,\psi\left(\left(\frac{x}{t}\right)^{1/a}\right)\frac{dt}{t^{\nu+1}},$$

where the integrals on the right-hand sides are convergent according to our assumption $-1/b < \nu < 1/a$. Thus, we obtain the representation (5.8) where first of all the term $K_{\nu,0}(a, b; x)$ is given by

$$K_{\nu,0}(a, b; x)$$

$$= \int_1^x \left\{t^{1/b} - \frac{1}{2}\right\}\left\{\left(\frac{x}{t}\right)^{1/a} - \frac{1}{2}\right\}\frac{dt}{t^{\nu+1}} - x^{-\nu}\int_1^\infty \left\{\left(\frac{x}{t}\right)^{1/b} - \frac{1}{2}\right\}\psi(t^{1/a})\frac{dt}{t^{1-\nu}}$$

$$- \int_1^\infty \left\{\left(\frac{x}{t}\right)^{1/a} - \frac{1}{2}\right\}\psi(t^{1/b})\frac{dt}{t^{\nu+1}}.$$

Calculating the first integral and using the well-known formula

$$\int_1^\infty t^{-s-1}\psi(t)\,dt = \frac{s+1}{2s(s-1)} - \frac{\zeta(s)}{s} \qquad (s > -1,\ s \neq 0, 1),$$

representation (5.9) follows at once.

Corollary. *We deduce a representation for the integral over* $D_{\nu,\mu}(a, b; x)$ *from* (5.8) *by integrating equation* (5.7):

$$\int_1^x D_{\nu,\mu}(a, b; t)\frac{dt}{t^{\nu+1}} = \int_1^x D_{\nu-\mu,0}(a, b; t)\frac{dt}{t^{\nu-\mu+1}} - \mu\int_1^x \frac{dt}{t^{\nu+1}}\int_1^t D_{\nu-\mu,0}(a, b; \tau)\frac{d\tau}{\tau^{1-\nu}}$$

$$= \int_1^x D_{\nu-\mu,0}(a, b; t)\frac{dt}{t^{\nu-\mu+1}} - \mu x^{-\nu}\int_1^x \frac{d\tau}{\tau^{1-\nu}}\int_1^\tau D_{\nu-\mu,0}(a, b; t)\frac{dt}{t^{\nu-\mu+1}}.$$

If $-1/b < \nu - \mu < 1/a$, (5.8) *now leads to*

$$\int_1^x D_{\nu,\mu}(a, b; t)\frac{dt}{t^{\nu+1}} = K_{\nu,\mu}(a, b; x) + \int_0^\infty \psi(t^{1/b})\,\psi\left(\left(\frac{x}{t}\right)^{1/a}\right)\frac{dt}{t^{\nu-\mu+1}} - \mu x^{-\nu}\int_1^x \tau^{\nu-1}\,d\tau\int_0^\infty \psi(t^{1/b})\,\psi\left(\left(\frac{x}{t}\right)^{1/a}\right)\frac{dt}{t^{\nu-\mu+1}} \tag{5.10}$$

where $K_{\nu,\mu}(a, b; x)$ *is defined by*

$$K_{\nu,\mu}(a, b; x) = \frac{a\zeta(b(1/a+\nu-\mu))}{1+a\nu}x^{\frac{1}{a}} + \frac{b\zeta(a(1/b+\mu-\nu))}{(1+b(\mu-\nu))(1+b\mu)}x^{\frac{1}{b}+\mu-\nu}$$
$$+\left\{\frac{a^2\mu\zeta(b(1/a+\nu-\mu))}{(1+a(\nu-\mu))(1+a\nu)} + \frac{b^2\mu\zeta(a(1/b+\mu-\nu))}{(1+b(\mu-\nu))(1+b\mu)} - \frac{\zeta(a(\mu-\nu))}{2(\mu-\nu)} - \frac{\mu\zeta(b(\nu-\mu))}{2\nu(\nu-\mu)}\right\}x^{-\nu} + \frac{\mu\zeta(b(\nu-\mu))}{2\nu(\nu-\mu)} \tag{5.11}$$

for $\nu - \mu \neq 1/b - 1/a, 0, 1/b$ and $\nu \neq -1/a$, $\mu \neq -1/b$. *Again, the excluded cases follow by considering the limiting values. In particular, we obtain*

$$K_{\nu,\mu}(a, b; x) = \frac{1}{b}x^{1/a}\log x + \frac{1}{b}\{(a+b)C - a - 1\}x^{1/a}$$
$$+\left\{(a+b)\left(1-\frac{1}{b}\right)C + \frac{a}{b}\left(1+\frac{1}{a}\right) - a - b\right.$$

$$+\frac{ab\zeta(1-a/b)}{2(a-b)}-\frac{a^2(b-1)\,\zeta(1-b/a)}{2(a-b)\,(a-1)}\Bigg\}\, x^{1/a-1}$$

$$+\frac{a^2(b-1)\,\zeta(1-b/a)}{2(a-b)\,(a-1)} \tag{5.12}$$

for $v = 1 - 1/a$, $\mu = 1 - 1/b$, $a < b$.

5.2.2. *Representations by infinite series*

It is well known that the integral

$$\int_0^\infty \sin\frac{2\pi x}{t}\,\sin 2\pi t\,\frac{dt}{t^{v+1}}$$

can be represented by cylinder functions. Here we need a generalization of these functions. Therefore, we begin with the definition of these functions by integrals of the above type.

Definition 5.1. *Let* $|v| < 1/a + 1/b$. *Let the function* $\varrho_v(a, b; x)$ *be defined by*

$$\varrho_v(a, b; x) = \int_0^\infty \sin 2\pi\left(\frac{x}{t}\right)^{1/a} \sin 2\pi t^{1/b}\,\frac{dt}{t^{v+1}} \tag{5.13}$$

for $x > 0$.

The next investigations are devoted to an asymptotic expansion of these functions, where we shall assume that $1 \leqq a \leqq b$, $-1/b \leqq v \leqq 1/a$. (5.13) may be written in the form

$$\varrho_v(a, b; x) = -\frac{b}{2}\,z^{-bv}\int_0^\infty \{\cos 2\pi z h(t) - \cos 2\pi z g(t)\}\,\frac{dt}{t^{bv+1}} \tag{5.14}$$

with

$$z = x^{\frac{1}{a+b}}, \qquad h(t) = t - t^{-b/a}, \qquad g(t) = t + t^{-b/a}.$$

We now formulate some lemmas.

Lemma 5.1. *Suppose that* $0 < \alpha < \beta < \infty$, $0 \leqq \gamma \leqq 1 + b/a$. *Then, as* $z \to \infty$, *the estimate*

$$\int_\alpha^\beta \cos 2\pi z g(t)\,\frac{dt}{t^\gamma} \ll \frac{1}{z} \tag{5.15}$$

holds uniformly in α *and* β. *The same is true if* $\cos 2\pi z g(t)$ *is replaced by* $\sin 2\pi z g(t)$.

Proof. Applying the second mean-value theorem, we obtain

$$\int_{\alpha}^{\beta} \cos 2\pi z g(t) \frac{dt}{t^{\gamma}} = \int_{\alpha}^{\beta} \frac{\frac{d}{dt} \sin 2\pi z g(t)}{2\pi z g'(t)\, t^{\gamma}} dt \ll \frac{1}{z}$$

uniformly in α and β since $g'(t)\, t^{\gamma} = t^{\gamma} + \frac{b}{a} t^{\gamma - b/a - 1} \geqq c$.

Lemma 5.2. *Let γ be a real number, and let t_0 be defined by $h'(t_0) = 0$ such that*

$$t_0 = \left(\frac{b}{a}\right)^{\frac{a}{a+b}}, \quad h(t_0) = \left(\frac{b}{a}\right)^{\frac{a}{a+b}} + \left(\frac{a}{b}\right)^{\frac{b}{a+b}}, \quad h''(t_0) = \left(\frac{b}{a}\right)^{\frac{b}{a+b}} + \left(\frac{a}{b}\right)^{\frac{a}{a+b}}.$$

Then, as $z \to \infty$,

$$\int_{t_0/2}^{2t_0} \cos 2\pi z h(t) \frac{dt}{t^{\gamma}} = \frac{t_0^{-\gamma}}{\sqrt{h''(t_0)\, z}} \cos 2\pi(z h(t_0) + 1/8) + O\left(\frac{1}{z}\right). \tag{5.16}$$

The same is true if $\cos 2\pi z h(t)$ *is replaced by* $\sin 2\pi z h(t)$.

Proof. We apply the method of stationary phase. The function $h(t)$ has one stationary point in $t_0/2 \leqq t \leqq 2t_0$, namely at $t = t_0$. We consider the integral

$$I = \int_{t_0}^{2t_0} e^{2\pi i z h(t)} \frac{dt}{t^{\gamma}}$$

and introduce a new variable u defined by $h(t) = h(t_0) + u^2$, where u is positive when $t_0 < t \leqq 2t_0$. We now obtain

$$I = e^{2\pi i z h(t_0)} \int_0^{u_0} e^{2\pi i z u^2} t^{-\gamma} \frac{dt}{du} du,$$

where $u_0 = \sqrt{h(2t_0) - h(t_0)}$. It is easily seen that

$$t^{-\gamma} \frac{dt}{du} = \frac{t_0^{-\gamma}}{\sqrt{\pi h''(t_0)}} + u\varphi(u),$$

where $\varphi(u)$ is differentiable in $0 \leqq u \leqq u_0$. The integral with respect to the second term $u\varphi(u)$ is of order $1/z$, which can be seen by integration by parts. Then

$$I = \frac{t_0^{-\gamma}}{\sqrt{\pi h''(t_0)}} e^{2\pi i z h(t_0)} \int_0^{u_0} e^{2\pi i z u^2} du + O\left(\frac{1}{z}\right).$$

Since

$$\int_0^{u_0} e^{2\pi i z u^2}\, du = \int_0^{\infty} e^{2\pi i z u^2}\, du + O\left(\frac{1}{z}\right) = \frac{1}{2}\sqrt{\frac{\pi}{z}}\, e^{\pi i/4} + O\left(\frac{1}{z}\right),$$

we obtain

$$I = \frac{t_0^{-\gamma}}{2\sqrt{h''(t_0)\, z}}\, e^{2\pi i z h(t_0) + \pi i/4} + O\left(\frac{1}{z}\right).$$

The same result holds for the integral over $t_0/2 \leqq t \leqq t_0$. Hence

$$\int_{t_0/2}^{2t_0} e^{2\pi i z h(t)} \frac{dt}{t^\gamma} = \frac{t_0^{-\gamma}}{\sqrt{h''(t_0)\, z}}\, e^{2\pi i z h(t_0) + \pi i/4} + O\left(\frac{1}{z}\right).$$

Clearly, we may replace i by —i. So (5.16) follows at once.

Lemma 5.3. *Let* t_0 *be defined as in Lemma 5.2. Suppose that* $0 < \alpha < \beta \leqq t_0/2$ *or* $2t_0 \leqq \alpha < \beta < \infty$, $0 \leqq \gamma \leqq 1 + b/a$. *Then, as* $z \to \infty$, *the estimate*

$$\int_\alpha^\beta \cos 2\pi z h(t) \frac{dt}{t^\gamma} \ll \frac{1}{z} \tag{5.17}$$

holds uniformly in α *and* β. *The same is true if* $\cos 2\pi z h(t)$ *is replaced by* $\sin 2\pi z h(t)$.

Proof. Applying the second mean-value theorem, we obtain

$$\int_\alpha^\beta \cos 2\pi z h(t) \frac{dt}{t^\gamma} = \int_\alpha^\beta \frac{\frac{d}{dt} \sin 2\pi z h(t)}{2\pi z h'(t)\, t} dt \ll \frac{1}{z}$$

uniformly in α and β, since $h'(t) \neq 0$ in the intervals under consideration and therefore

$$|h'(t)\, t^\gamma| = |t^\gamma - \frac{b}{a} t^{\gamma - b/a - 1}| \geqq c\,.$$

Lemmas 5.1—5.3 give an asymptotic representation of the function $\varrho_\nu(a, b; x)$ for $1 \leqq a \leqq b$, $-1/b \leqq \nu \leqq 1/a$ immediately. First of all we consider the integral of the representation (5.14) in the finite limits α and β, where $0 < \alpha < t_0/2$ and $2t_0 < \beta < \infty$. Here we can partly apply the results (5.15)—(5.17). Since this holds uniformly in α and β, we can take the limits $\alpha \to 0$, $\beta \to \infty$. So the following lemma is clear.

Lemma 5.4. *Let $1 \leqq a \leqq b$, $-1/b \leqq \nu \leqq 1/a$, and let t_0, $h(t_0)$, $h''(t_0)$ be defined as in Lemma 5.2. Then*

$$\varrho_\nu(a, b; x) = -\frac{bt_0^{-b\nu-1}}{2\sqrt{h''(t_0)}}\, x^{-\frac{2b\nu+1}{2(a+b)}} \cos 2\pi\left(x^{\frac{1}{a+b}} h(t_0) + \frac{1}{8}\right) + O\left(x^{-\frac{b\nu+1}{a+b}}\right). \tag{5.18}$$

Theorem 5.5. *Let $1 \leqq a \leqq b$, $-1/2b < \nu < 1/2a$, and let t_0, $h(t_0)$, $h''(t_0)$ be defined as in Lemma 5.2. Then*

$$\int_0^\infty \psi\left(\left(\frac{x}{t}\right)^{1/a}\right)\psi(t^{1/b})\frac{dt}{t^{\nu+1}} = \frac{1}{\pi^2}\sum_{m=1}^\infty\sum_{n=1}^\infty \frac{n^{b\nu-1}}{m}\,\varrho_\nu(a, b; m^a n^b x) \tag{5.19}$$

$$= R_\nu(a, b; x) + O\left(x^{-\frac{b\nu+1}{a+b}}\right), \tag{5.20}$$

where the main term is represented by

$$R_\nu(a, b; x) = -\frac{bt_0^{-b\nu-1}x^{-\frac{2b\nu+1}{2(a+b)}}}{2\pi^2\sqrt{h''(t_0)}}\sum_{m=1}^\infty\sum_{n=1}^\infty m^{a\alpha}n^{b\beta}\cos 2\pi\left((m^a n^b x)^{\frac{1}{a+b}}\, h(t_0) + \frac{1}{8}\right), \tag{5.21}$$

and the numbers α and β are defined by

$$\alpha = -\frac{2b\nu + 1}{2(a+b)} - \frac{1}{a}, \qquad \beta = \frac{2a\nu - 1}{2(a+b)} - \frac{1}{b}.$$

In particular, if a, b are integers, then

$$\int_0^\infty \psi\left(\left(\frac{x}{t}\right)^{1/a}\right)\psi(t^{1/b})\frac{dt}{t^{\nu+1}} = \frac{1}{\pi^2}\sum_{k=1}^\infty \sigma_{-1/a,\, \nu-1/b}(a, b; k)\,\varrho_\nu(a, b; kx), \tag{5.22}$$

and (5.20) holds with

$$R_\nu(a, b; x) = -\frac{bt_0^{-b\nu-1}x^{-\frac{2b\nu+1}{2(a+b)}}}{2\pi^2\sqrt{h''(t_0)}}\sum_{k=1}^\infty \sigma_{\alpha,\beta}(a, b; k)\cos 2\pi\left((kx)^{\frac{1}{a+b}}\, h(t_0) + \frac{1}{8}\right). \tag{5.23}$$

The series are absolutely and uniformly convergent with respect to x.

Proof. Let $0 < \delta < 1$. Then

$$\int_{\delta}^{1/\delta} \psi\left(\left(\frac{x}{t}\right)^{1/a}\right) \psi(t^{1/b}) \frac{dt}{t^{\nu+1}}$$

$$= -\frac{1}{\pi} \int_{\delta}^{1/\delta} \sum_{n=1}^{\infty} \frac{\sin 2\pi n t^{1/b}}{n} \psi\left(\left(\frac{x}{t}\right)^{1/a}\right) \frac{dt}{t^{\nu+1}}$$

$$= -\frac{1}{\pi} \sum_{n=1}^{\infty} \frac{1}{n} \int_{\delta}^{1/\delta} \sin 2\pi n t^{1/b} \psi\left(\left(\frac{x}{t}\right)^{1/a}\right) \frac{dt}{t^{\nu+1}}$$

$$= \frac{1}{\pi^2} \sum_{n=1}^{\infty} \sum_{m=1}^{\infty} \frac{1}{mn} \int_{\delta}^{1/\delta} \sin 2\pi m \left(\frac{x}{t}\right)^{1/a} \sin 2\pi n t^{1/b} \frac{dt}{t^{\nu+1}}.$$

Applying Lemmas 5.1—5.3 it is seen that the integral on the right-hand side is of order

$$(n^{b(2a\nu-1)} m^{-a(2b\nu+1)} x^{-2b\nu-1})^{\frac{1}{2(a+b)}}$$

which uniformly holds in δ. Because of $-1/2b < \nu < 1/2a$, the double sum converges absolutely and uniformly with respect to x. Hence, we can take the limit $\delta \to 0$ on both the sides. We now obtain equation (5.19) by means of (5.13). The asymptotic representation (5.20) where $R_\nu(a, b; x)$ is given by (5.21) follows from (5.18) and (5.19) immediately. The representations (5.21) and (5.22) and the properties of convergence are clear.

In case of $\mu = 0$ we obtain an asymptotic representation for the integral over $D_{\nu,\mu}(a, b; x)$ from (5.8) and (5.20) at once. However, if $\mu \neq 0$, we must use the representation (5.10). Then we get the following theorem.

Theorem 5.6. *Let $1 \leqq a \leqq b$, $-1/2b < \nu - \mu < 1/2a$. Then*

$$\int_{1}^{x} D_{\nu,\mu}(a, b; t) \frac{dt}{t^{\nu+1}}$$

$$= K_{\nu,\mu}(a, b; x) - Ax^{-\nu} + R_{\nu-\mu}(a, b; x) + O\left(x^{-\frac{b(\nu-\mu)+1}{a+b}} \log x\right), \tag{5.24}$$

where $K_{\nu,\mu}(a, b; x)$ is defined by (5.11), and $R_{\nu-\mu}(a, b; x)$, by (5.21) or (5.23). The constant A is given by

$$A = \mu \int_{1}^{\infty} \tau^{\nu-1}\, d\tau \int_{0}^{\infty} \psi\left(\left(\frac{\tau}{t}\right)^{1/a}\right) \psi(t^{1/b}) \frac{dt}{t^{\nu-\mu+1}}$$

for $a\nu + b\mu < 1$ and otherwise by $A = 0$.

Proof. It is seen from the representations (5.10) and (5.20) that we have to estimate only the integral

$$S = \mu x^{-\nu} \int_1^x \tau^{\nu-1}\, d\tau \int_0^\infty \psi\left(\left(\frac{\tau}{t}\right)^{1/a}\right) \psi(t^{1/b}) \frac{dt}{t^{\nu-\mu+1}} .$$

Using (5.20), we obtain

$$S = \mu x^{-\nu} \int_1^x \tau^{\nu-1} R_{\nu-\mu}(a, b; \tau)\, d\tau + O\left(x^{-\frac{b(\nu-\mu)+1}{a+b}} \log x\right)$$

for $a\nu + b\mu \geqq 1$. Applying the representation (5.21), we easily get by partial integration

$$S \ll x^{-\frac{b(\nu-\mu)+1}{a+b}} \log x .$$

In case of $a\nu + b\mu < 1$ we write

$$S = Ax^{-\nu} - \mu x^{-\nu} \int_x^\infty \tau^{\nu-1} R_{\nu-\mu}(a, b; \tau)\, d\tau + O\left(x^{-\frac{b(\nu-\mu)+1}{a+b}}\right),$$

and similarly we obtain

$$S = Ax^{-\nu} + O\left(x^{-\frac{b(\nu-\mu)+1}{a+b}}\right).$$

This proves representation (5.24).

5.2.3. *The Voronoi Identity*

In this section we are concerned with an identity which expresses $D(a, b; x)$ as an infinite series of generalized cylinder functions, provided that a, b are integers. We apply Theorems 5.4 and 5.5. Writing $D(a, b; x)$ and $K(a, b; x)$ instead of $D_{0,0}(a, b; x)$ and $K_{0,0}(a, b; x)$, we obtain from (5.8) and (5.22)

$$\int_1^x D(a, b; t) \frac{dt}{t} = K(a, b; x) + \frac{1}{\pi^2} \sum_{k=1}^\infty \sigma_{-1/a,\,-1/b}(a, b; k)\, \varrho_0(a, b; kx) . \quad (5.25)$$

We may hope to get a representation for $D(a, b; x)$, at least for non-integers x, if we differentiate this equation. Differentiating the infinite series term-by-term, we formally obtain

$$D(a, b; x) = xK'(a, b; x) + \frac{x}{\pi^2} \sum_{k=1}^\infty \sigma_{a', b'}(a, b; k)\, \varrho_0'(a, b; kx) ,$$

where $a' = 1 - 1/a$, $b' = 1 - 1/b$. This identity was first stated by G. Voronoi [1] in the special case of Dirichlet's divisor problem $a = b = 1$. W. Rogosinski [1] proved that in this case the series on the right-hand side converges for every x. More precisely, the series is uniformly convergent throughout any interval free from integral values of x. For integers x the series converges to the mean-value of $D(1, 1; x + 0)$ and $D(1, 1; x - 0)$. So the representation

$$\frac{D(1,1;x-0)+D(1,1;x+0)}{2} = xK'(1,1;x) + \frac{x}{\pi^2}\sum_{k=1}^{\infty} \sigma_{0,0}(1,1;k)\,\varrho_0'(1,1;kx)$$

is called *Voronoi's Identity*. Here we prove an analogous result for integers a, b, but the analysis is for $a < b$ a little more difficult. Moreover, we shall see that the series only converges for $a \leqq b < \frac{3}{2}a$. We begin with two lemmas.

Lemma 5.5. *Let t_0, $h(t_0)$, $h''(t_0)$ be defined as in Lemma 5.2. Then*

$$\varrho_0'(a, b; x) = \frac{\pi}{\sqrt{h''(t_0)}}\, x^{\frac{1}{2(a+b)}-1} \sin 2\pi\left(x^{\frac{1}{a+b}} h(t_0) + \frac{1}{8}\right) + O\left(\frac{1}{x}\right), \tag{5.26}$$

$$\varrho_0''(a, b; x) = \frac{2\pi^2 t_0}{b\sqrt{h''(t_0)}}\, x^{\frac{3}{2(a+b)}-2} \cos 2\pi\left(x^{\frac{1}{a+b}} h(t_0) + \frac{1}{8}\right) + O\left(x^{\frac{1}{a+b}-2}\right), \tag{5.27}$$

$$\varrho_0'''(a, b; x) = -\frac{4\pi^3 t_0^2}{b^2\sqrt{h''(t_0)}}\, x^{\frac{5}{2(a+b)}-3} \sin 2\pi\left(x^{\frac{1}{a+b}} h(t_0) + \frac{1}{8}\right) + O\left(x^{\frac{2}{a+b}-3}\right). \tag{5.28}$$

Proof. We obtain from (5.13)

$$\varrho_0'(a, b; x) = \frac{2\pi}{b} x^{1/b-1} \int_0^{\infty} \cos 2\pi(xt)^{1/b} \sin 2\pi t^{-1/a} \frac{dt}{t^{1-1/b}}$$

$$= \pi x^{\frac{1}{a+b}-1} \int_0^{\infty} \{\sin 2\pi z h(t) - \sin 2\pi z g(t)\}\, dt\,,$$

where z, $g(t)$, $h(t)$ are defined as in (5.14). The asymptotic representation (5.26) now follows from Lemmas 5.1—5.3 in the same way as we have proved (5.18).

To prove (5.27) we write

$$\varrho_0'(a, b; x) = \frac{2\pi}{bx} \int_0^{\infty} t^{1/b-1} \cos 2\pi t^{1/b} \sin 2\pi \left(\frac{x}{t}\right)^{1/a} dt\,.$$

Then

$$\varrho_0''(a, b; x) = \frac{(2\pi)^2}{ab} x^{\frac{1}{a}-2} \int_0^\infty t^{\frac{1}{b}-\frac{1}{a}-1} \cos 2\pi t^{\frac{1}{b}} \cos 2\pi \left(\frac{x}{t}\right)^{\frac{1}{a}} \mathrm{d}t - \frac{1}{x} \varrho_0'(a, b; x)$$

$$= \frac{2\pi^2}{a} x^{\frac{2}{a+b}-2} \int_0^\infty t^{-\frac{b}{a}} \{\cos 2\pi z h(t) + \cos 2\pi z g(t)\} \, \mathrm{d}t$$

$$+ O\left(x^{\frac{1}{2(a+b)}-2}\right).$$

(5.27) now again follows from Lemmas 5.1—5.3.

To prove (5.28) we write

$$\varrho_0''(a, b; x) = \frac{(2\pi)^2}{ab} x^{\frac{1}{b}-2} \int_0^\infty t^{\frac{1}{a}-\frac{1}{b}-1} \cos 2\pi t^{\frac{1}{a}} \cos 2\pi \left(\frac{x}{t}\right)^{\frac{1}{b}} \mathrm{d}t - \frac{1}{x} \varrho_0'(a, b; x).$$

Then

$$\varrho_0'''(a, b; x) = \frac{(2\pi)^2}{ab} x^{\frac{1}{b}-2} \frac{\mathrm{d}}{\mathrm{d}x} \int_0^\infty t^{\frac{1}{a}-\frac{1}{b}-1} \cos 2\pi t^{\frac{1}{a}} \cos 2\pi \left(\frac{x}{t}\right)^{\frac{1}{b}} \mathrm{d}t$$

$$+ \left(\frac{1}{b} - 1\right) \frac{1}{x^2} \varrho_0'(a, b; x) + \left(\frac{1}{b} - 2\right) \frac{1}{x} \varrho_0''(a, b; x).$$

If $a < b$, then

$$\varrho_0'''(a, b; x) = -\frac{(2\pi)^3}{ab^2} x^{\frac{2}{b}-3} \int_0^\infty t^{\frac{1}{a}-\frac{2}{b}-1} \cos 2\pi t^{\frac{1}{a}} \sin 2\pi \left(\frac{x}{t}\right)^{\frac{1}{b}} \mathrm{d}t + O\left(x^{\frac{3}{2(a+b)}-3}\right)$$

$$= \frac{4\pi^3}{b^2} x^{\frac{2}{a+b}-3} \int_0^\infty t^{-\frac{2a}{b}} \{\sin 2\pi z g(t) - \sin 2\pi z h(t)\} \, \mathrm{d}t + O\left(x^{\frac{3}{2(a+b)}-3}\right).$$

In this case (5.28) now follows as above.

It remains the case $a = b$, where we put $a = b = 1$ without loss of generality. Here we have

$$\varrho_0'''(1, 1; x) = \frac{4\pi^2}{x} \frac{\mathrm{d}}{\mathrm{d}x} \int_0^1 (\cos 2\pi t - 1) \cos \frac{2\pi x}{t} \frac{\mathrm{d}t}{t} + \int_0^1 \cos \frac{2\pi x}{t} \frac{\mathrm{d}t}{t}$$

$$+ \int_1^\infty \cos 2\pi t \cos \frac{2\pi x}{t} \frac{\mathrm{d}t}{t} + O(x^{-9/4})$$

$$= \frac{8\pi^3}{x} \int_0^1 (\cos 2\pi t - 1) \sin \frac{2\pi x}{t} \frac{dt}{t^2}$$

$$- \frac{8\pi^3}{x} \int_1^\infty \cos 2\pi t \sin \frac{2\pi x}{t} \frac{dt}{t^2} + O\left(\frac{1}{x^2}\right)$$

$$= \frac{8\pi^3}{x} \int_0^1 \cos \frac{2\pi}{t} \sin 2\pi x t \, dt + O\left(\frac{1}{x^2}\right)$$

$$= 4\pi^3 x^{-3/2} \int_0^{\sqrt{x}} \{\sin 2\pi z h(t) + \sin 2\pi z g(t)\} \, dt + O\left(\frac{1}{x^2}\right).$$

Now (5.28) follows from Lemmas 5.1—5.3 in this case, as well.

The reader will have no difficulty in proving the next lemma, which will be required in the proof of Theorem 5.7.

Lemma 5.6. *Let* $a' = 1 - 1/a$, $b' = 1 - 1/b$. *Then*

$$D_{a', b'}(a, b; x) = \frac{x}{ab} \log x + \frac{1}{ab} (C(a + b) - 1)\, x + O\left(x^{1 - \frac{1}{a+b}}\right), \tag{5.29}$$

where C is Euler's constant.

Theorem 5.7. *Let a, b be integers with* $1 \leqq a \leqq b < 3a/2$ *and let* $a' = 1 - 1/a$, $b' = 1 - 1/b$. *Then the series*

$$\sum_{k=1}^{\infty} \sigma_{a', b'}(a, b; k)\, \varrho_0'(a, b; kx) \tag{5.30}$$

is convergent for every positive value of x, uniformly convergent throughout any closed interval which does not contain an integer, and it is

$$\frac{D(a, b; x - 0) + D(a, b; x + 0)}{2} = H(a, b; x) + \frac{1}{4} + \frac{x}{\pi^2} \sum_{k=1}^{\infty} \sigma_{a', b'}(a, b; k)\, \varrho_0'(a, b; kx), \tag{5.31}$$

where $H(a, b; x)$ *is given by* (5.2).

Proof. We consider the partial sums

$$G_n(x) = \sum_{k=1}^{n} \sigma_{a',b'}(a, b; k)\, \varrho_0'(a, b; kx) = D_{a',b'}(a, b; n)\, \varrho_0'(a, b; nx)$$

$$- x \int_1^n D_{a',b'}(a, b; t)\, \varrho_0''(a, b; xt)\, dt$$

$$= D_{a',b'}(a, b; n)\, \varrho_0'(a, b; nx)$$

$$- n^{2-1/a} x \varrho_0''(a, b; nx) \int_1^n t^{1/a-2} D_{a',b'}(a, b; t)\, dt$$

$$+ x \int_1^n \frac{d}{dt}(t^{2-1/a} \varrho_0''(a, b; xt))\, dt \int_1^t D_{a',b'}(a, b; \tau)\, \tau^{1/a-2}\, d\tau .$$

We put

$$\int_1^x D_{a',b'}(a, b; \tau)\, \tau^{1/a-2}\, d\tau = K_{a',b'}(a, b; x) - A + R(x),$$

where A is the value of Theorem 5.6. Note that $A \neq 0$ only for $aa' + bb' < 1$. But this is only possible for $a = b = 1$. Thus we have $A = 0$ for $a < b$. Hence, (5.24) shows that

$$R(x) = R_{a'-b'}(a, b; x) + O\left(x^{-\frac{b(a'-b')+1}{a+b}} \log x\right).$$

From (5.23) we obtain

$$R(x) = c x^{\frac{2b-3a}{2a(a+b)}} \sum_{k=1}^{\infty} \sigma_{\alpha,\alpha}(a, b; k) \cos 2\pi \left((kx)^{\frac{1}{a+b}} h(t_0) + \frac{1}{8}\right)$$

$$+ O\left(x^{\frac{b-2a}{a(a+b)}} \log x\right) \tag{5.32}$$

with a certain constant c and $\alpha = -\dfrac{5}{2(a+b)}$.

Because of our condition $1 \leqq a \leqq b < \frac{3}{2} a$ the series is absolutely and uniformly convergent with respect to x. Now $G_n(x)$ becomes

$$G_n(x) = D_{a',b'}(a, b; n)\, \varrho_0'(a, b; nx)$$

$$- n^{2-1/a} x \varrho_0''(a, b; nx) \{K_{a',b'}(a, b; n) + R(n) - A\}$$

$$+ x \int_1^n K_{a',b'}(a, b; t) \frac{d}{dt}(t^{2-1/a} \varrho_0''(a, b; xt)\, dt$$

$$+ x \int_1^n (R(t) - A) \frac{d}{dt} (t^{2-1/a} \varrho_0''(a, b; xt))\, dt$$

$$= D_{a',b'}(a, b; n)\, \varrho_0'(a, b; nx) - n^{2-1/a} x R(n)\, \varrho_0''(a, b; nx)$$

$$+ (A - K_{a',b'}(a, b; 1))\, x \varrho_0''(a, b; x)$$

$$- x \int_1^n t^{2-1/a} K'_{a',b'}(a, b; t)\, \varrho_0''(a, b; xt)\, dt$$

$$+ x \int_1^n R(t) \frac{d}{dt} (t^{2-1/a} \varrho_0''(a, b; xt))\, dt$$

$$= \{D_{a',b'}(a, b; n) - n^{2-1/a} K'_{a',b'}(a, b; n)\}\, \varrho_0'(a, b; nx)$$

$$- n^{2-1/a} x R(n)\, \varrho_0''(a, b; nx) + (A - K_{a',b'}(a, b; 1))\, x \varrho_0''(a, b; x)$$

$$+ x K'_{a',b'}(a, b; 1)\, \varrho_0'(a, b; x)$$

$$+ x \int_1^n \frac{d}{dt} (t^{2-1/a} K'_{a',b'}(a, b; t))\, \varrho_0'(a, b; xt)\, dt$$

$$+ x \int_1^n R(t) \frac{d}{dt} (t^{2-1/a} \varrho_0''(a, b; xt))\, dt\,.$$

We now investigate the single terms in regard of their behaviour for $n \to \infty$.

It is seen from (5.12), (5.26) and (5.29) that the first term is of order

$$n^{1-\frac{1}{a+b}} |\varrho_0'(a, b; nx)| \ll n^{-\frac{1}{2(a+b)}} x^{\frac{1}{2(a+b)}-1}\,.$$

(5.27) and (5.30) show that the second term is of order

$$n^{-\frac{1}{a+b}} x^{\frac{3}{2(a+b)}-1}\,.$$

Hence, both terms uniformly tend to 0 in any closed interval, when $n \to \infty$.

The third and fourth terms are independent of n and continuous with respect to x.

The fifth term approaches a limit uniformly in any closed interval. For, we obtain by means of (5.12)

$$\int^n \frac{d}{dt} (t^{2-1/a} K'_{a',b'}(a, b; t))\, \varrho_0'(a, b; xt)\, dt$$

$$= \frac{1}{ab} \int_1^n (\log t + (a+b) C)\, \varrho_0'(a, b; xt)\, \mathrm{d}t$$

$$= \frac{\log n + (a+b) C}{abx} \varrho_0(a, b; xn) - \frac{(a+b) C}{abx} \varrho_0(a, b; x)$$

$$- \frac{1}{abx} \int_1^n \varrho_0(a, b; xt) \frac{\mathrm{d}t}{t},$$

and Lemma 5.4 proves the required property.

It remains to prove that the sixth term satisfies the assertion of the theorem. The asymptotic representations (5.27) and (5.28) show that

$$\frac{\mathrm{d}}{\mathrm{d}t} (t^{2-1/a} \varrho_0''(a, b; xt))$$

$$= c_1 x^{\frac{1}{a}-1} (xt)^{\frac{3a-2b}{2a(a+b)}-1} \sin 2\pi \left((xt)^{\frac{1}{a+b}} h(t_0) + \frac{1}{8} \right) + O\left(x^{\frac{1}{a}-1} (xt)^{\frac{a-b}{a(a+b)}-1} \right)$$

for $t \to \infty$ with a certain constant c_1. Putting

$$r(x) = (xt)^{\frac{1}{a+b}} h(t_0),$$

we obtain from (5.30)

$$R(t) \frac{\mathrm{d}}{\mathrm{d}t} (t^{2-1/a} \varrho_0''(a, b; xt))$$

$$= c_2 x^{\frac{5}{2(a+b)}-2} \frac{1}{t} \sum_{k=1}^{\infty} \sigma_{\alpha,\alpha}(a, b; k) \sin 2\pi \left(r(x) + \frac{1}{8} \right) \sin 2\pi \left(r(k) - \frac{1}{8} \right)$$

$$+ O\left(t^{-\frac{1}{2(a+b)}-1} \right)$$

with a certain constant c_2, where the error term depends on x. This means that we have to prove the assertion of the theorem for the integral

$$2 \int_1^{\infty} \sum_{k=1}^{\infty} \sigma_{\alpha,\alpha}(a, b; k) \sin 2\pi \left(r(x) + \frac{1}{8} \right) \sin 2\pi \left(r(k) - \frac{1}{8} \right) \frac{\mathrm{d}t}{\mathrm{d}}.$$

We use the functional equation

$$2 \sin 2\pi \left(r(x) + \frac{1}{8} \right) \sin 2\pi \left(r(k) - \frac{1}{8} \right)$$

$$= \sin 2\pi (r(k) - r(x)) - \cos 2\pi (r(k) + r(x)).$$

Obviously, the integral

$$\int_1^{\infty} \sum_{k=1}^{\infty} \sigma_{\alpha,\alpha}(a, b; k) \cos 2\pi(r(k) + r(x)) \frac{dt}{t}$$

is uniformly convergent for $x > 0$. Now we consider the function

$$F(x, y) = \int_1^{y} \sum_{k=1}^{\infty} \sigma_{\alpha,\alpha}(a, b; k) \sin 2\pi(r(k) - r(x)) \frac{dt}{t}$$

for $x > 0$ and $y > 1$. We have

$$F(x, y) = -\sum_{k \leqq x} \sigma_{\alpha,\alpha}(a, b; k) \int \sin 2\pi t^{\frac{1}{a+b}} h(t_0) \frac{dt}{t}$$

$$+ \sum_{k \geqq x} \sigma_{\alpha,\alpha}(a, b; k) \int \sin 2\pi t^{\frac{1}{a+b}} h(t_0) \frac{dt}{t},$$

$$IC(\textstyle\int): \left|k^{\frac{1}{a+b}} - x^{\frac{1}{a+b}}\right|^{a+b} \leqq t \leqq \left|k^{\frac{1}{a+b}} - x^{\frac{1}{a+b}}\right|^{a+b} y.$$

Since the integrals are uniformly bounded, the series is uniformly convergent with respect to $x > 0$ and $y > 1$ on the right-hand side. Hence, the limit $\lim_{y \to \infty} F(x, y) = F(x)$ exists for each fixed x, and the convergence is uniform in any closed interval which does not contain an integer. But if n is a positive integer, we obtain

$$F(x) = F(n) + \sigma_{\alpha,\alpha}(a, b; n) \int_N^{\infty} \sin 2\pi t^{\frac{1}{a+b}} h(t_0) \frac{dt}{t}$$

for $n - 1 < x < n$ and

$$F(x) = F(n) - \sigma_{\alpha,\alpha}(a, b; n) \int_N^{\infty} \sin 2\pi t^{\frac{1}{a+b}} h(t_0) \frac{dt}{t}$$

for $n < x < n + 1$, where N is given by $N = \left|n^{\frac{1}{a+b}} - x^{\frac{1}{a+b}}\right|^{a+b}$. Hence

$$F(n) = \frac{F(n-0) + F(n+0)}{2}.$$

We have now proved that the sequence $(G_n(x))$ and therefore the series (5.30) are convergent for every positive value of x, uniformly convergent throughout any closed interval which does not contain an integer. Thus, if x is not an integer, we can differentiate (5.25) term-by-term and obtain (5.31). Furthermore, we have proved that (5.31) also holds for integers x. This completes the proof of Theorem 5.7.

5.2.4. *The Ω-estimate*

We use Landau's method of Section 3.2.3 and apply Lemma 3.7. Let $\Delta(a, b; x)$ be given by (5.1). Then we put

$$\Delta_1(a, b; x) = \int_1^x \left(\Delta(a, b; t) - \frac{1}{4}\right) dt$$

$$= \frac{1 - x}{4} + x \int_1^x \Delta(a, b; t) \frac{dt}{t} - \int_1^x dy \int_1^y \Delta(a, b; t) \frac{dt}{t},$$

$$\Delta_2(a, b; x) = \int_1^x \Delta_1(a, b; t)\, dt .$$

Lemma 5.7. *Let* α *and* β *be defined by*

$$\alpha = -\frac{1}{2(a + b)} - \frac{1}{a}, \qquad \beta = -\frac{1}{2(a + b)} - \frac{1}{b}.$$

Then

$$\Delta_1(a, b; x) = -\frac{bx^{1-\frac{1}{2(a+b)}}}{2\pi^2 t_0 \sqrt{h''(t_0)}} \sum_{k=1}^{\infty} \sigma_{\alpha, \beta}(a, b; k) \cos 2\pi \left((kx)^{\frac{1}{a+b}} h(t_0) + \frac{1}{8}\right)$$

$$+ O\left(x^{1-\frac{1}{a+b}}\right) \tag{5.33}$$

In particular, the estimate

$$\Delta_1(a, b; x) \ll x^{1-\frac{1}{2(a+b)}} \tag{5.34}$$

holds.

Proof. Applying (5.1), we obtain

$$\int_1^x D(a, b; t) \frac{dt}{t} = \int_1^x H(a, b; t) \frac{dt}{t} + \int_1^x \Delta(a, b; t) \frac{dt}{t} .$$

(5.2), (5.8), (5.9) and (5.22) show that

$$\int_1^x D(a, b; t) \frac{dt}{t} = \int_0^x H(a, b; t) \frac{dt}{t} + \frac{1}{4} \log x + \frac{a + b}{4} \log 2\pi$$

$$+ \frac{1}{\pi^2} \sum_{k=1}^{\infty} \sigma_{-1/a, -1/b}(a, b; k)\, \varrho_0(a, b; kx) .$$

Hence

$$\int_1^x \Delta(a, b; t) \frac{dt}{t} = \frac{1}{4} \log x + \frac{a+b}{4} \log 2\pi + \int_0^1 H(a, b; t) \frac{dt}{t}$$

$$+ \frac{1}{\pi^2} \sum_{k=1}^{\infty} \sigma_{-1/a,\, -1/b}(a, b; k)\, \varrho_0(a, b; kx)\,,$$

$$\Delta_1(a, b;\ x)$$

$$= \frac{1}{\pi^2} \sum_{k=1}^{\infty} \sigma_{-1/a,\, -1/b}(a, b; k) \left\{ x\varrho_0(a, b; kx) - \int_1^x \varrho_0(a, b; kt)\, dt \right\} + O(1)\,. \tag{5.35}$$

Therefore, result (5.33) follows from (5.18).

Lemma 5.8.

$$\Delta_1(a, b; x) = \Omega\left(x^{1 - \frac{1}{2(a+b)}}\right). \tag{5.36}$$

Proof. Suppose on the contrary that

$$\Delta_1(a, b; x) = o\left(x^{1 - \frac{1}{2(a+b)}}\right).$$

Then (5.33) shows that

$$\sum_{k=1}^{\infty} \sigma_{\alpha,\, \beta}(a, b; k) \sin\left(xk^{\frac{1}{a+b}} - \frac{\pi}{4}\right) = o(1)\,. \tag{5.37}$$

However, we find with $\omega > 0$

$$\int_0^\omega \sin\left(x - \frac{\pi}{4}\right) \sum_{k=1}^{\infty} \sigma_{\alpha,\, \beta}(a, b; k) \sin\left(xk^{\frac{1}{a+b}} - \frac{\pi}{4}\right) dx$$

$$= \int_0^\omega \sin^2\left(x - \frac{\pi}{4}\right) dx + \sum_{k=2}^{\infty} \sigma_{\alpha,\, \beta}(a, b; k) \int_0^\omega \sin\left(x - \frac{\pi}{4}\right) \sin\left(xk^{\frac{1}{a+b}} - \frac{\pi}{4}\right) dx\,.$$

Because of

$$\int_0^\omega \sin^2\left(x - \frac{\pi}{4}\right) dx = \int_0^\omega \left(\frac{1}{2} - \frac{1}{2} \sin 2x\right) dx = \frac{\omega}{2} + O(1)\,,$$

$$2\int_0^\omega \sin\left(x-\frac{\pi}{4}\right)\sin\left(xk^{\frac{1}{a+b}}-\frac{\pi}{4}\right)dx$$

$$=\int_0^\omega \cos\left(x\left(k^{\frac{1}{a+b}}-1\right)\right)dx-\int_0^\omega \sin\left(x\left(k^{\frac{1}{a+b}}+1\right)\right)dx \ll 1$$

we obtain

$$\int_0^\omega \sin\left(x-\frac{\pi}{4}\right)\sum_{k=1}^\infty \sigma_{\alpha,\beta}(a,b;k)\sin\left(xk^{\frac{1}{a+b}}-\frac{\pi}{4}\right)dx=\frac{\omega}{2}+O(1).$$

From (5.37) we see that the integral must equal $o(\omega)$. This forms a contradiction.

Lemma 5.9.

$$\Delta_2(a,b;x)\ll x^{2-\frac{3}{2(a+b)}}. \tag{5.38}$$

Proof. Integrating equation (5.35), we obtain

$$\Delta_2(a,b;x)=\frac{1}{\pi^2}\sum_{k=1}^\infty \sigma_{-1/a,\,-1/b}(a,b;k)\,g(k;x)+O(x)$$

with

$$g(k;x)=\int_1^x y\varrho_0(a,b;ky)\,dy-\int_1^x dy\int_1^y \varrho_0(a,b;kt)\,dt$$

$$=\int_1^x (2y-x)\,\varrho_0(a,b;ky)\,dy.$$

We apply the representation (5.13) of the function $\varrho_0(a,b;ky)$. Supposing that $a<b$, we obtain by partial integrating

$$g(k,x)=b\int_1^{x^{1/b}}(2y^b-x)\,y^{b-1}\,dy\int_0^\infty \sin 2\pi\left(\frac{k}{t}\right)^{1/a}\sin 2\pi y t^{1/b}\,\frac{dt}{t}$$

$$=-\frac{b}{2\pi}x^{2-1/b}\int_0^\infty t^{-1-1/b}\sin 2\pi\left(\frac{k}{t}\right)^{1/a}\cos 2\pi(xt)^{1/b}\,dt$$

$$+\frac{b(2-x)}{2\pi}\int_0^\infty t^{-1-1/b}\sin 2\pi\left(\frac{k}{t}\right)^{1/a}\cos 2\pi t^{1/b}\,dt$$

$$+\frac{b}{2\pi}\int_1^{x^{1/b}}\{2(2b-1)\,y^b-(b-1)\,x\}\,y^{b-2}\,\mathrm{d}y$$

$$\times\int_0^\infty t^{-1-1/b}\sin 2\pi\left(\frac{k}{t}\right)^{1/a}\cos 2\pi(yt)^{1/b}\,\mathrm{d}t\,.$$

If we apply the functional equations for the trigonometric functions, we easily obtain by means of Lemmas 5.1—5.3

$$g(k,x)\ll x^2(kx)^{-\frac{3}{2(a+b)}}.$$

This gives at once the desired estimate (5.38), provided that $a<b$.

In case of $a=b=1$ we somewhat modify the proof. We get

$$g(k,x)=\int_1^x(2y-x)\,\mathrm{d}y\int_0^\infty\sin\frac{2\pi k}{t}\sin 2\pi yt\,\frac{\mathrm{d}t}{t}$$

$$=-\frac{x}{2\pi}\int_0^\infty\sin\frac{2\pi k}{t}(\cos 2\pi xt-1)\frac{\mathrm{d}t}{t^2}+\frac{2-x}{2\pi}\int_0^\infty\sin\frac{2\pi k}{t}(\cos 2\pi t-1)\frac{\mathrm{d}t}{t^2}$$

$$+\frac{1}{\pi}\int_1^x\mathrm{d}y\int_0^\infty\sin\frac{2\pi k}{t}(\cos 2\pi yt-1)\frac{\mathrm{d}t}{t^2}$$

$$=-\frac{x^2}{2\pi}\int_1^\infty\sin\frac{2\pi kx}{t}\cos 2\pi t\,\frac{\mathrm{d}t}{t^2}+\frac{2-x}{2\pi}\int_1^\infty\sin\frac{2\pi k}{t}\cos 2\pi t\,\frac{\mathrm{d}t}{t^2}$$

$$+\frac{1}{\pi}\int_1^x y\,\mathrm{d}y\int_1^\infty\sin\frac{2\pi ky}{t}\cos 2\pi t\,\frac{\mathrm{d}t}{t^2}+O\left(\frac{x}{k}\right).$$

We can now apply Lemmas 5.1—5.3, and we obtain $g(k,x)\ll x^2(kx)^{-3/4}$, such that (5.38) also follows in this case.

Theorem 5.8.

$$D(a,b;x)=H(a,b;x)+\Omega\left(x^{\frac{1}{2(a+b)}}\right). \tag{5.39}$$

Proof. We apply Lemma 3.7 with

$$f(x)=\begin{cases}\Delta(a,b;x)-\dfrac{1}{4} & \text{for}\quad x\geqq 1\,,\\[2mm] 0 & \text{for}\quad 0\leqq x<1\end{cases}$$

such that $f_1(x) = \Delta_1(a, b; x), f_2(x) = \Delta_2(a, b; x)$ for $x \geqq 1$. Assume that

$$\Delta(a, b; x) = o\left(x^{\frac{1}{2(a+b)}}\right).$$

Hence, we put $\alpha = 1/2(a + b)$. Because of (5.38) we put $\beta = 2 - 3/2(a + b)$. Then Lemma 3.7 shows that $f_1(x) = o\left(x^{1-\frac{1}{2(a+b)}}\right)$, which contradicts result (5.36). This proves the theorem.

5.3. IMPROVEMENTS OF THE O-ESTIMATES

Let us first consider Dirichlet's divisor problem.

Definition 5.2. *If*

$$D(x) = x \log x + (2C - 1)\, x + \Delta(x)\,,$$

we denote by ϑ *the value*

$$\vartheta = \inf\{\alpha : \Delta(x) \ll x^{\alpha}\}\,.$$

The basic estimates of this chapter lead to the inequality $1/4 \leqq \vartheta \leqq 1/3$. As has already been mentioned in the introduction the improvements in Dirichlet's divisor problem and in the circle problem are essentially the same. Also here J. G. van der Corput was the first who could prove $\vartheta < 1/3$ applying his method of exponential sums. Using three Weyl's steps, this method leads to $\vartheta \leqq 27/82$ in the circle and in the divisor problem. However, here we have the possibility to obtain better results by means of van der Corput's and Phillips' method of exponent pairs. This will be seen in Section 5.3.1.

But the simplest application of the theory of double exponential sums leads to essentially better estimates. As in the circle problem it is useful to apply Weyl's steps three times. Then one can obtain the following results. The simplest case in applying the method of double exponential sums is by making use of three Weyl's steps to one variable only. Then we obtain $\vartheta \leqq 19/58$, which will be proved in Section 5.3.2. Chih Tsung-tao and H.-E. Richert independently of each other applied Weyl's steps with respect to one variable twice and both the variables once and obtained $\vartheta \leqq 15/46$. But contrary to the circle problem in this case the corresponding Hessian presents a difficulty since it vanishes along certain curves. The result $\vartheta \leqq 13/40$, which can be obtained by using two Weyl's steps with respect to both the variables and one Weyl's step to one variable, has been proved by nobody. The estimate $\vartheta \leqq 12/37$ by applying Weyl's steps each time to both the variables was proved by G. Kolesnik. Further improvements were found by G. Kolesnik by a refinement of the method and extensive calculations.

In 1952 H.-E. Richert initiated the general problem of estimating $D(a, b; x)$ in connection with his investigations on the number of finite Abelian groups of a given order. He proved

$$\Delta(a, b; x) \ll \begin{cases} x^{\frac{2}{3}\frac{1}{a+b}} & \text{for} \quad b < 2a, \\ x^{\frac{2}{9a}} \log x & \text{for} \quad b = 2a, \\ x^{\frac{2}{5a+2b}} & \text{for} \quad b > 2a. \end{cases} \tag{5.40}$$

It is possible to improve this result, but we must consider many special cases depending on the values of a and b.

5.3.1. *Estimates by means of van der Corput's method*

The best general view of the order of the remainder $\Delta(a, b; x)$ for all numbers a, b with $1 \leqq a \leqq b$ is given by an application of Theorem 2.8.

Lemma 5.10. *Let* $N \geqq 1$, $K = 2^k$, $k = 2, 3, \ldots, u \geqq 1, v \geqq 1$. *Then*

$$\sum_{\substack{n^{u+v} \leqq x \\ N \leqq n \leqq 2N}} \psi\left(\left(\frac{x}{n^v}\right)^{1/u}\right) \ll (xN^{(K-k-1)u-v})^{\frac{1}{K-1}\cdot\frac{1}{u}}. \tag{5.41}$$

Proof. We apply Theorem 2.8 with $u = 2$, $a = N$, $b = 2N$ and $f(t) = (x/t^v)^{1/u}$. Clearly, this function satisfies the conditions of the theorem, and we have $\lambda = (x/N^v)^{1/u} \gg N$. Hence, (5.41) follows immediately from (2.19).

Theorem 5.9. *Let* $K = 2^k$, $k = 2, 3, \ldots$ *Then the estimate*

$$\Delta(a, b; x) \ll x^{\left(1-\frac{k-1}{k-1}\right)\frac{1}{a+b}} \log x \tag{5.42}$$

holds under the condition $(K - k - 1)\, a \geqq b$.

Proof. We can divide both the sums in the representation (5.3) of $\Delta(a, b; x)$ into nearly $\log x$ sums of the kind of (5.41) with $u + v = a + b$. Thus, if the condition of the theorem is satisfied, we can use the estimate $N^{u+v} \ll x$ in (5.41). This proves (5.42).

Note that the factor $\log x$ in (5.42) is only required in case of $(K - k - 1)\, a = b$.

Some special values in (5.42)

k	$1-\dfrac{k-1}{K-1}$	condition
2	$\frac{2}{3}$	$a \geqq b$ (that implies $a = b = 1$)
3	$\frac{5}{7}$	$4a \geqq b$
4	$\frac{4}{5}$	$11a \geqq b$
5	$\frac{27}{31}$	$26a \geqq b$

If we have equality in the condition of Theorem 5.9, then we cannot get better estimates in this way. But if we have a proper inequality, we can obtain a somewhat better result by a refinement of the proof. This is shown by the next theorem.

Theorem 5.10. *Let $K = 2^k$, $k = 3, 4, \ldots$ Then the estimate*

$$\Delta(a, b; x) \ll x^{\frac{K-2}{(K(k-2)+2)a+Kb}} \tag{5.43}$$

holds under the condition $(K-k-1)\,a > b > \left(\dfrac{K}{2} - k\right) a$.

Proof. We write with a suitably chosen value $z > 1$

$$\sum_{n^{a+b} \leq x} \psi\left(\left(\frac{x}{n^b}\right)^{1/a}\right) = \Sigma_1\, \psi\left(\left(\frac{x}{n^b}\right)^{1/a}\right) + \Sigma_2\, \psi\left(\left(\frac{x}{n^b}\right)^{1/a}\right),$$

$$SC(\Sigma_1)\colon n^{a+b} \leqq x, \qquad 1 \leqq z2^{-\nu-1} < n \leqq z2^{-\nu}, \qquad \nu = 0, 1, 2, \ldots;$$

$$SC(\Sigma_2)\colon n^{a+b} \leqq x, \qquad z2^{\nu} < n \leqq z2^{\nu+1} = x^{1/(a+b)}, \qquad \nu = 0, 1, 2, \ldots$$

Then (5.41) gives

$$\sum_{n^{a+b} \leq x} \psi\left(\left(\frac{x}{n^b}\right)^{1/a}\right) \ll \sum_{\nu=0}^{\infty} \left\{ (x(z2^{-\nu})^{(K-k-1)a-b})^{\frac{1}{K-1}\cdot\frac{1}{a}} + (x(z2^{\nu})^{(K/2-k)a-b})^{\frac{2}{K-2}\cdot\frac{1}{a}} \right\}$$

$$\ll (xz^{(K-k-1)a-b})^{\frac{1}{K-1}\cdot\frac{1}{a}} + (xz^{(K/2-k)\,a-b})^{\frac{2}{K-2}\cdot\frac{1}{a}}.$$

Both the terms are of the same order if we put $z = x^{\frac{K}{(K(k-2)+2)a+Kb}}$. Then

$$\sum_{n^{a+b} \leq x} \psi\left(\left(\frac{x}{n^b}\right)^{1/a}\right) \ll x^{\frac{K-2}{(K(k-2)+2)\,a+Kb}}.$$

Clarly, we obtain from (5.41) with $u = b$, $v = a$ and $k - 1$ for k

$$\sum_{n^{a+b} \leq x} \psi\left(\left(\frac{x}{n^a}\right)^{1/b}\right) \ll \sum_{v=0}^{\infty} \left(x \left(x^{\frac{1}{a+b}} 2^{-v}\right)^{(K/2-k)b-a}\right)^{\frac{2}{K-2} \cdot \frac{1}{b}}$$

$$\ll x^{\frac{K-2k+2}{K-2} \frac{1}{a+b}} \ll x^{\frac{K-2}{(K(k-2)+2)a+Kb}}.$$

This completes the proof of (5.43).

Some special values in (5.43)

k	$\frac{K-2}{(K(k-2)+2)a+Kb}$	condition
3	$\frac{3}{5a+4b}$	$4a > b > a$
4	$\frac{7}{17a+8b}$	$11a > b > 4a$
5	$\frac{15}{49a+16b}$	$26a > b > 11a$

Note that for these results the deep method of exponent pairs was not required. However, if we use this method, we have many possibilities to improve the results. It is clear that the functions under consideration satisfy all conditions of the method.

Lemma 5.11. *Let $N \geqq 1$, $u \geqq 1$, $v \geqq 1$, and let $(k, l) = A(\varkappa, \lambda)$ be any exponent pair. Then*

$$\sum_{\substack{n^{u+v} \leq x \\ N \leqq n \leqq 2N}} \psi\left(\left(\frac{x}{n^v}\right)^{1/u}\right) \ll (x^{2k} N^{(2l-1)u-2kv})^{1/u}. \tag{5.44}$$

Proof. We first consider the conditions of the method of exponent pairs. In Definition 2.1 we put $u = 2$, $a = N$, $b = 2N$. The function $f(t) = (x/t^v)^{1/u}$ possesses derivatives of all orders in $N \leqq t \leqq 2N$. Since $|f^{(v)}(t)| \asymp (x/N^v)^{1/u} N^{-v}$ $(v = 0, 1, \ldots)$, we put $z = \frac{1}{N}(x/N^v)^{1/u} > 1$ in (2.45). Hence, all conditions of the method are satisfied and we can use Theorem 2.14. So (5.44) follows immediately.

Theorem 5.11. *Let $(k, l) = A(\varkappa, \lambda)$ be any exponent pair. Then the estimate*

$$\Delta(a, b; x) \ll x^{(k+l-1/2)\frac{2}{a+b}} \log x \tag{5.45}$$

holds under the condition $(2l - 1)a \geqq 2kb$.

If otherwise $(2l - 1)a < 2kb$, *then*

$$\Delta(a, b; x) \ll x^{\frac{k}{(1-l)a+kb}} \log x. \tag{5.46}$$

Proof. In order to prove (5.45), we divide both the sums in the representation (5.3) of $\Delta(a, b; x)$ into nearly $\log x$ sums of the kind (5.44) with $u + v = a + b$. Because of $(2l - 1) u \geqq 2kv$ we can N^{u+v} replace by x. Thus, (5.45) follows from (5.44).

If $(2l - 1) a < 2kb$, on the one hand we obtain by (5.44) and on the other hand by trivial estimating

$$\sum_{n^{a+b} \leqq x} \psi\left(\left(\frac{x}{n^b}\right)^{1/a}\right) = \sum_{\substack{n^{a+b} \leqq x \\ n > z}} \psi\left(\left(\frac{x}{n^b}\right)^{1/a}\right) + \sum_{\substack{n^{a+b} \leqq x \\ n \leqq z}} \psi\left(\left(\frac{x}{n^b}\right)^{1/a}\right)$$
$$\ll \{(x^{2k} z^{(2l-1)a - 2kb})^{1/a} + z\} \log x .$$

If we put $z = x^{\frac{k}{(1-l)a+kb}}$, then

$$\sum_{n^{a+b} \leqq x} \psi\left(\left(\frac{x}{n^b}\right)^{1/a}\right) \ll x^{\frac{k}{(1-l)a+kb}} \log x .$$

If $(k, l) = A(\varkappa, \lambda)$, we have $2l - 1 \geqq 2k$ because of $\lambda \geqq \varkappa$. Thus we always have $(2l - 1) b \geqq 2ka$. Hence, (5.44) gives

$$\sum_{n^{a+b} \leqq x} \psi\left(\left(\frac{x}{n^a}\right)^{1/b}\right) \ll x^{(k+l-1/2)\frac{2}{a+b}} \log x \ll x^{\frac{k}{(1-l)a+kb}} \log x .$$

These results and (5.3) now lead immediately to (5.46).

Note that also here the factor $\log x$ is only required for equality in the condition of the theorem.

It is seen from (5.45) and (5.46) that we obtain Richert's result (5.40) by using the exponent pair $(k, l) = \left(\frac{2}{18}, \frac{13}{18}\right)$. Another interesting exponent pair is $(k, l) = \left(\frac{2}{40}, \frac{33}{40}\right)$. Here we get

$$\Delta(a, b; x) \ll \begin{cases} x^{\frac{3}{4}\cdot\frac{1}{a+b}} \log x & \text{for} \quad 13a \geqq 2b , \\ x^{\frac{2}{7a+2b}} \log x & \text{for} \quad 13a < 2b . \end{cases}$$

Some special exponent pairs in (5.45)

(k, l)	$2(k + l - 1/2)$	condition
$\left(\frac{11}{82}, \frac{57}{82}\right)$	$\frac{27}{41}$	$16a \geqq 11b$
$\left(\frac{33}{234}, \frac{161}{234}\right)$	$\frac{77}{117}$	$4a \geqq 3b$

$\left(\frac{75}{544}, \frac{376}{544}\right)$	$\frac{179}{272}$	$104a \geqq 75b$
$\left(\frac{97}{696}, \frac{480}{696}\right)$	$\frac{229}{348}$	$132a \geqq 97b$
$\left(\frac{141\,841}{1\,019\,718}, \frac{703\,527}{1\,019\,718}\right)$	$\frac{335\,509}{509\,859}$	$193\,668a \geqq 141\,841b$

Let us again consider the case $2a \geqq b$. We can get a slightly better estimate by using the convexity property of the exponent pairs. If (k, l) is any exponent pair, then $B(k, l) = (l - 1/2, k + 1/2)$ is another one. Applying the C-process, we obtain the exponent pair

$$\left(\frac{k+l-1/2}{2}, \frac{k+l+1/2}{2}\right).$$

This follows from Theorem 2.15 with $t = 1/2$. Now we form the new exponent pair

$$\begin{aligned}(\varkappa, \lambda) &= ABA\left(\frac{k+l-1/2}{2}, \frac{k+l+1/2}{2}\right)\\ &= \left(\frac{k+l+1/2}{6(k+l+7/6)}, \frac{1}{2} + \frac{2(k+l+1/2)}{6(k+l+7/6)}\right).\end{aligned}$$

This exponent pair has the property $2\lambda - 1 = 4\varkappa$. Hence, we obtain from (5.45) that the estimate

$$\Delta(a, b; x) \ll x^{\frac{\varrho}{a+b}} \log x\,, \qquad \varrho = \frac{k+l+1/2}{k+l+7/6}\,, \tag{5.47}$$

holds under the condition $2a \geqq b$. Note that $\varrho < 2/3$, if $k + l - 1/2 < 1/3$. If we use Rankin's exponent pair, we have

$$\varrho = \frac{1\,355\,227}{2\,035\,039} = 0{,}665946\,\ldots$$

5.3.2. *Estimates by means of double exponential sums*

Lemma 5.12. *Let $N \geqq 1$ and $u, v \geqq 1$. Then*

$$\sum_{\substack{n^{u+v} \leqq x\\ N \leqq n \leqq 2N}} \psi\left(\left(\frac{x}{n^v}\right)^{1/u}\right) \ll (x^8 N^{11u-8v})^{1/29u} \log x\,. \tag{5.48}$$

Proof. We apply Theorem 2.23, which belongs to the two-dimensional theory of exponential sums. There we put $u = 2$, $a = N$, $b = 2N$ and $f(t) = -(x/t^v)^{1/u}$. We must check the conditions of the theorem. It is easily seen that $\lambda = (x/N^v)^{1/u} \gg N$ and $|f^{(\nu)}(t)| \asymp \frac{\lambda}{N^\nu}$ ($\nu = 1, 2, 3, 4$). The function $\varphi(\tau)$ is defined by

$$f'(\varphi) = \frac{v}{u}\left(\frac{x}{\varphi^{u+v}}\right)^{1/u} = \tau$$

such that

$$\varphi(\tau) = \left(\frac{v^u x}{u^u \tau^u}\right)^{\frac{1}{u+v}}.$$

We have $\alpha \leqq \tau \leqq \beta$, where $\alpha = \min f'(t)$ and $\beta = \max f'(t)$, and we see that α, β, τ are of order λ/N. Hence

$$|\varphi^{(\nu)}(\tau)| \asymp \lambda\left(\frac{N}{\lambda}\right)^{\nu+1} \qquad (\nu = 3, 4)\,.$$

If we put $\tau = t_1/t_2$, it is easily seen that

$$t_2^4 \left|\frac{\partial^2}{\partial t_2^2} t_2^{-2} \varphi''\left(\frac{t_1}{t_2}\right)\right| \asymp \frac{N^3}{\lambda^2}\,.$$

If we put for the sake of simplicity $\varphi(\tau) = c\tau^{-\frac{u}{u+v}}$, $c = \left(\frac{v^u x}{u^u}\right)^{\frac{1}{u+v}}$, we obtain

$$\varphi''(\tau) = c_1 \tau^{-2-\frac{u}{u+v}}, \qquad c_1 = \frac{u}{u+v}\left(1 + \frac{u}{u+v}\right)c\,,$$

$$t_2^{-2}\varphi''\left(\frac{t_1'}{t_2}\right) = c_1 t_1'^{\,-2-\frac{u}{u+v}} t_2^{\frac{u}{u+v}}\,,$$

where t_1' is defined as in Theorem 2.23. It is seen that the second partial derivative of this function with respect to t_1 is positive, and with respect to t_2, negative. Therefore, the Hessian is negative, and we obtain the required estimate

$$t_2^8 \left| H\left(\int_0^1\int_0^1\int_0^1 t_2^{-2}\varphi''\left(\frac{t_1'}{t_2}\right) d\tau_1\, d\tau_2\, d\tau_3\right)\right|$$

$$\gg c_1^2 \left(\frac{a_2}{N}\right)^{-6-\frac{2u}{u+v}} a_2^{\frac{2u}{u+v}-2} a_2^8 \gg N^8 \lambda^{-6}\,.$$

Thus, all conditions of Theorem 2.23 are satisfied, and we deduce (5.48) from (2.75).

Tneorem 5.12. *If* $11a \geqq 8b$, *then*

$$\Delta(a, b; x) \ll x^{\frac{19}{29}\frac{1}{a+b}} \log^2 x\,. \tag{5.49}$$

If otherwise $11a < 8b$, *let* $(k, l) = A(\varkappa, \lambda)$ *be any exponent pair with* $(2l - 1)\, a \geqq 2kb$, $0 < 4 - 29k < 29l - 20$. *Then*

$$\Delta(a, b; x) \ll x^{\frac{8l-11k-4}{(29l-20)a+(4-29k)b}} \log^2 x\,. \tag{5.50}$$

Proof. In order to prove (5.49), we divide both the sums in the representation (5.3) into nearly $\log x$ sums of the kind (5.48) with $u + v = a + b$. Because of $11u \geqq 8v$ we can N^{u+v} replace by x. Thus, (5.49) follows from (5.48).

If $11a < 8b$ and $(2l - 1)\, a \geqq 2kb$, then on the one hand we obtain by (5.48), and on the other hand, by (5.44)

$$\sum_{n^{a+b}\leqq x} \psi\left(\left(\frac{x}{n^b}\right)^{1/a}\right) = \sum_{\substack{n^{a+b}\leqq x\\ n>z}} \psi\left(\left(\frac{x}{n^b}\right)^{1/a}\right) + \sum_{\substack{n^{a+b}\leqq x\\ n\leqq z}} \psi\left(\left(\frac{x}{n^b}\right)^{1/a}\right)$$

$$\ll \{(x^8 z^{11a-8b})^{1/29a} + (x^{2k} z^{(2l-1)a-2kb})^{1/a}\} \log^2 x\,.$$

If we put

$$z = x^{\frac{4-29k}{(29l-20)a+(4-29k)b}}\,,$$

provided that $29k < 4$ and $29l > 20$, then

$$\sum_{n^{a+b}\leqq x} \psi\left(\left(\frac{x}{n^b}\right)^{1/a}\right) \ll x^{\frac{8l-11k-4}{(29l-20)a+(4-29k)b}} \log^2 x\,.$$

Clearly, (5.48) gives

$$\sum_{n^{a+b}\leqq x} \psi\left(\left(\frac{x}{n^a}\right)^{1/b}\right) \ll x^{\frac{19}{29}\frac{1}{a+b}} \log^2 x \ll x^{\frac{8l-11k-4}{(29l-20)a+(4-29k)b}} \log^2 x$$

since $4 - 29k < 29l - 20$. These results and (5.3) now lead to (5.50).

Some special exponent pairs in (5.50).

(k, l)	$\dfrac{81 - 11k - 4}{(291 - 20)\, a + (4 - 29k)\, b}$	condition
$\left(\frac{2}{18}, \frac{13}{18}\right)$	$\frac{10}{17a + 14b}$	$2a \geqq b > \frac{11}{8}a$
$\left(\frac{1}{14}, \frac{11}{14}\right)$	$\frac{21}{39a + 27b}$	$4a \geqq b > \frac{11}{8}a$
$\left(\frac{2}{40}, \frac{33}{40}\right)$	$\frac{82}{157a + 102b}$	$\frac{13}{2}a \geqq b > \frac{11}{8}a$
$(0,1)$	$\frac{4}{9a + 4b}$	$b > \frac{11}{8}a$

5.4. A HISTORICAL OUTLINE OF THE DEVELOPMENT OF DIRICHLET'S DIVISOR PROBLEM

In the *history of O-estimates* we find the following values, where ϑ is given by Definition 5.2:

$\vartheta \leqq \frac{1}{2} = 0{,}5$ (P. G. L. Dirichlet [1], 1849),

$\vartheta \leqq \frac{1}{3} = 0{,}\bar{3}$ (G. Voronoi [1], 1903),

$\vartheta \leqq \frac{33}{100} = 0{,}33$ (J. G. van der Corput [5], 1922),

$\vartheta \leqq \frac{27}{82} = 0{,}3292\ldots$ (J. G. van der Corput [9], 1928),

$\vartheta \leqq \frac{15}{46} = 0{,}3260\ldots$ (Chih Tsung-tao [2], 1950), (H.-E. Richert [3], 1953),

$\vartheta \leqq \frac{12}{37} = 0{,}3243\ldots$ (G. Kolesnik [1], 1969),

$\vartheta \leqq \frac{346}{1067} = 0{,}3242\ldots$ (G. Kolesnik [2], 1973),

$\vartheta \leqq \frac{35}{108} = 0{,}3240\ldots$ (G. Kolesnik [6], 1982).

$\vartheta \leqq \frac{139}{429} = 0{,}32400\ldots$ (G. Kolesnik [7], 1985),

$\vartheta \leqq \frac{7}{22} = 0{,}3\overline{18}$ (H. Iwaniec, C. J. Mozzochi [1], 1987).

It should be remarked that in 1886 E. Pfeiffer [1] already made the attempt to prove the result

$$\Delta(x) \ll x^{1/3+\varepsilon} \qquad (\varepsilon > 0)\,.$$

But E. Landau [2] noted that the final step in the proof fails. Otherwise he showed that the ideas on which E. Pfeiffer established his proof are sound. Thus by applying Pfeiffer's method he proved the more precise result

$$\Delta(x) \ll x^{1/3} \log x$$

of G. Voronoi.

E. Landau [17] proved that $\vartheta \leqq \frac{37}{112}$ in 1926. At this time the result is weaker than van der Corput's $\vartheta \leqq \frac{33}{100}$, but the proof was much simpler. Analogously, the result

$\vartheta \leqq \frac{19}{58}$ coming from (5.49) is weaker as Chih's and Richert's result $\vartheta \leqq \frac{15}{46}$, but also here the proof is much simpler.

In the *history of Ω-estimates* we have first the trivial estimate $\vartheta \geqq 0$ and then

$$\vartheta \geqq \frac{1}{4} = 0{,}25\,, \qquad \text{(E. Landau [4], 1915; G. H. Hardy [2], [3], 1915).}$$

More precisely G. H. Hardy found

$$D(x) = x \log x + (2C - 1)\,x + \Omega_{\pm}(x^{1/4})\,.$$

Improvements of these results:

$$D(x) = x \log x + (2C - 1)\,x + \Omega_{-}(x^{1/4}\delta(x))\,.$$

$$\delta(x) \to \infty \text{ unspecified} \qquad \text{(A. E. Ingham [1], 1940),}$$

$$\delta(x) = (\log\log x)^{1/4}\,(\log\log\log x)^{5/4}$$

(K. S. Gangadharan [1], 1961),

$$\delta(x) = \exp\,(c(\log\log x)^{1/4}(\log\log\log x)^{-3/4})\,, \qquad c > 0$$

(K. Corrádi, I. Kátai [1], 1967).

$$D(x) = x \log x + (2C - 1)\,x + \Omega_{+}(x^{1/4}\varrho(x))\,.$$

$$\varrho(x) = (\log x)^{1/4} \log\log x \qquad \text{(G. H. Hardy [3], 1916),}$$

$$\varrho(x) = (\log x)^{1/4}\,(\log\log x)^{(3+\log 4)/4} \exp\,(-c(\log\log\log x)^{1/2})\,, \qquad c > 0\,,$$

(J. L. Hafner [1], 1981).

The first non-trivial *O*-estimate (5.40) in the general problem for $D(a, b; x)$ was proved by H.-E. Richert [1] in 1952. All the other estimates are new except the case $(a, b) = (1, 2)$, which is highly important for the problem of distribution of square-free numbers. If $\vartheta(1, 2)$ is defined by

$$\vartheta(1, 2) = \inf\,\{\alpha\colon \Delta(1, 2; x) \ll x^{\alpha}\}\,,$$

we have trivially $\vartheta(1, 2) \leqq 1/3$. Furthermore, we have the following values:

$$\vartheta(1, 2) \leqq \frac{2}{9} = 0{,}\overline{2} \qquad \text{(H.-E. Richert [1], 1952),}$$

$$\vartheta(1, 2) \leqq \frac{1355227}{6105117} = 0{,}22198\ldots \qquad \text{(R. A. Rankin [1], 1955),}$$

$$\vartheta(1, 2) \leqq \frac{109556}{494419} = 0{,}22158\ldots \qquad \text{(P. G. Schmidt [1], 1964).}$$

$$\vartheta(1,2) \leqq \frac{37}{167} = 0{,}22155\ldots$$

The estimate

$$\Delta(a, b; x) = \Omega_{\pm}\left(x^{\frac{1}{2}\frac{1}{a+b}}\right)$$

is a special case of a very general theorem due to E. Landau [12], 1924. A simple proof, based on Hardy's method, is given by A. Schierwagen [1], 1974. Furthermore, he [2] found the improvement in case of the upper sign

$$\Delta(a, b; x) = \Omega_{+}\left((x \log x)^{\frac{1}{2}\cdot\frac{1}{a+b}} \log \log x\right)$$

in 1978. Recently J. L. Hafner [3] proved the estimates

$$\Delta(a, b; x) = \Omega_{+}\left(x^{\frac{1}{2}\cdot\frac{1}{a+b}} \varrho(x)\right),$$

$$\varrho(x) = (\log x)^{\frac{b}{2(a+b)}} \log \log x$$

and

$$\Delta(a, b; x) = \Omega_{-}\left(x^{\frac{1}{2}\cdot\frac{1}{a+b}} \delta(x)\right),$$

$$\delta(x) = \exp\left(c(\log \log x)^{\frac{b}{2(a+b)}} (\log \log \log x)^{\frac{b}{2(a+b)}-1}\right).$$

NOTES ON CHAPTER 5

Section 5.2

The method presented in this section was originally developed by E. Landau [17] in 1926, where he considered the case of Dirichlet's divisor problem $a = b = 1$. The general case was introduced by E. Krätzel [5] in 1969.

Section 5.3

Theorems 5.9—5.12 present unpublished results of E. Krätzel.

The estimate $\vartheta(1,2) \leqq \frac{37}{167}$ follows from (5.45) by means of Huxley's and Watt's exponent pair $(37/334 + \varepsilon, 241/334 + \varepsilon)$.

Section 5.4

Note that the interesting asymptotic relation

$$\int_0^x \Delta^2(t)\, dt \sim c x^{3/2}$$

holds. For results in this direction see Tong Kwang-chang [3] and K. Chandrasekharan — R. Narasimhan [3].

Chapter 6

Many-dimensional multiplicative problems

Dirichlet's divisor problem discussed in Chapter 5 was generalized by A. Piltz [1], who considered the number $d_p(n)$ of decompositions of n into p factors instead of the divisor function $d(n)$. Moreover, we may consider the divisor function $d(\boldsymbol{a}; n)$ where $\boldsymbol{a} = (a_1, a_2, \dots, a_p)$ with positive integers $a_1, a_2, \dots, a_p$. This function counts the number of ways of expressing n as the product

$$n = n_1^{a_1} n_2^{a_2} \cdot \dots \cdot n_p^{a_p} .$$

We are now interested in the behaviour of the function

$$\sum_{n \leqq x} \mathrm{d}(\boldsymbol{a}; n)$$

for large x. But in reality this is a lattice point problem, for this function counts the number of lattice points under the surface

$$\xi_1^{a_1} \xi_2^{a_2} \cdot \dots \cdot \xi_p^{a_p} \leqq x$$

with $\xi_1 > 0$, $\xi_2 > 0, \dots, \xi_p > 0$. And then it is not required that the numbers $a_1, a_2, \dots, a_p$ are integers. Therefore we use the following definition.

Definition 6.1. *Let $a_1, a_2, \dots a_p$ be real numbers with $1 \leqq a_1 \leqq a_2 \leqq \dots \leqq a_p$, and let $\boldsymbol{a} = (a_1, a_2, \dots, a_p)$, $p \geqq 2$. Then $D(\boldsymbol{a}; x)$ denotes the number of lattice points under the surface*

$$\xi_1^{a_1} \xi_2^{a_2} \cdot \dots \cdot \xi_p^{a_p} = x$$

with $\xi_1, \xi_2, \dots, \xi_p > 0$. In particular, if $a_1 = a_2 = \dots = a_p = 1$, the notation $D(a; x) = D_p(x)$ is used.

If the numbers $a_1, a_2, \dots, a_p$ are all integers, we have

$$D(\boldsymbol{a}; x) = \sum_{n \leqq x} \mathrm{d}(\boldsymbol{a}; n)$$

and in particular

$$D_p(x) = \sum_{n \leqq x} \mathrm{d}_p(n) .$$

The problem of estimating $D_p(x)$ is called the *Piltz's divisor problem*,

The application of the method of exponential sums to the problem of estimating $D(\boldsymbol{a}; x)$ for large x does not give the best results in all cases. When the dimension p is greater than 3, one mostly obtains better results by applying the theory of the Riemann zeta-function. Therefore, we describe only the three-dimensional case in detail and give only some impressions in the other cases.

6.1. THE THREE-DIMENSIONAL PROBLEM

In this section a, b, c denote real numbers with $1 \leqq a \leqq b \leqq c$. Then the function $D(\boldsymbol{a}; x) = D(a, b, c; x)$ is defined by

$$D(a, b, c; x) = \#\{(n_1, n_2, n_3): n_1, n_2, n_3 \in \mathbf{Z}; n_1, n_2, n_3 \geqq 1; \\ n_1^a n_2^b n_3^c = x\}\,.$$

If in particular a, b, c are integers, we can introduce the divisor function

$$d(a, b, c; n) = \#\{(n_1, n_2, n_3): n_1, n_2, n_3 \in \mathbf{Z};\quad n_1, n_2, n_3 \geqq 1; \\ n_1^a n_2^b n_3^c = n\}\,,$$

and the counting function $D(a, b, c; x)$ is represented by

$$D(a, b, c; x) = \sum_{n \leqq x} \mathrm{d}(a, b, c; n)\,.$$

Especially in the last years many problems of number theory led to this function. Whereas the triplet $(a, b, c) = (1, 1, 1)$ describes Piltz's divisor problem, the triplet $(1, 2, 3)$ is of great importance in the theory of estimation of the number of non-isomorphic Abelian groups of a given finite order. However, in other applications such combinations as $(1, 2, 2)$, $(1, 2, 4)$, $(3, 4, 5)$ and other ones occur. Therefore, we should like to know general estimates of $D(a, b, c; x)$ for any combination (a, b, c). But there arise great difficulties when the differences between the numbers a, b, c are too large. Let us now make some remarks on this. We first write

$$D(a, b, c; x) = H(a, b, c; x) + \Delta(a, b, c; x)\,, \tag{6.1}$$

where $H(a, b, c; x)$ denotes the main term and $\Delta(a, b, c; x)$ the remainder. From the associated Dirichlet series

$$\sum_{n_1=1}^{\infty} \sum_{n_2=1}^{\infty} \sum_{n_3=1}^{\infty} \left(\frac{1}{n_1^a n_2^b n_3^c}\right)^s = \zeta(as)\,\zeta(bs)\,\zeta(cs)\,,$$

where $\zeta(s)$ denotes Riemann's zeta-function, we see at once that the main term must be of the form

$$H(a, b, c; x) = \zeta\left(\frac{b}{a}\right)\zeta\left(\frac{c}{a}\right)x^{1/a} + \zeta\left(\frac{a}{b}\right)\zeta\left(\frac{c}{b}\right)x^{1/b} + \zeta\left(\frac{a}{c}\right)\zeta\left(\frac{b}{c}\right)x^{1/c}, \tag{6.2}$$

provided that $a < b < c$. However, in cases of equalities $a = b$ or $b = c$ or both we can take the limit values. From the theory of Riemann zeta-function it is well known that

$$\zeta(s) = \frac{1}{s-1} + C + C_1(s-1) + O((s-1)^2) \qquad (s \to 1),$$

where C denotes Euler's constant

$$C = \lim_{N\to\infty}\left\{\sum_{n=1}^{N}\frac{1}{n} - \log N\right\}$$

and

$$C_1 = \lim_{N\to\infty}\left\{\sum_{n=1}^{N}\frac{\log n}{n} - \frac{1}{2}\log^2 N\right\}.$$

Then it is easy to prove that

$$H(a, a, c; x)$$
$$= \left\{\frac{1}{a}\zeta\left(\frac{c}{a}\right)\log x + (2C-1)\,\zeta\left(\frac{c}{a}\right) + \frac{c}{a}\zeta'\left(\frac{c}{a}\right)\right\} x^{1/a} + \zeta^2\left(\frac{a}{c}\right) x^{1/c} \qquad (6.3)$$

if $a = b < c$,

$$H(a, b, b; x)$$
$$= \zeta^2\left(\frac{b}{a}\right) x^{1/a} + \left\{\frac{1}{b}\zeta\left(\frac{a}{b}\right)\log x + (2C-1)\,\zeta\left(\frac{a}{b}\right) + \frac{a}{b}\zeta'\left(\frac{a}{b}\right)\right\} x^{1/b} \qquad (6.4)$$

if $a < b = c$, and

$$H(1, 1, 1; x) = \left\{\frac{1}{2}\log^2 x + (3C-1)\log x + 3C^2 - 3C + 3C_1 + 1\right\} x \qquad (6.5)$$

if $a = b = c = 1$.

In order to estimate the remainder, we can immediately use the classical result applying the theory of Dirichlet series. Thus we obtain from Landau's method (see Landau [6])

$$\Delta(a, b, c; x) \ll x^{1/2a}.$$

In case of Piltz's divisor problem $(a, b, c) = (1, 1, 1)$ this is the classical result. However, in the unsymmetrical case only the smallest number a arises, and the numbers b, c are omitted. Moreover, we shall see that in some cases this result is very bad in comparison to possible trivial estimates. Consequently, in order to prove better results, we have to find another way.

The simplest way is to trace back the three-dimensional problem to the two-dimensional one. In this way the first estimates, which were better than by means of Landau's method, were obtained by H.-E. Richert [1], W. Schwarz [1] and E. Krätzel [6]. But also this way has a great disadvantage. The numbers a, b, c do not enjoy the same rights. Indeed, with van der Corput's method one gets relatively good estimates for the remainder, however the conditions on the numbers a, b, c are by no means desirable.

P. G. Schmidt [2] gave a more symmetrical treatment of the most important unsymmetrical case $(a, b, c) = (1, 2, 3)$, and A. Ivić [5] and M. Vogts [1] showed that his method is also applicable to the other cases. In the first subsection we shall give a representation of $D(a, b, c; x)$ on this basis. In the next subsections we shall deal with the estimation of the remainder applying not only one-dimensional but also two-dimensional methods.

With respect to the Ω-estimates we refer to Section 6.2.4.

6.1.1. *The basic formula*

Theorem 6.1. *The main term $H(a, b, c; x)$ in expression* (6.1) *for $D(a, b, c; x)$ is represented by* (6.2) *and the limit values* (6.3)—(6.5), *respectively. The remainder is given by*

$$\Delta(a, b, c; x) = -\sum_{(u, v, w)\in\pi} S(u, v, w; x) + O\left(x^{\frac{1}{a+b+c}}\right), \tag{6.6}$$

where in the sum the triplet (u, v, w) runs over all permutations π of the numbers a, b, c. The function $S(u, v, w; x)$ is defined by

$$S(u, v, w; x) = \sum_{\substack{n^u m^{v+w}\leq x \\ n\leq m}} \psi\left(\left(\frac{x}{n^u m^v}\right)^{1/w}\right). \tag{6.7}$$

Proof. We write

$$D(a, b, c; x) = \sum_{(u, v, w)\in\pi} \sum_{\substack{n^u m^v k^w\leq x \\ n\leq m\leq k}}'' 1 + O\left(x^{\frac{1}{a+b+c}}\right),$$

where the sum gets factors 1/2 for the terms with equalities $n = m$ and $m = k$. The error term occurs, since the lattice points with $n = m = k$ are not counted correctly in the main term. Summing up over k, then we find

$$\begin{aligned}
&D(a, b, c; x)\\
&= \sum_{\pi} \sum_{\substack{n^u m^{v+w}\leq x \\ n\leq m}}' \left\{\left[\left(\frac{x}{n^u m^v}\right)^{1/w}\right] - m + \frac{1}{2}\right\} + O\left(x^{\frac{1}{a+b+c}}\right)\\
&= \sum_{\pi} \sum_{\substack{n^u m^{v+w}\leq x \\ n\leq m}}' \left\{\left(\frac{x}{n^u m^v}\right)^{1/w} - m\right\} + S + O\left(x^{\frac{1}{a+b+c}}\right)\\
&= \sum_{\pi} \sum_{\substack{n^u m^v \tau^w\leq x \\ n\leq m\leq \tau}}' \int d\tau + S + O\left(x^{\frac{1}{a+b+c}}\right),
\end{aligned}$$

where S denotes the sum

$$S = - \sum_{(u,\, v,\, w) \in \pi} S(u, v, w; x).$$

Summing up over m, we obtain

$$\begin{aligned} D(a, b, c; x) &= \sum_{\pi} \Bigg\{ \sum \int_{\substack{n^u \tau^{v+w} \leqq x \\ n \leqq \tau}} \left([\tau] - n + \frac{1}{2} \right) \mathrm{d}\tau \\ &\quad + \sum \int_{\substack{n^u \tau^{v+w} \geqq x \\ n^{u+v} \tau^w \leqq x}} \left(\left[\left(\frac{x}{n^u \tau^w} \right)^{1/v} \right] - n + \frac{1}{2} \right) \mathrm{d}\tau \Bigg\} + S + O\left(x^{\frac{1}{a+b+c}} \right) \\ &= \sum_{\pi} \Bigg\{ \sum \iint_{\substack{n^u t_2^v t_3^w \leqq x \\ n \leqq t_2 \leqq t_3}} \mathrm{d}t_2 \, \mathrm{d}t_3 - \sum \int_{\substack{n^u \tau^{v+w} \leqq x \\ n \leqq \tau}} \psi(\tau) \, \mathrm{d}\tau \\ &\quad - \sum \int_{\substack{n^u \tau^{v+w} \geqq x \\ n^{u+v} \tau^w \leqq x}} \psi\left(\left(\frac{x}{n^u \tau^w} \right)^{1/v} \right) \Bigg\} \mathrm{d}\tau + S + O\left(x^{\frac{1}{a+b+c}} \right). \end{aligned}$$

Since the integral over $\psi(\tau)$ is bounded, the second term is of order $x^{1/(u+v+w)} = x^{1/(a+b+c)}$. In the third term we use the substitution $n^u t^v \tau^w = x$. Then

$$\begin{aligned} \sum \int_{\substack{n^u \tau^{v+w} \geqq x \\ n^{u+v} \tau^w \leqq x}} \psi\left(\left(\frac{x}{n^u \tau^w} \right)^{1/v} \right) \mathrm{d}\tau &= \frac{v}{w} \sum \int_{\substack{n^u t^{v+w} \leqq x \\ n \leqq t}} \frac{1}{t} \left(\frac{x}{n^u t^v} \right)^{1/w} \psi(t) \, \mathrm{d}t \\ &= \frac{v}{w} \sum \int_{\substack{n^{u+v+w} \leqq x \\ n \leqq t < \infty}} \frac{1}{t} \left(\frac{x}{n^u t^v} \right)^{1/w} \psi(t) \, \mathrm{d}t - \frac{v}{w} \sum \int_{\substack{n^u t^{v+w} \leqq x \\ n^{u+v+w} \leqq x \\ n \leqq t < \infty}} \frac{1}{t} \left(\frac{x}{n^u t^v} \right)^{1/w} \psi(t) \, \mathrm{d}t \\ &= \frac{v}{w} \sum_{n \leqq t} \int \frac{1}{t} \left(\frac{x}{n^u t^v} \right)^{1/w} \psi(t) \, \mathrm{d}t + O\left(x^{\frac{1}{a+b+c}} \right). \end{aligned}$$

In the last step we have used partial integration and again the fact that the integral over $\psi(t)$ is bounded. Hence

$$\begin{aligned} D(a, b, c; x) &= \sum_{\pi} \Bigg\{ \sum \iint_{\substack{n^u t_2^v t_3^w \leqq x \\ n \leqq t_2 \leqq t_3}} \mathrm{d}t_2 \, \mathrm{d}t_3 - \frac{v}{w} \sum_{n \leqq t < \infty} \int \frac{1}{t} \left(\frac{x}{n^u t^v} \right)^{1/w} \psi(t) \, \mathrm{d}t \Bigg\} \\ &\quad + S + O\left(x^{\frac{1}{a+b+c}} \right). \end{aligned} \tag{6.8}$$

Finally we sum up over n and assume for the sake of simplicity that $a < b < c$. Then the first term on the right-hand side of (6.8) becomes

$$W_1 = \sum_{\pi} \sum_{\substack{n^u t_2^v t_3^w \leqq x \\ n \leqq t_2 \leqq t_3}} \iint dt_2\, dt_3$$

$$= \sum_{\pi} \left\{ \iint_{\substack{t_2^{u+v} t_3^w \leqq x \\ t_2 \leqq t_3}} [t_2]\, dt_2\, dt_3 + \iint_{\substack{t_2^{u+v} t_3^w \geqq x \\ t_2^v t_3^w \leqq x \\ t_2 \leqq t_3}} \left[\left(\frac{x}{t_2^v t_3^w}\right)^{1/u}\right] dt_2\, dt_3 \right\}$$

$$= \sum_{\pi} \left\{ \iiint_{\substack{t_1^u t_2^v t_3^w \leqq x \\ t_1 \leqq t_2 \leqq t_3}} dt_1\, dt_2\, dt_3 - \iint_{\substack{t_2^{u+v} t_3^w \leqq x \\ t_2 \leqq t_3}} \psi(t_2)\, dt_2\, dt_3 \right.$$

$$\left. = \iint_{\substack{t_2^{u+v} t_3^w \geqq x \\ t_2^v t_3^w \leqq x \\ t_2 \leqq t_3}} \psi\left(\left(\frac{x}{t_2^v t_3^w}\right)^{1/u}\right) dt_2\, dt_3 - \frac{1}{2} \iint_{\substack{t_2^v t_3^w \leqq x \\ t_2 \leqq t_3}} dt_2\, dt_3 \right\}.$$

We first compute the second and third term

$$- \iint_{\substack{t_2^{u+v} t_3^w \leqq x \\ t_2 \leqq t_3}} \psi(t_2)\, dt_2 = - \int_{t_2^{u+v+w} \leqq x} \left(\frac{x}{t_2^{u+v}}\right)^{1/w} \psi(t_2)\, dt_2 + O\left(x^{\frac{1}{a+b+c}}\right)$$

$$= - \int_1^{\infty} \left(\frac{x}{t^{u+v}}\right)^{1/w} \psi(t)\, dt + O\left(x^{\frac{1}{a+b+c}}\right),$$

$$- \iint_{\substack{t_2^{u+v} t_3^w \geqq x \\ t_2^v t_3^w \leqq x \\ t_2 \leqq t_3}} \psi\left(\left(\frac{x}{t_2^v t_3^w}\right)^{1/u}\right) dt_2\, dt_3 = - \frac{u}{w} \iint_{\substack{t_1^u t_2^{v+w} \leqq x \\ t_1 \leqq t_2}} \frac{1}{t_1} \left(\frac{x}{t_1^u t_2^v}\right)^{1/w} \psi(t_1)\, dt_1\, dt_2$$

$$= - \frac{u}{w-v} \int_1^{\infty} \frac{1}{t} \left(\frac{x}{t^u}\right)^{\frac{2}{v+w}} \psi(t)\, dt$$

$$+ \frac{u}{w-v} \int_1^{\infty} \left(\frac{x}{t^{u+v}}\right)^{1/w} \psi(t)\, dt$$

$$+ O\left(x^{\frac{1}{a+b+c}}\right).$$

If we consider all permutations or u, v, w, it is seen that the sum over the first integral vanishes. Hence

$$W_1 = \sum_{\pi} \iiint_{\substack{t_1^u t_2^v t_3^w \leqq x \\ t_1 \leqq t_2 \leqq t_3}} dt_1\, dt_2\, dt_3 - \frac{w-u-v}{w-v} \int_1^\infty \left(\frac{x}{t^{u+v}}\right)^{1/w} \psi(t)\, dt$$

$$- \frac{1}{2} \iint_{\substack{t_2^v t_3^w \leqq x \\ t_2 \leqq t_3}} dt_2\, dt_3 + O\left(x^{\frac{1}{a+b+c}}\right).$$

We apply Euler-Maclaurin's sum formula in the second term on the right-hand side of (6.8) and obtain

$$W_2 = -\sum_{\pi} \frac{v}{w} \sum_{n \leqq t} \int \frac{1}{t} \left(\frac{x}{n^u t^v}\right)^{1/w} \psi(t)\, dt$$

$$= -\sum_{\pi} \frac{v}{w} \left\{ \iint_{t_1 \leqq t_2} \frac{1}{t_2} \left(\frac{x}{t_1^u t_2^v}\right)^{1/w} \psi(t_2)\, dt_1\, dt_2 - \int_1^\infty \frac{1}{t} \left(\frac{x}{t^{u+v}}\right)^{1/w} \psi^2(t)\, dt \right.$$

$$\left. - \frac{u}{w} \iint_{t_1 \leqq t_2} \frac{1}{t_1 t_2} \left(\frac{x}{t_1^u t_2^v}\right)^{1/w} \psi(t_1)\, \psi(t_2)\, dt_1\, dt_2 + \frac{1}{2} \int_1^\infty \frac{1}{t} \left(\frac{x}{t^v}\right)^{1/w} \psi(t)\, dt \right\}.$$

First of all we compute the first and the second term:

$$-\frac{v}{w} \iint_{t_1 \leqq t_2} \frac{1}{t_2} \left(\frac{x}{t_1^u t_2^v}\right)^{1/w} \psi(t_2)\, dt_1\, dt_2$$

$$= -\frac{v}{w-u} \int_1^\infty \left(\frac{x}{t^{u+v}}\right)^{1/w} \psi(t)\, dt + \frac{v}{w-u} \int_1^\infty \frac{1}{t} \left(\frac{x}{t^v}\right)^{1/w} \psi(t)\, dt\, .$$

Because of $\psi^2(t) = 2 \int_1^t \psi(\tau)\, d\tau + \frac{1}{4}$ the second term becomes

$$\frac{v}{w} \int_1^\infty \frac{1}{t} \left(\frac{x}{t^{u+v}}\right)^{1/w} \psi^2(t)\, dt$$

$$= \frac{2v}{w} \iint_{\tau \leqq t} \frac{1}{t} \left(\frac{x}{t^{u+v}}\right)^{1/w} \psi(\tau)\, d\tau\, dt + \frac{v}{4w} \int_1^\infty \frac{1}{t} \left(\frac{x}{t^{u+v}}\right)^{1/w} dt$$

$$= \frac{2v}{u+v} \int_1^\infty \left(\frac{x}{\tau^{u+v}}\right)^{1/w} \psi(\tau)\, d\tau + \frac{v}{4(u+v)}\, x^{1/w}\, .$$

Hence

$$W_2 = \sum_{\pi}\left\{\left(\frac{-v}{w-u}+\frac{2v}{u+v}\right)\int_1^{\infty}\left(\frac{x}{t^{u+v}}\right)^{1/w}\psi(t)\,dt\right.$$

$$+\left(\frac{v}{w-u}-\frac{v}{2w}\right)\int_1^{\infty}\frac{1}{t}\left(\frac{x}{t^{v}}\right)^{1/w}\psi(t)\,dt$$

$$\left.+\frac{uv}{w_2}\iint_{t_1\leqq t_2}\frac{1}{t_1t_2}\left(\frac{x}{t_1^{u}t_2^{v}}\right)^{1/w}\psi(t_1)\,\psi(t_2)\,dt_1\,dt_2+\frac{v}{4(u+v)}\,x^{1/w}\right\}$$

$$=\sum_{\pi}\left\{\frac{w-u-v}{w-v}\int_1^{\infty}\left(\frac{x}{t^{u+v}}\right)^{1/w}\psi(t)\,dt\right.$$

$$+\left(\frac{v}{w-u}-\frac{v}{2w}\right)\int_1^{\infty}\frac{1}{t}\left(\frac{x}{t^{v}}\right)^{1/w}\psi(t)\,dt$$

$$\left.+\frac{uv}{2w^2}\int_1^{\infty}\int_1^{\infty}\frac{1}{t_1t_2}\left(\frac{x}{t_1^{u}t_2^{v}}\right)^{1/w}\psi(t_1)\,\psi(t_2)\,dt_1\,dt_2+\frac{1}{8}\,x^{1/w}\right\}.$$

Substituting these expressions for W_1 and W_2 into (6.8), then

$$D(a,b,c;x)=\sum_{\pi}\left\{\iiint_{\substack{t_1^{u}t_2^{v}t_3^{w}\leqq x\\ t_1\leqq t_2\leqq t_3}}dt_1\,dt_2\,dt_3-\frac{1}{2}\iint_{\substack{t_2^{v}t_3^{w}\leqq x\\ t_2\leqq t_3}}dt_2\,dt_3+\frac{1}{8}\,x^{1/w}\right.$$

$$+\left(\frac{v}{w-u}-\frac{v}{2w}\right)\int_1^{\infty}\frac{1}{t}\left(\frac{x}{t^{v}}\right)^{1/w}\psi(t)\,dt$$

$$\left.+\frac{uv}{2w^2}\int_1^{\infty}\int_1^{\infty}\frac{1}{t_1t_2}\left(\frac{x}{t_1^{u}t_2^{v}}\right)^{1/w}\psi(t_1)\,\psi(t_2)\,dt_1\,dt_2\right\}$$

$$+S+O\left(x^{\frac{1}{a+b+c}}\right)$$

$$=\frac{1}{2}\sum_{\pi}\left\{\frac{(w+u)(w+v)}{4(w-u)(w-v)}+\frac{v(w+u)}{w(w-u)}\int_1^{\infty}t^{-1-v/w}\psi(t)\,dt\right.$$

$$+ \frac{uv}{w^2} \int_1^\infty \int_1^\infty t_1^{-1-u/w} t_2^{-1-v/w} \psi(t_1)\, \psi(t_2)\, dt_1\, dt_2 \Big\} x^{1/w}$$

$$+ S + O\left(x^{\frac{1}{a+b+c}}\right)$$

$$= \frac{1}{2} \sum_{\pi} \left(\frac{w+u}{2(w-u)} + \frac{u}{w} \int_1^\infty t_1^{-1-u/w} \psi(t_1)\, dt_1 \right)$$

$$\times \left(\frac{w+v}{2(w-v)} + \frac{v}{w} \int_1^\infty t_2^{-1-v/w} \psi(t_2)\, dt_2 \right) x^{1/w}$$

$$+ S + O\left(x^{\frac{1}{a+b+c}}\right).$$

Applying the representation

$$\zeta(s) = \frac{s+1}{2(s-1)} - s \int_1^\infty t^{-1-s} \psi(t)\, dt \qquad (s > 0,\ s \neq 1)\,,$$

we get

$$D(a, b, c; x) = \frac{1}{2} \sum_{\pi} \zeta\left(\frac{u}{w}\right) \zeta\left(\frac{v}{w}\right) x^{1/w} + S + O\left(x^{\frac{1}{a+b+c}}\right)$$

$$= H(a, b, c; x) + S + O\left(x^{\frac{1}{a+b+c}}\right),$$

where $H(a, b, c; x)$ is given by (6.2). This completes the proof of (6.6), provided that $a < b < c$. But if two or three of these numbers are equal, similar proofs hold.

Remark. Even the trivial estimate of the sums $S(u, v, w; x)$ gives a proper result. It is seen that

$$S(u, v, w; x) \ll \sum_{\substack{n^u m^{v+w} \leq x \\ n \leq m}} 1 \ll \sum_{n^{u+v+w} \leq x} \left(\frac{x}{n^u}\right)^{\frac{1}{v+w}}.$$

Supposing $c < a + b$, the inequality $u < v + w$ holds for all permutations of (u, v, w). Hence

$$S(u, v, w; x) \ll x^{\frac{2}{u+v+w}} = x^{\frac{2}{a+b+c}}.$$

Thus we obtain the *trivial estimate*

$$\Delta(a, b, c; x) \ll x^{\frac{2}{a+b+c}}.$$

The required condition $c < a + b$ arises quite naturally. It ensures that the error term is smaller than the main term. Otherwise the problem is in reality a two-dimensional one.

6.1.2. *Estimates by means of van der Corput's method I*

In this section we give estimates of $\Delta(a, b, c; x)$ applying van der Corput's method to one of the variables in the sums $S(u, v, w; x)$, whereas we use only a trivial estimate with respect to the other variable. It will be seen that the results are nevertheless better than we may think. First of all this is the case, when the differences between the numbers a, b, c are comparatively large.

In the next lemmas of this chapter we always use the following notation. *Let D denote the domain*

$$D = \{(t_1, t_2)\colon t_1^{v+w} t_2^u \leqq x,\ t_2 \leqq t_1,\ M \leqq t_1 \leqq 2M,\ N \leqq t_2 \leqq 2N\} \tag{6.9}$$

with $M, N \geqq 1$ and $S(u, v, w; M, N; x)$ the sum

$$S(u, v, w; M, N; x) = \sum_{(m, n) \in D} \psi\left(\left(\frac{x}{n^u m^v}\right)^{1/w}\right). \tag{6.10}$$

Lemma 6.1. *If (k, l) is any exponent pair with $(k, l) = A(\varkappa, \lambda)$, the estimate*

$$S(u, v, w; M, N; x) \ll (x^{2k} M^{(2l-1)w - 2kv} N^{w - 2ku})^{1/w} \tag{6.11}$$

holds.

Proof. We apply the method of exponent pairs with respect to the sum over m in (6.10). In Definition 2.1 of exponent pairs we put $u = 2$, $a = M$. There the function

$$f(t) = \left(\frac{x}{n^u t^v}\right)^{1/w}$$

possesses derivatives of all order. The derivatives and (2.45) show that we can put

$$z = \frac{1}{M}\left(\frac{x}{n^u M^v}\right)^{1/w}.$$

Then Theorem 2.14 leads to the estimate

$$S(u, v, w; M, N; x) \ll \sum_{N \leqq n \leqq 2N} M^{2(k+l-1/2)} \frac{1}{M^{2k}} \left(\frac{x}{n^u M^v}\right)^{2k/w}$$

$$\ll (x^{2k} M^{(2l-1)w - 2kv} N^{w - 2ku})^{1/w},$$

which proves (6.11).

Theorem 6.2. *If* $(k, l) = A(\varkappa, \lambda)$ *is any exponent pair, the estimate*

$$\Delta(a, b, c; x) \ll x^{\frac{2(k+l)}{a+b+c}} \log^2 x \tag{6.12}$$

holds under the conditions $la \geqq k(b + c)$, $a + b \geqq 2(k + l - 1/2)\, c$.

Proof. We rewrite the estimate (6.11) in the form

$$S(u, v, w; M, N; x) \ll \left(x^{2k}(M^{v+w}N^u)^y \left(\frac{N}{M}\right)^z\right)^{1/w},$$

$$(u + v + w)\, y = 2lw - 2k(u + v), \quad (u + v + w)\, z = (v + w - 2(k + l - 1/2)\, u)\, w .$$

Under the conditions of the theorem we always have $y \geqq 0$, $z \geqq 0$. If we use the inequalities $M^{v+w}N^u \ll x$, $N \ll M$, we get

$$S(u, v, w; M, N; x) \ll x^{\frac{2(k+l)}{a+b+c}} .$$

We can divide the region of summation in the representation (6.7) of the function $S(u, v, w; x)$ into $O(\log^2 x)$ subregions of type (6.9). Therefore, we obtain

$$S(u, v, w; x) \ll x^{\frac{2(k+l)}{a+b+c}} \log^2 x ,$$

and from (6.6) result (6.12) follows immediately.

Note that the second condition $a + b + c \geqq 2(k + l)\, c$ again arises quite naturally. For, if we cannot satisfy this condition, the problem is in reality a two-dimensional one.

We obtain *the best general view of the order of the remainder* $\Delta(a, b, c; x)$ for all numbers with $1 \leqq a \leqq b \leqq c$ when we use the exponent pair

$$A^{k-1}\left(\frac{1}{2}, \frac{1}{2}\right) = \left(\frac{1}{2(K-1)}, \frac{2K-k-2}{2(K-1)}\right)$$

with $K = 2^k$ and $k = 2, 3, \ldots$ *Then the estimate*

$$\Delta(a, b, c; x) \ll x^{\left(2-\frac{k-1}{K-1}\right)\frac{1}{a+b+c}} \log^2 x$$

holds under the conditions $(2K - k - 2)\, a \geqq b + c$, $(K - 1)(a + b) \geqq (K - k)\, c$.

We notice five other special exponent pairs in (6.12):

(k, l)	$2(k + l)$	conditions
$\left(\frac{2}{40}, \frac{33}{40}\right)$	$\frac{7}{4}$	$33a \geqq 2(b + c), \quad 4(a + b) \geqq 3c$
$\left(\frac{2}{18}, \frac{13}{18}\right)$	$\frac{5}{3}$	$13a \geqq 2(b + c), \quad 3(a + b) \geqq 2c$
$\left(\frac{11}{82}, \frac{57}{82}\right)$	$\frac{68}{41}$	$57a \geqq 11(b + c), \quad 41(a + b) \geqq 27c$
$\left(\frac{33}{234}, \frac{161}{234}\right)$	$\frac{194}{117}$	$161a \geqq 33(b + c), \quad 117(a + b) \geqq 77c$
$\left(\frac{97}{696}, \frac{480}{696}\right)$	$\frac{577}{348}$	$480a \geqq 97(b + c), \quad 348(a + b) \geqq 229c$

6.1.3. *Estimates by means of van der Corput's method II*

In this section we give estimates, where we use the transformation formula for one of the variables in the sums $S(u, v, w; x)$ and then we apply the method of exponent pairs to the other variable. That means, we give an application of Theorem 2.20.

Lemma 6.2. *If* (k, l) *is any exponent pair, the estimate*

$$S(u, v, w; M, N; x) \ll (x^{2k+1} M^{(2k+1)(w-v)} N^{(2l+1)w-(2k+1)} u)^{\frac{1}{2k+3} \cdot \frac{1}{w}} \log x \quad (6.13)$$

holds.

Proof. We apply Theorem 2.20 with respect to the domain (6.9) and the sum (6.10). We put there $u_1 = u_2 = 2, a_1 = M, b_1 = 2M, a_2 = N, b_2 = 2N$. The function $f(t_1, t_2)$ is given by

$$f(t_1, t_2) = -\left(\frac{x}{t_1^v t_2^u}\right)^{1/w}.$$

Putting $\lambda = (x/M^v N^u)^{1/w}$, it is easily seen that $\lambda \gg M \gg N$ and

$$\left|\frac{\partial^\nu}{\partial t_1^\nu} f(t_1, t_2)\right| \asymp \frac{\lambda}{M^\nu} \qquad (\nu = 1, 2, 3).$$

The function

$$f_{t_1 t_1} f_{t_1 t_1 t_2} - f_{t_1 t_2} f_{t_1 t_1 t_1} = \frac{uv^2}{w}\left(1 + \frac{v}{w^3}\right) \frac{f^2}{t_1^4 t_2}$$

is positive in D. For the parts of the curve of boundary we have the possibilities t_2 = const and $t_1 = \varrho(t_2)$ with

$$\varrho(t_2) = \text{const}\,, \qquad t_2\,, \qquad \left(\frac{x}{t_2^u}\right)^{\frac{1}{v+w}}.$$

Now

$$\frac{\mathrm{d}}{\mathrm{d}t}(f_{t_1}(\varrho(t), t)) = \frac{v}{w}\frac{\mathrm{d}}{\mathrm{d}t}\left(\frac{x}{t^u\varrho^{v+w}}\right)^{1/w}.$$

In the third case this expression is identically zero. In the first two cases we get

$$\left|\frac{\mathrm{d}}{\mathrm{d}t}(f_{t_1}(\varrho(t), t))\right| \gg \frac{1}{\varrho^t}\left(\frac{x}{t^u\varrho^v}\right)^{1/w} \gg \frac{\lambda}{MN}.$$

The function $\varphi(y_1, t_2)$ is defined by

$$f_{t_1}(\varphi, t_2) = \frac{v}{w}\left(\frac{x}{\varphi^{v+w}t_2^u}\right)^{1/w} = y_1$$

such that

$$\varphi(y_1, t_2) = \left(\frac{v^w x}{w^w y_1^w t_2^u}\right)^{\frac{1}{v+w}}.$$

Therefore, we obtain with a certain constant c

$$f_1(m_1, t_2) = f(\varphi, t_2) - m_1\varphi(m_1, t_2) = c(m_1^v x t_2^{-u})^{\frac{1}{v+w}}.$$

It is clear that $f_1(m_1, t_2)$ satisfies the conditions of the method of exponent pairs with respect to the variable t_2, and that $|f_1(m_1, t_2)| \asymp \lambda$. Consequently, we can apply inequality (2.69) which proves (6.13)

Theorem 6.3. *If (k, l) is any exponent pair, the estimate*

$$\Delta(a, b, c; x) \ll x^{\left(2-\frac{3-2l}{3+2k}\right)\frac{1}{a+b+c}}\log^3 x \tag{6.14}$$

holds under the condition $2(k + l + 1)\,a \geqq (2k + 1)\,(b + c)$.

Proof. We rewrite the estimate (6.13) in the form

$$S(u, v, w; M, N; x) \ll \left(x^{2k+1}(M^{v+w}N^u)^y\left(\frac{N}{M}\right)^z\right)^{\frac{1}{(2k+3)w}}\log x,$$

$$(u + v + w)\,y = 2(k + l + 1)\,w - (2k + 1)\,(u + v)\,,$$

$$(u + v + w)\,z = ((2l + 1)\,(v + w) - 2(2k + 1)\,u)\,w\,.$$

We have $y \geqq 0$, $z \geqq 0$ for all combinations of u, v, w if

$$2(k+l+1)\,a \geqq (2k+1)\,(b+c)\,, \qquad (2l+1)\,(a+b) \geqq 2(2k+1)\,c\,.$$

Now it is easily seen that the second inequality follows from the first:

$$\begin{aligned}2(2k+1)\,c &\leqq 4(k+l+1)\,a - 2(2k+1)\,b\\ &= (2l+1)\,(a+b) + (4k+2l+3)\,(a-b) \leqq (2l+1)\,(a+b)\,.\end{aligned}$$

Analogous to the proof of Theorem 6.2 we obtain (6.14) immediately.

We now consider some special exponent pairs. We first use again the exponent pairs

$$A^{k-1}\left(\frac{1}{2}, \frac{1}{2}\right) = \left(\frac{1}{2(K-1)}, \frac{2K-k-2}{2(K-1)}\right)$$

with $K = 2^k$, $k = 2, 3, \ldots$ *Then the estimate*

$$\Delta(a, b, c; x) \ll x^{\left(2-\frac{K+k-1}{3K-2}\right)\frac{1}{a+b+c}} \log^3 x$$

holds under the condition $(4K-k-3)\,a \geqq K(b+c)$.

We notice four other special exponent pairs in (6.14):

(k, l)	$2 - \dfrac{3-2l}{3+2k}$	condition
$\left(\frac{2}{18}, \frac{13}{18}\right)$	$\frac{44}{29}$	$3a \geqq b + c$
$\left(\frac{1}{6}, \frac{4}{6}\right)$	$\frac{3}{2}$	$11a \geqq 4(b + c)$
$\left(\frac{2}{7}, \frac{4}{7}\right)$	$\frac{37}{25}$	$26a \geqq 11(b + c)$
$\left(\frac{13}{40}, \frac{22}{40}\right)$	$\frac{108}{73}$	$25a \geqq 11(b + c)$

In order to obtain a small exponent, we can also proceed in the following manner. If (k, l) is an exponent pair, then the B-process shows that $(l - 1/2, k + 1/2)$ is also an exponent pair. Now applying the C-process, we obtain the pair $(\alpha, \alpha + 1/2)$ with $\alpha = (k + l - 1/2)/2$. Finally we get

$$BA\left(\alpha, \alpha + \frac{1}{2}\right) = \left(\frac{2\alpha+1}{4(\alpha+1)}, \frac{2(2\alpha+1)}{4(\alpha+1)}\right).$$

Then the estimate

$$\Delta(a, b, c; x) \ll x^{\left(2-\frac{2\alpha+4}{8\alpha+7}\right)\frac{1}{a+b+c}} \log^2 x$$

holds under the condition $(10\alpha + 7)\,a \geqq (4\alpha + 3)\,(b + c)$.

We obtain *the smallest known value* α by means of Ranking's exponent pair

$$(k, l) = \left(\frac{141\,841}{1\,019\,718}, \frac{703\,527}{1\,019\,718}\right).$$

Then

$$\alpha = \frac{335\,509}{2\,039\,436} = 0{,}1645106\ldots, \qquad 2 - \frac{2\alpha + 4}{8\alpha + 7} = \frac{12\,545\,743}{8\,480\,062} = 1{,}47944\,,$$

$$\frac{10\alpha + 7}{4\alpha + 3} = \frac{8\,815\,571}{3\,730\,172} = 2{,}3633\ldots$$

It is easily seen that Lemma 6.2 leads to a result similar to that of (5.6). Using the exponent pair $(k, l) = (1/2, 1/2)$, we obtain from (6.9) and (6.13)

$$S(u, v, w; M, N; x) \ll (xM^{w-v}N^{w-u})^{1/2w} \ll (xM^{a-v'}N^{a-u'})^{1/2a}$$
$$\ll (xN^{2a-b-c})^{1/2a} \ll x^{\frac{2}{2a+b+c}}$$

for each permutation (u, v, w), provided that $N^{2a+b+c} \geqq x$. Otherwise, if $N^{2a+b+c} \leqq x$, we apply Theorem 2.2 with respect to the variable m in (6.10) and we trivially estimate the sum over n. Assuming that $c < 2a + b$, then

$$S(u, v, w; M, N; x) \ll N(xN^{-u}M^{w-v})^{1/3w} \ll N(xN^{-u'}M^{a-v'})^{1/3a}$$
$$\ll N(xN^{-u'})^{\frac{1}{2a-v'}} \ll x^{\frac{2}{2a+b+c}},$$

provided that $M^{2a+v'} \geqq xN^{-u'}$. Otherwise trivial estimation of (6.10) yields

$$S(u, v, w; M, N; x) \ll MN \ll (xN^{-u'})^{\frac{1}{2a+v'}}\, N x^{\frac{2}{2a+b+c}}.$$

Hence

$$S(u, v, w; M, N; x) \ll x^{\frac{2}{2a+b+c}}$$

holds for each permutation (u, v, w) and for $c < 2a + b$. Hence, we have the *general estimation*

$$\Delta(a, b, c; x) \ll x^{\frac{2}{2a+b+c}} \log^2 x\,.$$

6.1.4. *Estimates by means of Titchmarsh's method*

The estimates of this section are based on Theorem 2.22.

Lemma 6.3. *Let* $K = 2^k$, $k = 2, 3, \ldots$ *Let* $S(u, v, w; M, N; x)$ *denote the sum* (6.10) *with respect to the domain*

$$D = \{(t_1, t_2)\colon t_1^{v+w} t_2^{u} \leqq x,\ t_2 \leqq t_1,\ M \leqq t_1 \leqq \alpha M,\ N \leqq t_2 \leqq 2N\}\,,$$

where $\alpha = 2$ *for* $k = 2$ *and* $\alpha > 1$ *such that*

$$(k-1)\,w + u + v > (\alpha - 1)\,(k-2)\,u((k-2)\,w + v)$$

for $k > 2$. *Then*

$$S(u, v, w; M, N; x) \ll (x^4 M^{(3K-4k)\,w-4v} N^{(3K-8)\,w-4u})^{\frac{1}{(3K-4)\,w}} \log x\,. \qquad (6.15)$$

Proof. In Theorem 2.22 we put $u_1 = \alpha$, $u_2 = 2$, $a_1 = M$, $b_1 = \alpha M$, $a_2 = N$, $b_2 = 2N$,

$$f(t_1, t_2) = -\left(\frac{x}{t_1^v t_2^u}\right)^{1/w}.$$

The required derivatives are given by

$$\frac{\partial^\nu f}{\partial t_1^\nu} = (-1)^{\nu-1} \prod_{r=0}^{\nu-1} \left(r + \frac{v}{w}\right) \frac{f}{t_1^\nu} \qquad (\nu = 1, 2, \ldots, k)\,,$$

$$\frac{\partial^k f}{\partial t_1^{k-2}\,\partial t_2^2} = (-1)^{k-1}\,\frac{u}{w}\left(1 + \frac{u}{w}\right) \prod_{r=0}^{k-3} \left(r + \frac{v}{w}\right) \frac{f}{t_1^{k-2} t_2^2}\,,$$

$$\frac{\partial^k f}{\partial t_1^{k-1}\,\partial t_2} = (-1)^{k-1}\,\frac{u}{w} \prod_{r=0}^{k-2} \left(r + \frac{v}{w}\right) \frac{f}{t_1^{k-1} t_2}\,.$$

Further we have to consider the Hessian

$$H = H\left(\int_0^1 \ldots \int_0^1 \frac{\partial^{k-2} f(t_1', t_2)}{\partial t_1^{k-2}}\,\mathrm{d}\tau_1 \cdot \ldots \cdot \mathrm{d}\tau_{k-2}\right),$$

where t_1' is defined by

$$t_1' = t_1 + h_1\tau_1 + \ldots + h_{k-2}\tau_{k-2}\,,$$

$$1 \leqq h_1, h_2, \ldots, h_{k-2} \leqq (\alpha - 1)\,M$$

for $k > 2$. In case of $k = 2$ we have $t_1' = t_1$ and $H = H(f(t_1, t_2))$. Expanding this determinant for $k > 2$, it is seen that there arise two groups of integrals. In the first group we use the above notation t_1', in the second one we write

$$t_1'' = t_1 + h_1\tau_1' + \ldots + h_{k-2}\tau_{k-2}'\,.$$

Furthermore, we use the notations $\int g(t_1', t_2)$ and $\int g(t_1'', t_2)$ for the corresponding integrals and functions g as abbreviations. We obtain without trouble

$$H = \frac{u}{w}\left(k - 2 + \frac{v}{w}\right) \prod_{r=0}^{k-3} \left(r + \frac{v}{w}\right)^2 \int \frac{f(t_1', t_2)}{t_1'^k} \int \frac{f(t_1'', t_2)}{t_1''^{k-1}}\,\frac{\omega(t_1', t_1'')}{t_2^2}\,,$$

$$\omega(t_1', t_1'') = \left(k - 1 + \frac{v}{w}\right)\left(1 + \frac{u}{w}\right) t_1'' - \left(k - 2 + \frac{v}{w}\right)\frac{u}{w}\, t_1'$$

$$\geqq \left(k - 1 + \frac{v}{w}\right)\left(1 + \frac{u}{w}\right) t_1 - \left(k - 2 + \frac{v}{w}\right)\frac{u}{w}\,(t_1 + (\alpha - 1)(k - 2)M)$$

$$\geqq M\left\{k - 1 + \frac{u + v}{w} - (\alpha - 1)(k - 2)\frac{u}{w}\left(k - 2 + \frac{v}{w}\right)\right\} > 0\,.$$

Clearly, this property also holds in case of $k = 2$. Putting

$$\lambda = \left(\frac{x}{M^v N^u}\right)^{1/w},$$

it is easily seen that $\lambda \gg M \gg N$ and that the conditions of Theorem 2.22 are satisfied with respect to the derivatives and Hessian.

For the parts of the curve of boundary we have the possibilities $t_2 = \text{const}$ or $t_1 = \varrho(t_2)$ with

$$\varrho(t_2) = \text{const}\,, \qquad t_2\,, \qquad \left(\frac{x}{t_2^u}\right)^{\frac{1}{v+w}}.$$

Only the last possibility is of interest:

$$\varrho''(t_2) = \frac{u}{v + w}\left(1 + \frac{u}{v + w}\right)\frac{1}{t_2^2}\left(\frac{x}{t_2^u}\right)^{\frac{1}{v+w}} \ll \frac{M}{N^2}\,.$$

Therefore, all conditions of Theorem 2.22 are satisfied, and (6.15) follows from (2.73).

Theorem 6.4. *If* $K = 2^k$, $k = 2, 3, \ldots$, *the estimate*

$$\Delta(a, b, c; x) \ll x^{\left(2 - \frac{4(k-1)}{3K-4}\right)\frac{1}{a+b+c}} \log^3 x \tag{6.16}$$

holds under the conditions $(3K - 2k - 4)\,a \geqq 2(b + c)$, $(3K - 8)(a + b) \geqq (3K - 4k + 4)\,c$.

Proof. We rewrite the stimate (6.15) in the form

$$S(u, v, w; M, N; x) \ll \left(x^4 (M^{v+w} N^u)^y \left(\frac{N}{M}\right)^z\right)^{\frac{1}{(3K-4)w}} \log x\,,$$

$$(u + v + w)\,y = 2((3K - 2k - 4)\,w - 2(u + v))\,,$$

$$(u + v + w)\,z = w((3K - 8)(v + w) - (3K - 4k + 4)\,u)\,.$$

The conditions of the theorem show that we have $y \geqq 0$, $z \geqq 0$ for all combinations of u, v, w. Analogously to the proof of Theorem 6.2 (6.16) now follows.

We notice the results for $k = 2, 3, 4, 5$:

k	$2 - \frac{4(k-1)}{3K-4}$	conditions
2	$\frac{3}{2}$	$2a \geqq b + c,\ a + b \geqq 2c (a = b = c = 1)$
3	$\frac{8}{5}$	$7a \geqq b + c,\ a + b \geqq c$
4	$\frac{19}{11}$	$18a \geqq b + c,\ 10(a + b) \geqq 9c$
5	$\frac{42}{23}$	$41a \geqq b + c,\ 11(a + b) \geqq 10c$

6.1.5. *Estimates by transformation of double exponential sums*

The estimates of this section are based on Theorems 2.26 and 2.27. That means, we work with the transformation formula of double exponential sums, and then we use one or two Weyl's steps. We again consider the sum (6.10), but in applying Theorem 2.25 it is useful to change the notations in (6.9). Thus we consider the domain

$$D = \{(t_1, t_2): t_1^u t_2^{v+w} \leqq x,\ t_1 \leqq t_2,\ N \leqq t_1 \leqq 2N,\ M \leqq t_2 \leqq 2M\}$$

and the function

$$f(t_1, t_2) = -\left(\frac{x}{t_1^u t_2^v}\right)^{1/w}.$$

Lemma 6.4.

$$S(u, v, w; M, N; x) \ll (x^2 M^{3w-2v} N^{4w-2u})^{1/6w} \log x\,, \tag{6.17}$$

$$S(u, v, w; M, N; x) \ll (x^7 M^{8w-7v} N^{10w-7u})^{1/17w} \log x \tag{6.18}$$

Proof. We apply Theorem 2.26 for the proof of (6.17) and Theorem 2.27 for the proof of (6.18). That means that we have to check all the conditions of Theorem 2.25 with respect to $k = 3$ and $k = 4$. There we put

$$u_1 = u_2 = 2\,,\quad a_1 = N\,,\ b_1 = 2N\,,\quad a_2 = M\,,\ b_2 = 2M\,,$$

$$\lambda = \left(\frac{x}{N^u M^v}\right)^{1/w},$$

such that $\lambda \gg M \gg N$.

Now we have to check all the conditions of Theorem 2.24. It is easily seen that $c_1 \asymp N$, $c_2 \asymp M$ and

$$\lambda_1 = \frac{\lambda}{N^2}, \quad \lambda' = \frac{\lambda}{N^3}, \quad c_1\lambda' \ll \lambda_1, \quad \lambda_2 = \frac{\lambda}{M^2}, \quad \Lambda = \frac{\lambda^2}{M^2N^2}.$$

For the parts of the curve of boundary we have the possibilities $t_2 = \text{const}$ or $t_1 = \varrho(t_2)$ with

$$\varrho(t_2) = \text{const}, \qquad t_2, \qquad \left(\frac{x}{t_2^{v+w}}\right)^{1/u}.$$

Then

$$\left|\frac{\mathrm{d}}{\mathrm{d}t} f_{t_1}(\varrho(t), t)\right| = \frac{u}{w}\left|\frac{\mathrm{d}}{\mathrm{d}t}\left(\frac{x}{\varrho^{u+w}t^v}\right)^{1/w}\right| \gg \frac{1}{\varrho t}\left(\frac{x}{\varrho^u t^v}\right)^{1/w} \gg \frac{\lambda}{MN}.$$

D_1 is the image of D under the mapping

$$y_1 = f_{t_1}(t_1, t_2) = \frac{u}{w}\left(\frac{x}{t_1^{u+w}t_2^v}\right)^{1/w}, \qquad y_2 = t_2.$$

The function $\varphi_1(y_1, y_2)$ is defined by

$$\varphi_1(y_1, y_2) = \left(\frac{u^w x}{w^w y_1^w y_2^v}\right)^{\frac{1}{u+w}}$$

and we find

$$f_1(y_1, y_2) = d_1\left(\frac{x y_1^u}{y_2^v}\right)^{\frac{1}{u+w}}$$

with a certain constant $d_1 < 0$. Of course, we have $\lambda'' = \lambda M^{-3}$ such that $c_2\lambda_1\lambda'' \ll \Lambda$. Obviously, the function

$$\frac{\partial}{\partial y_2}\frac{1}{\sqrt{|f_{t_1t_1}(\varphi_1, y_2)|}}$$

is monotonic with respect to y_2, and we can put $G_2 = N/M\sqrt{\lambda}$.

D_2 is the image of D under the mapping

$$y_1 = f_{t_1}(t_1, t_2) = \frac{u}{w}\left(\frac{x}{t_1^{u+w}t_2^v}\right)^{1/w},$$

$$y_2 = f_{t_2}(t_1, t_2) = \frac{v}{w}\left(\frac{x}{t_1^u t_2^{v+w}}\right)^{1/w}.$$

The functions $\varphi_1(y_1, y_2)$ and $\varphi_2(y_1, y_2)$ are defined by

$$\varphi_1(y_1, y_2) = \left(\frac{u^{v+w} y_2^v x}{v^v w^w y_1^{v+w}}\right)^{\frac{1}{u+v+w}},$$

$$\varphi_2(y_1, y_2) = \left(\frac{v^{u+w} y_1^u x}{u^u w^w y_2^{u+w}}\right)^{\frac{1}{u+v+w}}$$

and we find

$$f_2(y_1, y_2) = d_2(x y_1^u y_2^v)^{\frac{1}{u+v+w}}$$

with a certain constant $d_2 < 0$.

It remains to check the conditions of Theorem 2.25 for $k = 3, 4$. When we apply this theorem with y_1, y_2 instead of t_1, t_2, we can take the same λ, but now the numbers a_1, a_2 are of order λ/N, λ/M. Since $f_2(y_1, y_2) \asymp \lambda$, the condition of the theorem are obviously satisfied with respect to the derivatives of this function.

Moreover, it is easily seen that the Hessians

$$H\left(\int_0^1\int_0^1 \frac{\partial^{k-2}}{\partial y_1^{k-2}} f_2(y_1', y_2)\, d\tau_1\, d\tau_{k-2}\right),$$

$y_1' = y_1 + h_1\tau_1 + h_{k-2}\tau_{k-2}$, $1 \leqq h_1, h_{k-2} \ll \lambda/N$, are always negative for $k = 3$, 4 and therefore satisfy the required conditions. For the parts of the curve of boundary of D_2 we have the possibilities $y_2 =$ const or $y_1 = \varrho(y_2)$ with

$$\varrho(y_2) = \frac{v}{u}\, y_2\,, \qquad \gamma\left(\frac{y_2^{u+w}}{x}\right)^{\frac{1}{u}}, \qquad \delta(x y_2^v)^{\frac{1}{v+w}}.$$

Clearly, in the first case there is $y_1'' = 0$. In the last two cases we have

$$|y_1''| \asymp \frac{y_1}{y_2^2} \asymp \frac{M^2}{\lambda N}.$$

Hence, all conditions of Theorem 2.25 are satisfied.

Finally, it is seen from

$$H(f(\varphi_1, \varphi_2)) = d_3(y_1^{v+w} y_2^{u+w} x^{-1})^{\frac{2}{u+v+w}}$$

where d_3 is a certain negative constant that the first partial derivatives of this function with respect to y_1, y_2 are of fixed sign.

Thus, having checked all the required conditions, the estimates (6.17) and (6.18) now follow from (2.83) and (2.84).

Theorem 6.5. *The estimate*

$$\Delta(a, b, c; x) \ll x^{\frac{3}{2}\frac{1}{a+b+c}} \log^3 x \tag{6.19}$$

holds under the conditions $7a \geqq 2(b + c)$, $4(a + b) \geqq 5c$.

Proof. (6.17) shows that

$$S(u, v, w; M, N; x) \ll \left(x^2 (M^{v+w} N^u)^y \left(\frac{N}{M}\right)^z \right)^{1/6w} \log x ,$$

$$(u + v + w) y = 7w - 2(u + v) , \quad (u + v + w) z = w(4(v + w) - 5u) .$$

The conditions of the theorem show that we have $y \geqq 0$, $z \geqq 0$ for all combinations of u, v, w. Analogous to the proof of Theorem 6.2, (6.19) now follows.

Theorem 6.6. *The estimate*

$$\Delta(a, b, c; x) \ll x^{\frac{25}{17}\frac{1}{a+b+c}} \log^3 x \tag{6.20}$$

holds under the conditions $18a \geqq 7(b + c)$, $2(a + b) \geqq 3c$.

Proof. (6.18) shows that

$$S(u, v, w; M, N; x) \ll \left(x^7 (M^{v+w} N^u)^y \left(\frac{N}{M}\right)^z \right)^{1/17w} \log x ,$$

$$(u + v + w) y = 18w - 7(u + v) , \quad (u + v + w) z = 5w(2(v + w) - 3u) .$$

Again, we have $y \geqq 0$, $z \geqq 0$, and (6.20) follows immediately.

6.1.6. *The divisor problem of Piltz*

We now investigate the special case $a = b = c = 1$ which denotes the divisor problem of Piltz. We write

$$D_3(x) = xP_3(\log x) + \Delta_3(x) ,$$

where the main term $xP_3(\log x)$ is defined by (6.5). We now define:

Definition 6.2.

$$\vartheta_3 = \inf \{\alpha : \Delta_3(x) \ll x^\alpha\} .$$

We know the trivial inequality $0 \leqq \vartheta_3 \leqq 2/3$. The following improvements of the upper bound are obtained from previous theorems.

$$\vartheta_3 \leqq \frac{5}{9}, \ \frac{68}{123}, \ \frac{194}{351}, \ \frac{577}{1044} \qquad \text{(Theorem 6.2)} ,$$

$$\vartheta_3 \leqq \frac{1}{2}, \quad \frac{37}{75}, \quad \frac{36}{73}, \quad \frac{12545743}{25440186} \qquad \text{(Theorem 6.3)},$$

$$\vartheta_3 \leqq \frac{25}{51} \qquad \text{(Theorem 6.6)}.$$

We shall prove in this section $\vartheta_3 \leqq 15/31$ by means of I. M. Vinogradov's method which he used in the sphere problem. This value cannot be found in any paper but it demonstrates the power of this method in a simple way. Further improvements were obtained by refinements of the investigations, but thereby many technical difficulties arose.

Lemma 6.5. *Let m be a positive integer. Let D denote the domain*

$$D = \{(t_1, t_2)\colon t_1 t_2^2 \leqq x,\ t_1 \leqq t_2,\ N \leqq t_1 \leqq 2N,\ M \leqq t_2 \leqq 2M\}$$

and D_2 the domain

$$D_2 = \{(y_1, y_2)\colon m \leqq y_2 \leqq y_1,\ N \leqq (xmy_1^{-2}y_2)^{1/3} \leqq 2N,$$
$$M \leqq (xmy_1y_2^{-2})^{1/3} \leqq 2M\}.$$

Then

$$\sum_{(n_1, n_2) \in D} \mathrm{e}\left(-\frac{mx}{n_1 n_2}\right) = \frac{-im}{\sqrt{3}} \sum_{(m_1, m_2) \in D_2} \left(\frac{x}{m^2 m_1^2 m_2^2}\right)^{1/3} \mathrm{e}\,(-3(xmm_1m_2)^{1/3})$$
$$+ O(M \log (mx)) + O\left(m \sqrt{\frac{x}{MN}} \log (mx)\right). \quad (6.21)$$

Proof. We apply Theorem 2.24 to the domain D and the function $f(t_1, t_2) = -mx/t_1 t_2$. We can use the whole first part of the proof of Lemma 6.4 with the special values $u = v = w = 1$. Clearly, D_2 is the image of D under the mapping

$$y_1 = \frac{mx}{t_1^2 t_2}, \quad y_2 = \frac{mx}{t_1 t_2^2}.$$

Further, we obtain

$$H(f(\varphi_1, \varphi_2)) = \frac{3}{m^2}(m^2 y_1^2 y_2^2 x^{-1})^{2/3},$$

$$f_2(\varphi_1, \varphi_2) = -3(xmy_1y_2)^{1/3}.$$

It is easily seen that we can use the values

$$c_1 = N, \quad c_2 = M, \qquad \lambda_1 = \frac{m\lambda}{N^2}, \quad \lambda' = \frac{m\lambda}{N^3}, \qquad \Lambda = \left(\frac{m\lambda}{MN}\right)^2,$$

$$G = 1, \quad G_1 = 0, \qquad \gamma_1 \ll \frac{m\lambda}{N}, \quad \gamma_2 \ll \frac{m\lambda}{M}, \qquad \lambda'' = \frac{m\lambda}{M^3}, \qquad G_2 = \frac{N}{M\sqrt{m\lambda}}$$

in Theorem 2.24. Thereby it is $\lambda = mx/MN$ such that $\lambda \gg M \gg N$ is satisfied. (2.79) shows that

$$\Sigma\, T(\varrho(n_2), n_2) \ll \sqrt{m\lambda} + M \log(mx)\,.$$

Moreover, we get

$$\Delta \ll M + \sqrt{m\lambda}\log(mx)\,, \quad G_2\,\Delta_2 \ll \sqrt{m\lambda}\,.$$

Applying the transformation formula (2.78), equation (6.21) now follows.

Lemma 6.6. *Let D denote the domain of Lemma* 6.5. *Then*

$$\sum_{(n_1, n_2) \in D} \psi\left(\frac{x}{n_1 n_2}\right) \ll (x^3 N^{-2})^{1/5} (\log x)^{18/5} + x^{2/5} + M \log^2 x\,. \tag{6.22}$$

Proof. The proof is similar to that of Theorem 4.2. We consider a sum of type

$$S = \sum_{m=1}^{\infty} c_m \sum_{(m_1, m_2) \in D} \mathrm{e}\left(-\frac{mx}{n_1 n_2}\right),$$

where

$$c_m \ll K_m = \min\left(\frac{z^2}{m^3}, \frac{1}{m}\right), \qquad z > 1\,.$$

Applying the transformation formula (6.21), we obtain

$$S = \frac{-i}{\sqrt{3}} \sum_{m=1}^{\infty} m c_m \sum_{(m_1, m_2) \in D_2} \left(\frac{x}{m^2 m_1^2 m_2^2}\right)^{1/3} \mathrm{e}(-3(xmm_1m_2)^{1/3})$$
$$+ O(M \log^2 (zx)) + O\left(z \sqrt{\frac{x}{MN}} \log(zx)\right). \tag{6.23}$$

We first consider the sum

$$S_H = \sum_{m=H}^{2H} m c_m \sum_{(m_1, m_2) \in D_2} \left(\frac{x}{m^2 m_1^2 m_2^2}\right)^{1/3} \mathrm{e}(-3(xmm_1m_2)^{1/3})\,,$$

where $H \geqq 1$. It is clear from the definition of D_2 that this domain is contained in a rectangle, where the lengths of sides are of order Hx/MN^2 and Hx/M^2N. We take the sum over m_1, m_2 such that $m_1 - m_2 \geqq 0$ and $m_2 - m \geqq 0$. Assume that A, B are the smallest integers such that $m_1 - m_2 \leqq A$, $m_2 - m \leqq B$. Then

$$A \ll Hx/MN^2\,, \quad B \ll Hx/M^2N\,.$$

Let D_3 denote the image of the rectangle

$$\{(t_1, t_2) \colon N \leqq t_1 \leqq 2N,\ M \leqq t_2 \leqq 2M\}$$

under the mapping

$$y_1 = \frac{mx}{t_1^2 t_2}, \quad y_2 = \frac{mx}{y_1 t_2^2},$$

that is

$$D_3 = \{(y_1, y_2) : N \leqq (xmy_1^{-2} y_2)^{1/3} \leqq 2N,\ M \leqq (xmy_1 y_2^{-2})^{1/3} \leqq 2M\}.$$

Now

$$S_H = \frac{1}{4AB} \sum_{r=1}^{2A} \sum_{s=1}^{2B} \sum_{k_1=0}^{A} \sum_{k_2=0}^{B} \sum_{m=H}^{2H} mc_m$$

$$\times \sum_{(m_1, m_2) \in D_3} \left(\frac{x}{m^2 m_1^2 m_2^2}\right)^{1/3} \mathrm{e}(-3(smm_1 m_2)^{1/3})$$

$$\times \mathrm{e}\left(\frac{r}{2A}(m_1 - m_2 - k_1) + \frac{s}{2B}(m_2 - m - k_2)\right).$$

If $\varrho(n)$ denotes the number of solutions of $n = mm_2$ under the considered restrictions, we get

$$S_H \ll \frac{HK_H}{AB} \sum_{r=1}^{2A} \sum_{s=1}^{2B} \left| \sum_{k_1=0}^{A} \sum_{k_2=0}^{B} \mathrm{e}\left(\frac{rk_1}{2A} + \frac{sk_2}{2B}\right) \right|$$

$$\times \sum_n \left(\frac{x}{n^2}\right)^{1/3} \varrho(n) \left| \sum_{m_1} m_1^{-2/3} \mathrm{e}(-\mu m_1 - 3(xm_1 n)^{1/3}) \right|.$$

Here μ denotes that value $r/2A$, where the sum takes on its maximum. m_1 runs over values of order Hx/MN^2, and n takes values of order H^2x/M^2N. Now

$$S_H \ll \left(\frac{M^4 N^2}{Hx}\right)^{1/3} K_H \log^2(Hx) \sum_n \varrho(n)$$

$$\times \left| \sum_{m_1} m_1^{-2/3} \mathrm{e}(-\mu m_1 - 3(xm_1 n)^{1/3}) \right|.$$

Applying Schwarz's inequality and using the fact that

$$\sum_n \varrho^2(n) \ll \sum_n d^2(n) \ll \frac{H^2 x}{M^2 N} \log^3(Hx),$$

we obtain

$$S_H^2 \ll (H^4 x M^2 N)^{1/3} K_H^2 \log^7(Hx)$$

$$\times \sum_n \left| \sum_{m_1} m_1^{-2/3} \mathrm{e}(-\mu m_1 - 3(xm_1 n)^{1/3}) \right|^2$$

$$\ll (H^4 x M^2 N)^{1/3} K_H^2 \log^7(Hx) \sum_u \sum_v (uv)^{-2/3} \sum_n \mathrm{e}^{2\pi i f(n)},$$

where

$$f(t) = -\mu(u-v) - 3(xt)^{1/3}(u^{1/3} - v^{1/3}),$$

and u, v run over values of order Hx/MN^2. If $u = v$, we get an error term of order $\frac{H^3x}{M} K_H^2 \log^7(Hx)$.

We now estimate the sum $T = \sum_n e^{2\pi i f(n)}$ with van der Corput's Theorem 2.1 supposing that $u > v$. Because of

$$f''(t) = \frac{2}{3}(xt^{-5})^{1/3}(u^{1/3} - v^{1/3})$$

we have

$$|f''(t)| \gg \frac{M^4N^3}{H^4x^2}(u-v).$$

Thus we obtain from (2.6)

$$T \ll \sqrt{N(u-v)} + \frac{H^2x}{M^2N\sqrt{N(u-v)}}.$$

A similar result holds if $v > u$. Hence

$$\begin{aligned} S_H^2 &\ll (H^4xM^2N)^{1/3} K_H^2 \log^7(Hx) \\ &\quad \times \sum_{u \neq v} (uv)^{-2/3} \left\{ \sqrt{N|u-v|} + \frac{H^2x}{M^2N\sqrt{N|u-v|}} \right\} + \frac{H^3x}{M} K_H^2 \log^7(Hx) \\ &\ll (H^4xM^2N)^{1/3} K_H^2 \log^7(Hx) \left(\frac{Hx}{MN^2}\right)^{2/3} \left\{ \sqrt{\frac{Hx}{MN}} + \sqrt{\frac{H^3x}{M^3N}} \right\} \\ &\quad + \frac{H^3x}{M} K_H^2 \log^7(Hx). \end{aligned}$$

We now assume that $H \ll M$. Moreover, since $N^3 \ll x$, we obtain

$$S_H \ll \left(\frac{H^5x^3}{MN^3}\right)^{1/4} K_H (\log x)^{7/2}.$$

In case $H \gg M$ we use trivial estimation. Suppose that $1 < z < M \leqq \sqrt{x}$. Then we obtain from (6.23)

$$S \ll \left(\frac{zx^3}{MN^3}\right)^{1/4} (\log x)^{9/2} + \sum_{m \geqq M} \left(\frac{z}{m}\right)^2 \frac{x}{MN} + M \log^2 x + z\sqrt{\frac{x}{MN}} \log x.$$

Now (1.17) shows that

$$\sum_{(n_1, n_2) \in D} \psi\left(\frac{x}{n_1 n_2}\right) \ll \frac{MN}{z} + \left(\frac{zx^3}{MN^3}\right)^{1/4} (\log x)^{9/2} + \frac{z^2 x}{M^2 N}$$

$$+ M \log^2 x + z \sqrt{\frac{x}{MN}} \log x .$$

The first two terms are of the same order if we put

$$z = (M^5 N^7 x^{-3})^{1/5} (\log x)^{-18/5} .$$

Then

$$\sum_{(n_1, n_2) \in D} \psi\left(\frac{x}{n_1 n_2}\right) \ll (x^3 N^{-2})^{1/5} (\log x)^{18/5} + (x^{-1} N^9)^{1/5}$$

$$+ M \log^2 x + (M^5 N^9 x^{-1})^{1/10} .$$

Because of $M^2 N \ll x$ and $N^3 \ll x$ (6.22) now follows.

Theorem 6.7.

$$\Delta_3(x) \ll x^{15/31} \log^4 x . \tag{6.24}$$

Proof. Theorem 6.1 shows that

$$\Delta_3(x) = -6 \sum \psi\left(\frac{x}{n_1 n_2}\right) + O(x^{1/3}),$$

$$SC(\Sigma):\ n_1 n_1^2 \leqq x,\ n_1 \leqq n_2 .$$

Let $z > 1$. Then we split up the sum into two parts

$$\Delta_3(x) = -6(\Sigma_1 + \Sigma_2)\ \psi\left(\frac{x}{n_1 n_2}\right) + O(x^{1/3}) ,$$

$$SC(\Sigma_1):\ n_1 n_2^2 \leqq x,\ n_1 \leqq n_2,\ n_1 > z ,$$

$$SC(\Sigma_2):\ n_1 n_2^2 \leqq x,\ n_1 \leqq n_2,\ n_1 \leqq z .$$

We estimate the first sum by means of Lemma 6.6. We obtain from (6.22), since $M^2 N \ll x$,

$$\Sigma_1\ \psi\left(\frac{x}{n_1 n_2}\right) \ll \max \{(x^3 N^{-2})^{1/5} + (xN^{-1})^{1/2}\} \log^6 x + x^{2/5} \log^2 x$$

$$\ll \{(x^3 z^{-2})^{1/5} + (xz^{-1})^{1/2}\} \log^6 x + x^{2/5} \log^2 x .$$

We use Lemma 6.4 for the estimation of the second sum. Then (6.18) gives

$$\Sigma_2\ \psi\left(\frac{x}{n_1 n_2}\right) \ll \max\ (x^7 M N^3)^{1/17} \log^3 x \ll \max\ (x^3 N)^{5/34} \log^3 x$$

$$\ll (x^3 z)^{5/34} \log^3 x .$$

If we choose z such that, apart from logarithm factors, the leading terms are of the same order, that is $z = x^{9/31} \log^5 x$, then (6.24) follows immediately.

6.1.7. *A historical outline of the development of Piltz's divisor problem for $p = 3$*

The history of O-estimates:

$\vartheta_3 \leqq \frac{2}{3} = 0{,}\overline{6}$ (A. Piltz [1], 1881),

$\vartheta_3 \leqq \frac{1}{2} = 0{,}5$ (E. Landau [1], 1912),

$\vartheta_3 \leqq \frac{43}{87} = 0{,}4942 \ldots$ (A. Walfisz [1], 1926),

$\vartheta_3 \leqq \frac{37}{75} = 0{,}49\overline{3}$ (F. V. Atkinson [1], 1941),

$\vartheta_3 \leqq 0{,}4931 \ldots$ (R. A. Rankin [1], 1955),

$\vartheta_3 \leqq \frac{14}{29} = 0{,}4827 \ldots$ (Yuh Ming-i [2], 1958),

$\vartheta_3 \leqq \frac{25}{52} = 0{,}4807 \ldots$ (Yin Wen-Lin [2], 1959),

$\vartheta_3 \leqq \frac{10}{21} = 0{,}4761 \ldots$ (Yin Wen-Lin [3], 1959),

$\vartheta_3 \leqq \frac{8}{17} = 0{,}4705 \ldots$ (Wu Fang-Yuh Ming-i [2], 1962),

$\vartheta_3 \leqq \frac{5}{11} = 0{,}\overline{45}$ (Chen Jing-run [6], [7], [8], 1964),

$\vartheta_3 \leqq \frac{127}{282} = 0{,}4503 \ldots$ (Li Zhong-Fu—Yin Wen-Lin [1], 1981),

$\vartheta_3 \leqq \frac{43}{96} = 0{,}4479 \ldots$ (G. Kolesnik [4], 1981).

In the other direction we have

$\Delta_3(x) = \Omega_{\pm}(x^{1/3})$ (G. H. Hardy [3], 1916),

$\Delta_3(x) = \Omega_{+}((x \log x)^{1/3} (\log \log x)^2)$ (G. Szegö—A. Walfisz [1], 1927),

$\Delta_3(x) = \Omega_{+}\left((x \log x)^{1/3} (\log \log x)^{\log 3 + 4/3} e^{-A\sqrt{\log \log \log x}}\right)$ $(A > 0)$ (J. L. Hafner [2], 1982).

6.2 MANY-DIMENSIONAL PROBLEMS

In this section we consider the function $D(\boldsymbol{a}; x)$, where $\boldsymbol{a} = (a_1, a_2, \dots, a_p)$, $1 \leqq a_1 \leqq a_2 \leqq \dots \leqq a_p$, $p \geqq 4$. So, as defined in Definition 6.1, $D(\boldsymbol{a}; x)$ counts the number of lattice points under the surface

$$\xi_1^{a_1} \xi_2^{a_2} \cdot \dots \cdot \xi_p^{a_p} = x$$

with $\xi_1 > 0$, $\xi_2 > 0, \dots, \xi_p > 0$. It is possible to obtain results which are quite similar to those of the three-dimensional case. For this purpose we must apply a generalization of Theorem 6.1 obtained by M. Vogts [2]. However, the technical difficulties will be enormously increasing. We do not develop such a theory, since on the other hand the application of deep results of the theory of Riemann zeta-function leads to better estimates in general as already mentioned in the introduction to this chapter. We can state as a rule of thumb the following fact: If the differences between the exponents a_k are "small", then apply the theory of Riemann zeta-function. But if the differences between the exponents a_k are "large", then use the theory of exponential sums. Therefore we prove only a few simple, but useful theorems in the following section.

6.2.1. *O-estimates*

We write

$$D(\boldsymbol{a}; x) = H(\boldsymbol{a}; x) + \Delta(\boldsymbol{a}; x) \tag{6.25}$$

such that $H(\boldsymbol{a}; x)$ denotes the main term and $\Delta(\boldsymbol{a}; x)$ the remainder. The associated Dirichlet series $\prod_{\mu=1}^{p} \zeta(a_\mu s)$ where $\zeta(s)$ again denotes Riemann's zeta-function has poles at the points $s = 1/a_\mu$. Therefore, if all poles are simple, the main term must be of the form

$$H(\boldsymbol{a}; x) = \sum_{\nu=1}^{p} \operatorname*{Res}_{s=1/a_\nu} \prod_{\mu=1}^{p} \zeta(a_\mu s) \frac{x^s}{s}$$

$$= \sum_{\nu=1}^{p} \prod_{\substack{\mu=1 \\ \mu \neq \nu}}^{p} \zeta\left(\frac{a_\mu}{a_\nu}\right) x^{1/a_\nu}. \tag{6.26}$$

In case of some equalities of the exponents there are multiple poles. Then we get a representation of $H(\boldsymbol{a}; x)$ by taking the corresponding limiting values in (6.26). In case of Piltz's divisor problem $a_1 = a_2 = \dots = a_p = 1$ we write

$$D_p(x) = H_p(x) + \Delta_p(x), \tag{6.27}$$

where the main term is given by $H_p(x) = xP_p(\log x)$, and $P_p(y)$ denotes a well-defined polynomial of degree $p - 1$.

In the following theorems we assume for the sake of simplicity that the numbers $a_1, a_2, \dots, a_p$ are positive integers such that

$$D(\boldsymbol{a}; x) = \sum_{n \leqq x} d(\boldsymbol{a}; n) .$$

But it is easy to extend the theorems to non-integers x. First of all we need a simple lemma.

Lemma 6.7. *Let* $p \geqq 1$ *and* $\Delta(\boldsymbol{a}; x)$ *be defined by* (6.25). *If* $\Delta(\boldsymbol{a}; x) \ll x^{\alpha_p}$, $\alpha_1 = 0$, $\alpha_p < \dfrac{1}{a_p}$, *then*

$$\sum_{n \leqq x} d(\boldsymbol{a}; n)\, n^{-s} = \sum_{\nu=1}^{p} \frac{1}{1 - a_\nu s} \prod_{\substack{\mu=1 \\ \mu \neq \nu}}^{p} \zeta\left(\frac{a_\mu}{a_\nu}\right) x^{1/a_\nu - s} + \prod_{\nu=1}^{p} \zeta(a_\nu s) + O(x^{\alpha_p - s}) \quad (6.28)$$

holds for $s > \alpha_p$. *For* $s = 1/a_\nu$ *take the limiting values.*

Proof. We obtain by partial summation and (6.26)

$$\sum_{n \leqq x} d(\boldsymbol{a}; n)\, n^{-s} = D(\boldsymbol{a}; x)\, x^{-s} + s \int_1^x D(\boldsymbol{a}; t)\, t^{-s-1}\, \mathrm{d}t$$

$$= \sum_{\nu=1}^{p} \frac{1}{1 - a_\nu s} \prod_{\substack{\mu=1 \\ \mu \neq \nu}}^{p} \zeta\left(\frac{a_\mu}{a_\nu}\right) x^{1/a_\nu - s}$$

$$- \sum_{\nu=1}^{p} \frac{a_\nu s}{1 - a_\nu s} \prod_{\substack{\mu=1 \\ \mu \neq \nu}}^{p} \zeta\left(\frac{a_\mu}{a_\nu}\right) + s \int_1^\infty \Delta(\boldsymbol{a}; t)\, t^{-s-1}\, \mathrm{d}t + O(x^{\alpha_p - s}) . \quad (6.29)$$

Assuming that $s > 1$, we have, as $x \to \infty$,

$$\prod_{\nu=1}^{p} \zeta(a_\nu s) = - \sum_{\nu=1}^{p} \frac{a_\nu s}{1 - a_\nu s} \prod_{\substack{\mu=1 \\ \mu \neq \nu}}^{p} \zeta\left(\frac{a_\mu}{a_\nu}\right) + s \int_1^\infty \Delta(\boldsymbol{a}; t)\, t^{-s-1}\, \mathrm{d}t .$$

By analytic continuation this equation also holds for $\mathrm{Re}(s) > \alpha_p$. Substituting this into (6.29), the asymptotic representation (6.28) follows.

Theorem 6.8. *Let* $1 \leqq q \leqq p - 1$. *Suppose that*

$$\Delta(a_1, \dots, a_q; x) \ll x^{\alpha_q} , \quad \alpha_1 = 0 , \quad \alpha_q < \frac{1}{a_p} ,$$

$$\Delta(a_{q+1}, \dots, a_p; x) \ll x^{\beta_q} , \quad \beta_{p-1} = 0 , \quad \beta_q < \frac{1}{a_p} .$$

Then

$$\Delta(\boldsymbol{a}; x) \ll x^{\alpha_p}$$

holds, where

$$\alpha_p = \frac{1 - a_1 a_{q+1} \alpha_q \beta_q}{a_1 + a_{q+1} - a_1 a_{q+1}(\alpha_q + \beta_q)}. \tag{6.30}$$

Proof. Let $1 < y < x$. Then

$$D(\boldsymbol{a}; x) = \sum_{mn \leq x} d(a_1, \dots, a_q; m)\, d(a_{q+1}, \dots, a_p; n) = T_1 + T_2 + T_3\,,$$

where T_1, T_2, T_3 are given by

$$T_1 = \sum_{m \leq y} d(a_1, \dots, a_q; m) \sum_{n \leq x/m} d(a_{q+1}, \dots, a_p; n)\,,$$

$$T_2 = \sum_{n < x/y} d(a_{q+1}, \dots, a_p; n) \sum_{m \leq x/n} d(a_1, \dots, a_q; m)\,,$$

$$T_3 = -\sum_{m \leq y} d(a_1, \dots, a_q; m) \sum_{n < x/y} d(a_{q+1}, \dots, a_p; n)\,.$$

We assume for the sake of simplicity that $a_1 < a_2 < \dots < a_p$, and apply Lemma 6.7. Then

$$T_1 = \sum_{m \leq y} d(a_1, \dots, a_q; m)$$
$$\times \left\{ \sum_{\nu = q+1}^{p} \prod_{\substack{\varrho = q+1 \\ \varrho \neq \nu}}^{p} \zeta\left(\frac{a_\varrho}{a_\nu}\right) \left(\frac{x}{m}\right)^{1/a_\nu} + O\left(\left(\frac{x}{m}\right)^{\beta_q}\right) \right\}$$
$$= \sum_{\nu = q+1}^{p} \left\{ \sum_{\mu = 1}^{q} \frac{a_\nu}{a_\nu - a_\mu} \prod_{\substack{\varrho = q+1 \\ \varrho \neq \nu}}^{p} \zeta\left(\frac{a_\varrho}{a_\nu}\right) \prod_{\substack{r = 1 \\ r \neq \mu}}^{q} \zeta\left(\frac{a_r}{a_\mu}\right) \left(\frac{x}{y}\right)^{1/a_\nu} y^{1/a_\mu} \right.$$
$$\left. + \prod_{\substack{\varrho = 1 \\ \varrho \neq \nu}}^{p} \zeta\left(\frac{a_\varrho}{a_\nu}\right) x^{1/a_\nu} \right\} + O\left(\left(\frac{x}{y}\right)^{1/a_{q+1}} y^{\alpha_q}\right) + O\left(\left(\frac{x}{y}\right)^{\beta_q} y^{1/a_1}\right).$$

Analogously

$$T_2 = \sum_{\nu = 1}^{q} \left\{ \sum_{\mu = q+1}^{p} \frac{a_\nu}{a_\nu - a_\mu} \prod_{\substack{\varrho = 1 \\ \varrho \neq \nu}}^{q} \zeta\left(\frac{a_\varrho}{a_\nu}\right) \prod_{\substack{r = q+1 \\ r \neq \mu}}^{p} \zeta\left(\frac{a_r}{a_\mu}\right) \left(\frac{x}{y}\right)^{1/a_\mu} y^{1/a_\nu} \right.$$
$$\left. + \prod_{\substack{\varrho = 1 \\ \varrho \neq \nu}}^{p} \zeta\left(\frac{a_\varrho}{a_\nu}\right) x^{1/a_\nu} \right\} + O\left(\left(\frac{x}{y}\right)^{1/a_{q+1}} y^{\alpha_q}\right) + O\left(\left(\frac{x}{y}\right)^{\beta_q} y^{1/a_1}\right),$$

$$T_3 = -\left\{ \sum_{\mu=1}^{q} \prod_{\substack{r=1 \\ r \neq \mu}}^{q} \zeta\left(\frac{a_r}{a_\mu}\right) y^{1/a_\mu} + O(y^{\alpha_q}) \right\}$$

$$\times \left\{ \sum_{\nu=q+1}^{p} \prod_{\substack{\varrho=q+1 \\ \varrho \neq \nu}}^{p} \zeta\left(\frac{a_\varrho}{a_\nu}\right) \left(\frac{x}{y}\right)^{1/a_\nu} + O\left(\left(\frac{x}{y}\right)^{\beta_q}\right) \right\}$$

$$= -\sum_{\mu=1}^{q} \prod_{\nu=q+1}^{p} \prod_{\substack{r=1 \\ r \neq \mu}}^{q} \zeta\left(\frac{a_r}{a_\mu}\right) \prod_{\substack{\varrho=q+1 \\ \varrho \neq \nu}}^{p} \zeta\left(\frac{a_\varrho}{a_\nu}\right) \left(\frac{x}{y}\right)^{1/a_\nu} y^{1/a_\mu}$$

$$+ O\left(\left(\frac{x}{y}\right)^{1/a_{q+1}} y^{\alpha_q}\right) + O\left(\left(\frac{x}{y}\right)^{\beta_q} y^{1/a_1}\right).$$

We now obtain equation (6.25) from these representations of T_1, T_2, T_3, where the remainder is estimated by

$$\Delta(\boldsymbol{a}; x) \ll \left(\frac{x}{y}\right)^{1/a_{q+1}} y^{\alpha_q} + \left(\frac{x}{y}\right)^{\beta_q} y^{1/a_1}$$

Both the terms are of the same order if

$$y = x^\gamma; \quad \gamma = \frac{a_1(1 - a_{q+1}\beta_q)}{a_1 + a_{q+1} - a_1 a_{q+1}(\alpha_q + \beta_q)}.$$

This proves (6.30). If some of the coefficients are equal, a similar proof holds.

Let A_p be defined by $A_p = a_1 + a_2 + \ldots + a_p$. Then it is easy to prove the trivial estimate

$$\alpha_p \leqq \frac{A_{p-1}}{A_p} \cdot \frac{1}{a_1}. \tag{6.31}$$

Clearly, by (5.4), (6.31) is correct for $p = 2$, Applying the induction argument and using Theorem 6.8 with $q = p - 1$, we then obtain from (6.30)

$$\alpha_p = \frac{1}{a_1 + a_p - a_1 a_p \alpha_{p-1}} \leqq \frac{A_{p-1}}{(a_1 + a_p) A_{p-1} - a_p A_{p-2}}$$

$$= \frac{A_{p-1}}{a_1 A_{p-1} + a_p a_{p-1}} \leqq \frac{A_{p-1}}{A_p} \cdot \frac{1}{a_1}.$$

Thus (6.31) is proved by induction.

Note that

$$\alpha_p \leqq \left(1 - \frac{1}{p}\right) \frac{1}{a_1}.$$

We now proceed in proving a non-trivial estimate.

Theorem 6.9. *The estimate*

$$\Delta(\boldsymbol{a}; x) \ll x^{\gamma_p+\varepsilon}, \qquad \gamma_p = \frac{2(p-1)}{2p+1} \cdot \frac{1}{a_1} \tag{6.32}$$

holds for $p \geqq 1$ *and every* $\varepsilon > 0$.

Proof. Clearly, we have $\gamma_1 = 0$ and, by (5.6),

$$\Delta(a_1, a_2; x) \ll x^{\frac{1}{2a_1+a_2}} \log x$$

such that we can put $\gamma_2 = 2/5a_1$. In Section 6.1.3 we proved that

$$\Delta(a_1, a_2, a_3; x) \ll x^{\frac{2}{2a_1+a_2+a_3}} \log^2 x\,. \tag{6.33}$$

Hence, $\gamma_3 = 4/7a_1$ is a suitable value. We now apply Theorem 6.8 and prove that $\alpha_p \leqq \gamma_p + \varepsilon$. As we have seen, this is correct for $p = 1, 2, 3$. Now we use the induction argument and apply (6.30) with $q = p - 3$. Then

$$\alpha_p = \frac{1 - a_1 a_{p-2} \alpha_{p-3} \beta_{p-3}}{a_1 + a_{p-2} - a_1 a_{p-2}(\alpha_{p-3} + \beta_{p-3})}\,.$$

(6.33) shows that

$$\Delta(a_1, a_2, a_3; x) \ll x^{1/2a_1+\varepsilon}\,.$$

Therefore, we put $\beta_{p-3} = 1/2a_{p-2} + \varepsilon_1$. Then

$$\alpha_p = \frac{2 - a_1 \alpha_{p-3}}{a_1 + 2a_{p-2}(1 - a_1 \alpha_{p-3})} + \varepsilon_2$$

$$\leqq \frac{2 - \dfrac{2(p-4)}{2p-5}}{a_1 + 2a_{p-2}\left(1 - \dfrac{2(p-4)}{2p-5}\right)} + \varepsilon_3 \leqq \frac{2(p-1)}{2p+1} \cdot \frac{1}{a_1}\,,$$

which proves the theorem.

Theorem 6.9 enables us to prove better estimates for the error term $\Delta(\boldsymbol{a}; x)$ by applying an interation process. This is based on the following theorem.

Theorem 6.10. *The estimate*

$$\Delta(\boldsymbol{a}; x) \ll x^{\gamma_p+\varepsilon}, \quad \gamma_p = \frac{2p-1}{(2q+1)\,a_1 + (2(p-q)+1)\,a_{q+1}}\,, \tag{6.34}$$

holds for $q \geqq 1$, $p - q \geqq 1$ *and every* $\varepsilon > 0$.

Proof. It is seen from (6.32) that we can use (6.30) with

$$\alpha_q = \frac{2(q-1)}{2q+1} \cdot \frac{1}{a_1} + \varepsilon_1 , \quad \beta_q = \frac{2(p-q-1)}{2(p-q)+1} \cdot \frac{1}{a_{q+1}} + \varepsilon_2 .$$

A simple calculation gives $\alpha_p = \gamma_p + \varepsilon$, where γ_p is the value of (6.34).

Finally, we consider the case $p = 4$. Here we will give a refinement of Theorem 6.8 in case $q = p - 1 = 3$. We need some lemmas.

Lemma 6.8. *Let* $1 \leqq y \leqq x$. *Suppose that*

$$\Delta(a_1, a_2, a_3; x) \ll x^{\alpha_3} , \quad \alpha_3 < \frac{1}{a_4} .$$

Then

$$\Delta(a_1,\ldots, a_4; x) = - \sum_{n \leqq x/y} d(a_1, a_2, a_3; n)\, \psi\left(\left(\frac{x}{n}\right)^{1/a_4}\right)$$
$$+ O\left(y^{1/a_4}\left(\frac{x}{y}\right)^{\alpha_3}\right) + O\left(y^{-1/a_4}\left(\frac{x}{y}\right)^{1/a_1}\right). \tag{6.35}$$

Proof.

$$D(a_1,\ldots, a_4; x) = \sum_{nm^{a_4} \leqq x} d(a_1, a_2, a_3; n)$$
$$= \sum_{\substack{nm^{a_4} \leqq x \\ m^{a_4} \leqq y}} d(a_1, a_2, a_3; n) + \sum_{\substack{nm^{a_4} \leqq x \\ y < m^{a_4} \leqq x}} d(a_1, a_2, a_3; n)$$
$$= \sum_{m^{a_4} \leqq y} D\left(a_1, a_2, a_3; \frac{x}{m^{a_4}}\right)$$
$$+ \sum_{n \leqq x/y} d(a_1, a_2, a_3; n) \left\{\left[\left(\frac{x}{n}\right)^{1/a_4}\right] - [y^{1/a_4}]\right\}. \tag{6.36}$$

Applying the asymptotic representation

$$\sum_{n \leqq y} n^{-z} = \zeta(z) + \frac{y^{1-z}}{1-z} - y^{-z}\psi(y) + O(y^{-z-1})$$

$(0 < z \neq 1)$, we obtain

$$\sum_{m^{a_4} \leqq y} D\left(a_1, a_2, a_3; \frac{x}{m^{a_4}}\right)$$
$$= \sum_{m^{a_4} \leqq y} \left\{\sum_{\nu=1}^{3} \prod_{\substack{\varrho=1 \\ \varrho \neq \nu}}^{3} \zeta\left(\frac{a_\varrho}{a_\nu}\right)\left(\frac{x}{m^{a_4}}\right)^{1/a_\nu} + O\left(\left(\frac{x}{m^{a_4}}\right)^{\alpha_3}\right)\right\}$$

$$= \sum_{\nu=1}^{3} \left\{ \prod_{\substack{\varrho=1 \\ \varrho \neq \nu}}^{4} \zeta\left(\frac{a_\varrho}{a_\nu}\right) x^{1/a_\nu} + \frac{a_\nu}{a_\nu - a_4} \prod_{\substack{\varrho=1 \\ \varrho \neq \nu}}^{3} \zeta\left(\frac{a_\varrho}{a_\nu}\right) y^{1/a_4} \left(\frac{x}{y}\right)^{1/a_\nu} \right.$$

$$\left. - \prod_{\substack{\varrho=1 \\ \varrho \neq \nu}}^{3} \zeta\left(\frac{a_\varrho}{a_\nu}\right) \left(\frac{x}{y}\right)^{1/a_\nu} \psi(y^{1/a_4}) \right\}$$

$$+ O\left(y^{1/a_4} \left(\frac{x}{y}\right)^{\alpha_3}\right) + O\left(y^{-1/a_4} \left(\frac{x}{y}\right)^{1/a_4}\right). \tag{6.37}$$

Clearly, we must assume that $a_1 < a_2 < a_3 < a_4$. But, if we have some equalities between the numbers a_1, a_2, a_3, a_4, a similar proof holds. Now we use Lemma 6.7 for the second sum in (6.36). Then

$$\sum_{n \leqq x/y} d(a_1, a_2, a_3; n) \left\{ \left[\left(\frac{x}{y}\right)^{1/a_4}\right] - [y^{1/a_4}] \right\}$$

$$= \sum_{n \leqq x/y} d(a_1, a_2, a_3; n) \left\{ \left(\frac{x}{n}\right)^{1/a_4} - \psi\left(\left(\frac{x}{n}\right)^{1/a_4}\right) \right\}$$

$$- \{y^{1/a_4} - \psi(y^{1/a_4})\}\ D\left(a_1, a_2, a_3; \frac{x}{y}\right)$$

$$= \sum_{\nu=1}^{3} \frac{a_\nu}{a_4 - a_\nu} \prod_{\substack{\varrho=1 \\ \varrho \neq \nu}}^{3} \zeta\left(\frac{a_\varrho}{a_\nu}\right) y^{1/a_4} \left(\frac{x}{y}\right)^{1/a_\nu}$$

$$+ \prod_{\substack{\varrho=1 \\ \varrho \neq \nu}}^{3} \zeta\left(\frac{a_\varrho}{a_\nu}\right) \left(\frac{x}{y}\right)^{1/a_\nu} \psi(y^{1/a_4}) + \prod_{\varrho=1}^{3} \zeta\left(\frac{a_\varrho}{a_4}\right) x^{1/a_4}$$

$$- \sum_{n \leqq x/y} d(a_1, a_2, a_3; n)\ \psi\left(\left(\frac{x}{n}\right)^{1/a_4}\right) + O\left(y^{1/a_4} \left(\frac{x}{y}\right)^{\alpha_3}\right). \tag{6.38}$$

Substituting (6.37) and (6.38) into (6.36), we obtain (6.25), where the remainder $\Delta(\boldsymbol{a}; x) = \Delta(a_1, \ldots, a_4; x)$ is represented by (6.35). Thus the lemma is proved.

We shall estimate the remaining sum of (6.35) by van der Corput's method of exponent pairs. The application of this method for a sum $\Sigma\, \psi(f(n))$ requires the property $f'(t) > 1$ which is not always satisfied. Therefore, we first apply Theorem 1.5 and then Theorem 2.14. So we prove the following lemma.

Lemma 6.9. *Let* $1 \leqq y < z \leqq x^{1/b}$, $a > 0$, $b > 0$. *If* $(k, l) = A(\varkappa, \lambda)$ *is any exponent pair and if*

$$(2l - 1)\, a > 2kb\,, \quad (2l - 1)\, b > 2ka\,, \tag{6.39}$$

then

$$\sum_{y<n\leqq z} \psi\left(\left(\frac{x}{n^b}\right)^{1/a}\right) \ll x^{\frac{2(k+l-1/2)}{a+b}} \log x + z\left(\frac{z^b}{x}\right)^{1/a} \log x \,. \tag{6.40}$$

Proof. First suppose that $z^{a+b} \leqq x$. We then divide the interval $(y, z]$ into $O(\log x)$ subintervals of type $(N, N']$ with $N < N' \leqq 2N$. Then we obtain from Theorem 2.14

$$\sum_{N<n\leqq N'} \psi\left(\left(\frac{x}{n^b}\right)^{1/a}\right) \ll N^{2(k+l-1/2)}\left(\frac{x}{N^{a+b}}\right)^{2k/a} = N^{2l-1}\left(\frac{x}{N^b}\right)^{2k/a}$$

$$\ll z^{2l-1}\left(\frac{x}{z^b}\right)^{2k/a} \ll x^{\frac{2(k+l-1/2)}{a+b}} .$$

The last inequality follows from the first condition of (6.39). This proves (6.40) in this case.

Now let $z^{a+b} > x$. If $y^{a+b} < x$, (6.40) holds for the interval $(y, x^{1/(a+b)}]$. Therefore it remains to consider an interval $x^{1/(a+b)} \leqq y < n \leqq z$. We again divide this interval into $O(\log x)$ subintervals of type $(N, N']$ with $N < N' \leqq 2N$, where now $N \geqq x^{1/(a+b)}$. Here we apply Theorem 1.5 with

$$f(t) = \left(\frac{x}{t^b}\right)^{1/a}, \qquad f^{-1}(t) = \left(\frac{x}{t^a}\right)^{1/b} .$$

Let $M = (x/N^b)^{1/a}$ and $M' = (x/N'^b)^{1/a}$. Then we get from (1.9)

$$\sum_{N<n\leqq N'} \psi\left(\left(\frac{x}{n^b}\right)^{1/a}\right) = \sum_{M'\leqq m<M} \psi\left(\left(\frac{x}{m^a}\right)^{1/b}\right) - \frac{b}{a}\int_N^{N'} \left(\frac{x}{t^{a+b}}\right)^{1/a} \psi(t)\,\mathrm{d}t$$

$$+ \frac{a}{b}\int_{M'}^{M} \left(\frac{x}{t^{a+b}}\right)^{1/b} \psi(t)\,\mathrm{d}t + O(1) .$$

Since $M^{a+b} \leqq x$, the sum on the right-hand side can be estimated by applying Theorem 2.14 as before. Using partial integration, we see that the first integral is of order 1 and the second of order $(xM^{-a-b})^{1/b}$. Hence

$$\sum_{N<n\leqq N'} \psi\left(\left(\frac{x}{n^b}\right)^{1/a}\right) \ll M^{2(k+l-1/2)}\left(\frac{x}{M^{a+b}}\right)^{2k/b} + \left(\frac{x}{M^{a+b}}\right)^{1/b}$$

$$\ll \left(\frac{x}{N^b}\right)^{(2l-1)/a} N^{2k} + N\left(\frac{N^b}{x}\right)^{1/a}$$

$$\ll x^{\frac{2(k+l-1/2)}{a+b}} + z\left(\frac{z^b}{x}\right)^{1/a} .$$

Here the last inequality follows from the second condition of (6.39). Now (6.40) also is proved in this case.

Theorem 6.11. *Suppose that*

$$\Delta(a_1, a_2, a_3; x) \ll x^{\alpha_3}, \quad \alpha_3 < \frac{1}{a_4}.$$

Let $(k, l) = A(\varkappa, \lambda)$ be any exponent pair with

$$(2l - 1)\, a_1 > 2ka_4, \tag{6.41}$$

$$a_3(2(k + l)\, a_2 + a_4) < (a_1 + a_2)\,(a_2 + a_4). \tag{6.42}$$

Let y be defined by

$$\left(\frac{x}{y}\right)^{\frac{2}{a_1+a_2+a_3}} (x^{a_1} y^{a_2+a_3})^{\frac{2(k+l-1/2)}{(a_1+a_4)(a_1+a_2+a_3)}} = y^{\frac{1}{a_4}} \left(\frac{x}{y}\right)^{\alpha_3} \tag{6.43}$$

and assume that

$$y^{a_1+a_2+a_3+a_4} \geqq x^{a_4}. \tag{6.44}$$

Then

$$\Delta(a_1, \dots, a_4; x) \ll y^{1/a_4} \left(\frac{x}{y}\right)^{\alpha_3} \log x + y^{-1/a_4} \left(\frac{x}{y}\right)^{1/a_1} \log^3 x \tag{6.45}$$

In particular, if additionally

$$2\left(k + l + \frac{1}{2}\right) a_1 \geqq a_2 + a_3, \qquad a_1 + a_2 \geqq 2\left(k + l - \frac{1}{2}\right) a_3,$$

then

$$\Delta(a_1, \dots, a_4; x) \ll x^{\frac{2(k+l+1/2)}{a_1+a_2+a_3+a_4} + \varepsilon} \tag{6.46}$$

for every $\varepsilon > 0$.

Proof. Applying Lemma 6.8, we write for the sum of the right-hand side of (6.35)

$$S = \sum_{n \leqq x/y} d(a_1, a_2, a_3; n)\, \psi\left(\left(\frac{x}{n}\right)^{1/a_4}\right) = \sum_{\pi} \sum_{\substack{n_1^u n_2^v n_3^w \leqq x/y \\ n_1 \leqq n_2 \leqq n_3}} \psi\left(\left(\frac{x}{n_1^u n_2^v n_3^w}\right)^{1/a_4}\right).$$

The first sum is taken over all permutations $\pi = \pi(u, v, w)$ of the numbers a_1, a_2, a_3. In the inner sum we may take the equality signs as we like it, only such that the equation remains correct altogether. We now apply Lemma 6.9 to the sum over n_3. There we put $a = a_4$, $b = w$, $y = n_2 - 1$ or n_2, $z^w = x/y n_1^u n_2^v$. Because of (6.41) condition (6.39) is satisfied. Thus, we obtain from (6.40)

$$S \ll \sum_{\pi} \sum_{\substack{n_1^u n_2^{v+w} \leqq x/y \\ n_1 \leqq n_2}} \left\{ \left(\frac{x}{n_1^u n_2^v}\right)^{\frac{2(k+l-1/2)}{w+a_4}} + \left(\frac{x}{n_1^u n_1^v y}\right)^{\frac{1}{w}} y^{-\frac{1}{a_4}} \right\} \log x.$$

Since $2(k+l-1/2) \leqq 1$, we have $2(k+l-1/2)\, v < w + a_4$. Hence

$$S \ll \sum_{\pi} \sum_{n^{u+v+w} \leqq x/y} \left(\frac{x}{n^u y}\right)^{\frac{1}{v+w}\left(1+\frac{2(k+l-1/2)w}{a+a_4}\right)} y^{\frac{2(k+l-1/2)}{w+a_4}} \log x$$

$$+ \left(\frac{x}{y}\right)^{1/a_1} y^{-1/a_4} \log^3 x \,.$$

It is easily seen that

$$\frac{u}{v+w}\left(1+\frac{2(k+l-1/2)\, w}{w+a_4}\right) \leqq \frac{a_3}{a_1+a_2}\left(1+\frac{2(k+l-1/2)\, a_2}{a_2+a_4}\right) < 1\,,$$

where the last inequality follows from condition (6.42). Thus, since $u + v + w = a_1 + a_2 + a_3$,

$$S \ll \left(\frac{x}{y}\right)^{\frac{1}{a_1+a_2+a_3}\left(2+\frac{2(k+l-1/2)w}{w+a_4}\right)} y^{\frac{2(k+l-1/2)}{w+a_4}} \log x + \left(\frac{x}{y}\right)^{1/a_1} y^{-1/a_4} \log^3 x$$

$$= \left(\frac{x}{y}\right)^{\frac{2}{a_1+a_2+a_3}} (x^w y^{u+v})^{\frac{2(k+l-1/2)}{(w+a_4)(a_1+a_2+a_3)}} \log x + \left(\frac{x}{y}\right)^{1/a_1} y^{-1/a_4} \log^3 x \,.$$

Condition (6.44) ensures that

$$(x^w y^{u+v})^{\frac{1}{w+a_4}} \leqq (x^{a_1} y^{a_2+a_3})^{\frac{1}{a_1+a_4}} \,.$$

Using condition (6.43), we obtain

$$S \ll y^{1/a_4} \left(\frac{x}{y}\right)^{\alpha_3} \log x + y^{-1/a_4} \left(\frac{x}{y}\right)^{1/a_1} \log^3 x \,.$$

Result (6.45) now follows from this estimation and (6.35).

In order to obtain (6.46), we apply Theorem 6.2, which requires the conditions

$$la_1 \geqq k(a_2 + a_3)\,, \qquad a_1 + a_2 \geqq 2\left(k + l - \frac{1}{2}\right) a_3 \,.$$

The first condition is always satisfied, because (6.41) shows that

$$la_1 \geqq k(a_2 + a_3)\,, \quad 2ka_4 \geqq k(a_2 + a_3)\,.$$

We see from (6.12) that we can put

$$\alpha_3 = \frac{2(k+l)}{a_1 + a_2 + a_3} + \varepsilon \,.$$

If we put $y^{a_1+a_2+a_3+a_4} = x^{a_4}$, we obtain (6.46) from (6.45), where the condition $2(k+l+1/2)\, a_1 \geqq a_2 + a_3$ ensures that the second term of (6.45) is less than the first. Now the proof of the theorem is complete.

6.2.2. *The Ω-estimates*

We present the method of G. H. Hardy which in principle does not differ from the corresponding one used in the circle problem. It is based on analytical properties of the function

$$F(\boldsymbol{a}; s) = \sum_{n_1=1}^{\infty} \dots \sum_{n_p=1}^{\infty} \mathrm{e}^{-s(n_1^{a_1} \cdot \dots \cdot n_p^{a_p})^{1/A_p}}, \tag{6.47}$$

where $p \geqq 2$, $A_p = a_1 + a_2 + \dots + a_p$. Obviously, this function is holomorphic for $\mathrm{Re}(s) > 0$. In a first lemma we study this function in a neighbourhood of the origin. Then we show that it is possible to give an analytic continuation over the whole plane with the exception of an infinity of isolated singular points. The Ω-theorem is then an immediate consequence of these properties.

Lemma 6.10. *Let $\zeta(s)$ denote the Riemann zeta-function. If*

$$0 < |s| < q = 2\pi A_p \left(\prod_{\nu=1}^{p} a_\nu^{a_\nu} \right)^{-1/A_p},$$

the function (6.47) *is represented by*

$$F(\boldsymbol{a}; s) = K(\boldsymbol{a}; s) + P(\boldsymbol{a}; s), \tag{6.48}$$

$$K(\boldsymbol{a}; s) = \sum_{\mu=1}^{p} \Gamma\left(1 + \frac{A_p}{a_\mu}\right) \prod_{\substack{\nu=1 \\ \nu \neq \mu}}^{p} \zeta\left(\frac{a_\nu}{a_\mu}\right) s^{-A_p/a_\mu}, \tag{6.49}$$

$$P(\boldsymbol{a}; s) = \sum_{n=0}^{\infty} \frac{(-s)^n}{n!} \prod_{\nu=1}^{p} \zeta\left(\frac{-a_\nu n}{A_p}\right). \tag{6.50}$$

The representation for $K(\boldsymbol{a}; s)$ holds for $1 \leqq a_1 < a_2 < \dots < a_p$. In the cases of equalities take the limit values. The power series $P(\boldsymbol{a}; s)$ is convergent for $|s| < q$ and represents a branch of an analytic function.

Proof. Assume that $s > 0$. From the well-known formula

$$\mathrm{e}^{-s} = \frac{1}{2\pi \mathrm{i}} \int_{c-\mathrm{i}\infty}^{c+\mathrm{i}\infty} s^{-z} \Gamma(z) \, \mathrm{d}z \qquad (c > 0)$$

we deduce

$$F(\boldsymbol{a}; s) = \sum_{n_1=1}^{\infty} \dots \sum_{n_p=1}^{\infty} \frac{1}{2\pi \mathrm{i}} \int_{c-\mathrm{i}\infty}^{c+\mathrm{i}\infty} s^{-z} (n_1^{a_1} \cdot \dots \cdot n_p^{a_p})^{-z/A_p} \Gamma(z) \, \mathrm{d}z$$

$$= \frac{1}{2\pi \mathrm{i}} \int_{c-\mathrm{i}\infty}^{c+\mathrm{i}\infty} s^{-z} \Gamma(z) \prod_{\nu=1}^{p} \zeta\left(\frac{a_\nu z}{A_p}\right) \mathrm{d}z,$$

where now $c > A_p/a_1$. The integrand has simple poles at the points $z = 0, -1, -2, \ldots$ and, assuming that $1 \leqq a_1 < a_2 < \ldots < a_p$, at $z = A_p/a_\nu$. Then, by Cauchy's theorem,

$$F(\boldsymbol{a}; s) = K(\boldsymbol{a}; s) + \sum_{n=0}^{N} \frac{(-s)^n}{n!} \prod_{\nu=1}^{p} \zeta\left(\frac{-a_\nu n}{A_p}\right) + R_N,$$

where R_N is given by

$$R_N = \frac{1}{2\pi\,\mathrm{i}} \int_{-N-1/2-\mathrm{i}\infty}^{-N-1/2+\mathrm{i}\infty} s^{-z}\Gamma(z) \prod_{\nu=1}^{p} \zeta\left(\frac{a_\nu z}{A_p}\right) \mathrm{d}z\,.$$

Supposing $0 < s < q$, then it will be shown that $R_N \to 0$ as $N \to \infty$. Thereby we apply the functional equations

$$\Gamma(s)\,\Gamma(1-s) = \frac{\pi}{\sin \pi s},$$

$$2(2\pi)^{-s}\,\Gamma(s)\,\zeta(s) \cos\frac{\pi s}{2} = \zeta(1-s)\,. \tag{6.51}$$

If $z = -N - 1/2 - \mathrm{i}y$, we have

$$|\Gamma(z)| = \frac{\pi}{\cosh \pi y |\Gamma(N + 3/2 - \mathrm{i}y)|},$$

$$\left|\zeta\left(\frac{a_\nu z}{A_p}\right)\right| = \left|2(2\pi)^{-1+a_\nu z/A_p} \cos\frac{\pi}{2}\left(1 - \frac{a_\nu z}{A_p}\right) \Gamma\left(1 - \frac{a_\nu z}{A_p}\right) \cdot \zeta\left(1 - \frac{a_\nu z}{A_p}\right)\right|$$

$$\ll (2\pi)^{-a_\nu N/A_p}\, \mathrm{e}^{\pi a_\nu |y|/2A_p}\, \Gamma\left(1 - \frac{a_\nu z}{A_p}\right).$$

Hence

$$s^{-z}\Gamma(z) \prod_{\nu=1}^{p} \zeta\left(\frac{a_\nu z}{A_p}\right)$$

$$\ll \left(\frac{s}{2\pi}\right)^N \mathrm{e}^{-\pi|y|/2} (N + |y|)^{p-1} \left| \frac{\prod_{\nu=1}^{p} \Gamma\left(\frac{a_\nu(N + 1/2 - \mathrm{i}y)}{A_p}\right)}{\Gamma(N + 1/2 - \mathrm{i}y)} \right|.$$

Applying the well-known asymptotic representation

$$\Gamma(x) = \sqrt{2\pi}\, x^{x-1/2}\, \mathrm{e}^{-x} \left(1 + O\left(\frac{1}{x}\right)\right) \tag{6.52}$$

$(x \to \infty,\ |\mathrm{arc}\ x| < \pi)$, we obtain

$$s^{-z}\Gamma(z) \prod_{\nu=1}^{p} \zeta\left(\frac{a_\nu z}{A_p}\right) \ll \left(\frac{s}{2\pi A_p} \prod_{\nu=1}^{p} a_\nu^{a_\nu/A_p}\right)^N \mathrm{e}^{-\pi|y|/2} (N + |y|)^{\frac{p-1}{2}}\,.$$

Consequently

$$|R_N| \ll \left(\frac{S}{2\pi A_p}\prod_{\nu=1}^{p} a_\nu^{a_\nu/A_p}\right)^N \int_{-\infty}^{+\infty} (N+|y|)^{\frac{p-1}{2}} \mathrm{e}^{-\pi|y|/2}\,\mathrm{d}y \to 0\,.$$

The representations (6.48)—(6.50) now follow at once. The analytical property of the power series (6.50) is trivial.

The behaviour of the function $F(\boldsymbol{a}; s)$ in the circle $|s| < q$ is now known. But we have to investigate the singularities on the imaginary axis for the proof of the Ω-theorem. For this purpose we give another representation for $P(\boldsymbol{a}; s)$ by means of Riemann's functional equation (6.51).

Lemma 6.11. *If* $|s| < q$, *then*

$$P(\boldsymbol{a}; s) = (-2)^{-p} + \left(\frac{\mathrm{i}}{2\pi}\right)^p M\left(\boldsymbol{a}; \frac{-\mathrm{i}s}{2\pi}\right) + \left(\frac{-\mathrm{i}}{2\pi}\right)^p M\left(\boldsymbol{a}; \frac{\mathrm{i}s}{2\pi}\right)$$

$$+ \left(\frac{\mathrm{i}}{2\pi}\right)^p \Sigma(-1)^r M\left(\boldsymbol{a}; \mathrm{e}\left(\frac{B_r}{4A_p}\right)\right), \tag{6.53}$$

$SC(\Sigma)$: *Let* B_r *be defined by* $B_r = \pm a_1 \pm a_2 \pm \ldots \pm a_p$, *where there is the minus sign for r times exactly. Then sum over r from* 1 *up to* $p-1$ *and for each fixed r over all combinations of signs.*

If $|z| < q/2\pi$, $M(\boldsymbol{a}; z)$ *is defined by*

$$M(\boldsymbol{a}; z) = \sum_{n_1=1}^{\infty} \cdots \sum_{n_p=1}^{\infty} \frac{L(\boldsymbol{a}; (n_1^{a_1} \cdot \ldots \cdot n_p^{a_p})^{-1/A_p} z)}{n_1 \cdot \ldots \cdot n_p}, \tag{6.54}$$

$$L(\boldsymbol{a}; z) = \sum_{n=2}^{\infty} \prod_{\nu=1}^{p} \Gamma\left(1 + \frac{a_\nu(n-1)}{A_p}\right) \frac{z^{n-1}}{(n-1)!}\,. \tag{6.55}$$

Proof. Since $\zeta(0) = -1/2$, it follows from (6.50) that

$$P(\boldsymbol{a}; s) = (-2)^{-p} + \sum_{n=2}^{\infty} \frac{(-s)^{n-1}}{(n-1)!} \prod_{\nu=1}^{p} \zeta\left(-\frac{a_\nu(n-1)}{A_p}\right).$$

Applying the functional equation (6.51), we get

$$\prod_{\nu=1}^{p} \zeta\left(-\frac{a_\nu(n-1)}{A_p}\right)$$

$$= \frac{2^p}{(2\pi)^{p+n-1}} \prod_{\nu=1}^{p} \cos\frac{\pi}{2}\left(1 + \frac{a_\nu(n-1)}{A_p}\right) \Gamma\left(1 + \frac{a_\nu(n-1)}{A_p}\right) \zeta\left(1 + \frac{a_\nu(n-1)}{A_p}\right).$$

Now

$$2^p \prod_{\nu=1}^{p} \cos \frac{\pi}{2}\left(1 + \frac{a_\nu(n-1)}{A_p}\right)$$

$$= \mathrm{i}^p \prod_{\nu=1}^{p} \left\{ \mathrm{e}\left(\frac{a_\nu(n-1)}{4A_p}\right) - \mathrm{e}\left(-\frac{a_\nu(n-1)}{4A_p}\right)\right\}$$

$$= \mathrm{i}^{p+n-1} + (-\mathrm{i})^{p+n-1} + \mathrm{i}^p \sum (-1)^r \, \mathrm{e}\left(\frac{B_r(n-1)}{4A_p}\right)$$

and

$$\prod_{\nu=1}^{p} \zeta\left(1 + \frac{a_\nu(n-1)}{A_p}\right) = \sum_{n_1=1}^{\infty} \cdots \sum_{n_p=1}^{\infty} \frac{(n_1^{a_1} \cdot \ldots \cdot n_p^{a_p})^{-(n-1)/A_p}}{n_1 \cdot \ldots \cdot n_p}.$$

Substituting these expressions, representation (6.53) follows at once.

Lemma 6.12. *The function $F(\boldsymbol{a}; s)$ is holomorphic at all points of the imaginary axis except the origin and isolated algebraic infinities of order $(p+1)/2$, At the origin the function $F(\boldsymbol{a}; s) - K(\boldsymbol{a}; s)$ is holomorphic.*

Proof. Obviously, the behaviour of $F(\boldsymbol{a}; s)$ at the origin follows from the representations (6.48) to (6.50). Further, the representation (6.53) for $P(\boldsymbol{a}; s)$ shows that we have to consider the function $L(\boldsymbol{a}; z)$ on the real axis. From (6.52) we obtain

$$\frac{1}{(n-1)!} \prod_{\nu=1}^{p} \Gamma\left(1 + \frac{a_\nu(n-1)}{A_p}\right)$$

$$= (2\pi)^{\frac{p-1}{2}} n^{\frac{1}{2}-n} \mathrm{e}^{1-p} \prod_{\nu=1}^{p} \left(1 + \frac{a_\nu(n-1)}{A_p}\right)^{\frac{1}{2}+\frac{a_\nu(n-1)}{A_p}} \left\{1 + O\left(\frac{1}{n}\right)\right\}$$

$$= (2\pi n)^{\frac{p-1}{2}} \mathrm{e}^{1-p} \left(1 - \frac{1}{n}\right)^n$$

$$\times \prod_{\nu=1}^{p} \left(\frac{a_\nu}{A_p}\right)^{\frac{1}{2}+\frac{a_\nu(n-1)}{A_p}} \left(1 + \frac{A_p}{a_\nu(n-1)}\right)^{\frac{a_\nu(n-1)}{A_p}} \left\{1 + O\left(\frac{1}{n}\right)\right\}$$

$$= (2\pi(n-1))^{\frac{p-1}{2}} \prod_{\nu=1}^{p} \left(\frac{a_\nu}{A_p}\right)^{\frac{1}{2}+\frac{a_\nu(n-1)}{A_p}} \left\{1 + O\left(\frac{1}{n}\right)\right\}.$$

Putting $c = \prod_{\nu=1}^{p} \left(\frac{a_\nu}{A_p}\right)^{a_\nu/A_p}$, $d = \prod_{\nu=1}^{p} \left(\frac{a_\nu}{A_p}\right)^{1/2}$ and using the well-known fact that

$$(n-1)^{\frac{p-1}{2}} = \Gamma\left(\frac{p+1}{2}\right)(-1)^{n-1}\binom{-\frac{p+1}{2}}{n-1}\left\{1 + O\left(\frac{1}{n}\right)\right\},$$

then

$$\frac{1}{(n-1)!}\prod_{\nu=1}^{p}\Gamma\left(1+\frac{a_\nu(n-1)}{A_p}\right)$$
$$= (2\pi)^{\frac{p-1}{2}}\Gamma\left(\frac{p+1}{2}\right) d(-c)^{n-1}\binom{-\frac{p+1}{2}}{n-1}\left\{1 + O\left(\frac{1}{n}\right)\right\}.$$

Hence

$$L(\boldsymbol{a}; z) \sim (2\pi)^{\frac{p-1}{2}}\Gamma\left(\frac{p+1}{2}\right) d\sum_{n=2}^{\infty}\binom{-\frac{p+1}{2}}{n-1}(-cz)^{n-1}$$
$$\sim (2\pi)^{\frac{p-1}{2}}\left(\frac{p+1}{2}\right) d(1-cz)^{-\frac{p+1}{2}}$$

for $0 < z \to 1/c$. This shows that $L(\boldsymbol{a}; z)$ has its only singularity at the point $z = 1/c$ which is an algebraic infinity of order $(p+1)/2$. Then it is seen from (6.53) to (6.55) that $F(\boldsymbol{a}; s)$ is holomorphic at all points of the imaginary axis except the origin and the points

$$s = \pm\frac{2\pi i}{c}(n_1^{a_1}\cdot\ldots\cdot n_p^{a_p})^{1/A_p}$$

which are algebraic infinities of order $(p+1)/2$. This proves the lemma.

Theorem 6.12.

$$D(\boldsymbol{a}; x) = H(\boldsymbol{a}; x) + \Omega_\pm\left(x^{\frac{p-1}{2}\frac{1}{A_p}}\right).$$

Proof. Let $\Delta(\boldsymbol{a}; x)$ be defined by

$$D(\boldsymbol{a}; x) = H(\boldsymbol{a}; x) + \Delta(\boldsymbol{a}; x).$$

Suppose that $\Delta(\boldsymbol{a}; t^{A_p}) \leq Kt^{\frac{p-1}{2}}$ for $K > 0$ and for all sufficiently large values of t. Then we put

$$g(t) = \Delta(\boldsymbol{a}; t^{A_p}) - Kt^{\frac{p-1}{2}},$$

$$G(s) = \frac{1}{s}P(\boldsymbol{a}; s) - K_1 s^{-\frac{p+1}{2}}, \qquad K_1 = K\Gamma\left(\frac{p+1}{2}\right),$$

such that

$$G(s) = \int_0^\infty g(t)\, e^{-st}\, dt .$$

Lemma 6.12 shows that $G(s)$ has isolated, algebraic infinities of order $(p + 1)/2$ on the imaginary axis and that

$$G(s) \sim -K_1 s^{-(p+1)/2} \quad \text{for } s \to 0 .$$

Therefore, it exists a real value $\tau \neq 0$ such that

$$G(\sigma + i\tau) \sim c\sigma^{-(p+1)/2}$$

with a certain $c \neq 0$, $\sigma > 0$ and $\sigma \to 0$. We have $g(t) \leqq 0$ for $t \geqq t_0$ under the above assumption and

$$|G(\sigma + i\tau)| \leqq \int_{t_0}^\infty |g(t)|\, e^{-\sigma t}\, dt + O(1)$$

$$\leqq -G(\sigma) + O(1) = K_1 \sigma^{-(p+1)/2} + O(1) .$$

Hence, because these results are contradictory if $K_1 < |c|$, we proved that

$$\Delta(\boldsymbol{a}; t^{A_p}) = \Omega_+\left(t^{\frac{p-1}{2}}\right).$$

The proof in the order direction is analogous.

6.2.3. *A historical outline of the development of Piltz's divisor problem for p ≧ 4*

Definition 6.3.

$$\vartheta_p = \inf\{\alpha : \Delta_p(x) \ll x^\alpha\} .$$

The history of O-estimates

$\vartheta_p \leqq \dfrac{p-1}{p}$ (A. Piltz [1], 1881),

$\vartheta_p \leqq \dfrac{p-1}{p+1}$ (E. Landau [1], 1912),

$\vartheta_p \leqq \dfrac{p-1}{p+2}$ (G. H. Hardy — J. E. Littlewood [2], 1922),

$\vartheta_7 \leqq \dfrac{71}{107}, \quad \vartheta_8 \leqq \dfrac{41}{59}, \quad \vartheta_9 \leqq \dfrac{31}{43}, \quad \vartheta_{10} \leqq \dfrac{26}{35}, \quad \vartheta_{11} \leqq \dfrac{19}{25}$

(Tong Kwang-chang [1], 1953),

$$\left.\begin{aligned} &\vartheta_p \leqq \frac{3p-4}{4p} \quad \text{for} \quad 4 \leqq p \leqq 8\,, \\ &\vartheta_p \leqq \frac{p-3}{p} \quad \text{for} \quad p > 8 \end{aligned}\right\} \quad \text{(D. R. Heath-Brown [1], 1978),}$$

$$\left.\begin{aligned} &\vartheta_9 \leqq \frac{35}{54}, \quad \vartheta_{10} \leqq \frac{41}{60}, \quad \vartheta_{11} \leqq \frac{7}{10}, \\ &\vartheta_p \leqq \frac{p-2}{p+2} \quad \text{for} \quad 12 \leqq p \leqq 25\,, \\ &\vartheta_p \leqq \frac{p-1}{p+4} \quad \text{for} \quad 26 \leqq p \leqq 50\,, \\ &\vartheta_p \leqq \frac{31p-98}{32p} \quad \text{for} \quad 51 \leqq p \leqq 57\,, \\ &\vartheta_p \leqq \frac{7p-34}{7p} \quad \text{for} \quad p \geqq 58 \end{aligned}\right\} \quad \text{(A. Ivić [4], 1980)}\,.$$

If p is very large, the estimate

$$\Delta_p(x) \ll x^{1-cp^{-2/3}}(c_1 \log x)^p\,, \qquad c, c_1 > 0\,,$$

uniformly holds in p. This was proved by H.-E. Richert [5] in a slightly weaker form ($\log x$ replaced by x^ε) in 1960. The stronger result is due to A. K. Karacuba [2], 1972.

In the other direction we have

$$\Delta_p(x) = \Omega_\pm(x^{(p-1)/2p}) \qquad \text{(G. H. Hardy [3], 1916)},$$

$$\Delta_p(x) = \Omega_\pm((x \log x)^{(p-1)/2p}(\log\log x)^{p-1}) \qquad \text{(G. Szegö – A. Walfisz [1], 1927)}.$$

In 1956 Tong Kwang-chang [3] found some improvements of the result of G. Szegö and A. Walfisz by considering p in residue classes module 4. The best result obtained by J. L. Hafner [2] in 1982 is:

$$\Delta_p(x) = \Omega_\pm\big((x \log x)^{(p-1)/2p} (\log\log x)^A \, e^{-B\sqrt{\log\log\log x}}\big),$$

$$A = \frac{p-1}{2p}(p \log p - p + 1) + p - 1\,, \qquad B > 0\,.$$

The general many-dimensional lattice point problem was initiated by the problems of counting the number of powerful integers and the number of non-isomorphic Abelian groups of a given order, which are described in Section 7. The first results are already contained in a very general theorem of E. Landau [1], 1912. E. Krätzel [6], 1969, was the first who investigated the general three-dimensional problem. M. Vogts [1] and A. Ivić [5] found new interesting results in 1981 by generalizing the work of P. G. Schmidt [2] who investigated the special problem of estimating $D(1, 2, 3; x)$ in 1968. M. Vogts [2] extended the method to p-dimensional problems in 1985. The best results for large p were obtained by E. Krätzel [15], 1983: The estimate

$$\Delta(\boldsymbol{a}; x) \ll x^{(p-c)/A_p+\varepsilon}, \qquad \varepsilon > 0,$$

holds under the condition

$$(p - c)\, a_p < A_p \leqq 2a_1 q \frac{p - c}{2q - c}$$

with

$$c = \frac{4q}{q + 2}, \qquad q = 2, 3, 4, 5, 6, \quad \text{for} \quad p \geqq 2q,$$

$$c \geqq 1, \quad q = c2^{c-1} + 2, \quad \text{for} \quad p \geqq 2q.$$

The estimate

$$\Delta(\boldsymbol{a}; x) = \Omega_{\pm}(x^{(p-1)/2A_p})$$

is contained in a very general theorem of E. Landau [12], 1924.

NOTES ON CHAPTER 6

Section 6.1

Theorem 6.1 is due to A. Ivić [5] and independently to M. Vogts [1], where their methods are based on the method of P. G. Schmidt [2]. The proof presented here is a simpler version.

Theorem 6.2. was proved by M. Vogts [1] and independently by A. Ivić [5].

Theorems 6.5 and 6.6 are to be found in E. Krätzel [17]. Theorems 6.3 and 6.4 are unpublished results of the author.

Further on, A. Ivić and P. Shiu [1] obtained an estimate of $\Delta(a, b, c; x)$, but which is based on the incorrect work of B. R. Srinivasan [3].

Section 6.2

Theorems 6.8—6.11 are generalizations of results obtained by E. Krätzel [9]. The proof of Theorems 6.12 follows the lines of Hardy's [3] work.

Chapter 7

Some applications to special multiplicative problems

7.1. POWERFUL NUMBERS

This section is concerned with the distribution of powerful numbers. Thereby a positive integer is said to be powerful if it contains only powers of primes as factors. More precisely:

Definition 7.1. *Let $k \geqq 2$ be an integer. A natural number n_k is said to be powerful of type k if $n_k = 1$ or if each prime factor of n_k divides it at least to the k-th power.*

Powerful numbers of type k are also called *k-full numbers.* In particular, if $k = 2$ or $k = 3$, we say *square-full* or *cube-full numbers.* The symbol n_k denotes a powerful number of type k throughout this section. The canonical representation of $n_k > 1$ is then given by

$$n_k = \prod_{i=1}^{r} p_i^{\nu_i}, \qquad \nu_i \geqq k \quad \text{for} \quad i = 1, 2, \dots, r\,.$$

Hence, every powerful number n_k can be uniquely represented by

$$n_k = a_0^k a_1^{k+1} \cdot \dots \cdot a_{k-1}^{2k-1},$$

where $a_1, a_2, \dots, a_{k-1}$ are square-free numbers and $(a_i, a_j) = 1$ for $1 \leqq i < j \leqq k-1$. Thereby a number is said to be *square-free* if it has no squared factor. We put

$$f_k(n) = \begin{cases} 1 & \text{for} \quad n = n_k\,, \\ 0 & \text{for} \quad n \neq n_k\,. \end{cases}$$

Let $N_k(x)$ denote the number of k-full integers not exceeding x. Then

$$N_k(x) = \sum_{n_k \leqq x} 1 = \sum_{n \leqq x} f_k(n) = \sum\nolimits_1 |\mu(a_1 \cdot \dots \cdot a_{k-1})|\,,$$

$$SC(\Sigma_1)\colon a_0^k a_1^{k+1} \cdot \dots \cdot a_{k-1}^{2k-1} \leqq x\,,$$

where $\mu(n)$ denotes the Möbius function. The investigation of the distribution of powerful numbers was initiated in 1935, when P. Erdös and G. Szekeres proved in an elementary way that

$$N_k(x) = \gamma x^{1/k} + O(x^{1/(k+1)})\,,$$

where γ is a certain positive constant. Considering the Dirichlet series

$$F_k(s) = \sum_{n_k=1}^{\infty} \frac{1}{n_k^s} = \sum_{n=1}^{\infty} \frac{f_k(n)}{n^s}$$

it is easy to see that the problem of estimating the number of powerful integers is connected with divisor problems in a direct manner. Clearly, the Dirichlet series is absolutely convergent for $\mathrm{Re}(s) > 1/k$. Since the arithmetical function $f_k(n)$ is multiplicative, it follows that

$$F_k(s) = \prod_p \left(1 + \sum_{\nu=k}^{\infty} p^{-\nu s}\right) = \prod_p \left(1 + \frac{p^{-ks}}{1 - p^{-s}}\right).$$

The case $k = 2$ is simple. We have

$$F_2(s) = \prod_p \frac{1 - p^{-6s}}{(1 - p^{-2s})(1 - p^{-3s})}.$$

Using the product representation of Riemann's zeta-function

$$\zeta(s) = \prod_p (1 - p^{-s})^{-1},$$

we obtain

$$F_2(s) = \frac{\zeta(2s)\,\zeta(3s)}{\zeta(6s)} \tag{7.1}$$

and therefore

$$F_2(s) = \sum_{a=1}^{\infty} \sum_{b=1}^{\infty} \sum_{m=1}^{\infty} \frac{\mu(m)}{a^{2s} b^{3s} m^{6s}} = \sum_{n=1}^{\infty} \sum_{m=1}^{\infty} \frac{\mathrm{d}(2, 3; n)\,\mu(m)}{(nm^6)^s}.$$

Hence

$$N_2(x) = \sum_{nm^6 \leq x} d(2, 3; n)\,\mu(m). \tag{7.2}$$

The representation of $F_k(s)$ is a little more complicated for $k > 2$. We get

$$\begin{aligned} F_k(s) &= \prod_p \frac{1 - p^{-s} + p^{-ks}}{1 - p^{-ks}} (1 + p^{-s} + p^{-2s} + \ldots + p^{-(k-1)s}) \\ &= \prod_p \frac{1 + p^{-(k+1)s} + p^{-(k+2)s} + \ldots + p^{-(2k-1)s}}{1 - p^{-ks}} \\ &= \prod_p \frac{1 - p^{-(2k+2)s} + b_1 p^{-(2k+3)s} + \ldots + b_q p^{-(2k+2+q)s}}{(1 - p^{-ks})(1 - p^{-(k+1)s}) \cdot \ldots \cdot (1 - p^{-(2k-1)s})}, \end{aligned}$$

where $b_1, b_2, \dots, b_q$ are some constants and $q = \frac{3}{2}(k^2 - k - 2)$. In what follows $c_r(n)$ denotes a certain arithmetical function whose associated Dirichlet series $\sum_{n=1}^{\infty} \frac{c_r(n)}{n^s}$ is absolutely convergent for $\mathrm{Re}(s) > 1/r$. Then

$$F_k(s) = \prod_{v=k}^{2k-1} \zeta(vs) \frac{1}{\zeta((2k+2)s)} \sum_{n=1}^{\infty} \frac{c_{2k+3}(n)}{n^s} \tag{7.3}$$

and

$$N_k(x) = \sum_{nn'm^{2k+2} \leq x} d(k, k+1, \dots, 2k-1; n)\, c_{2k+3}(n')\, \mu(m)\,. \tag{7.4}$$

Hence, it turns out that the problem of estimating the number of k-full integers not exceeding x is connected with the divisor problem for $D(k, k+1, \dots, 2k-1; x)$. Having an estimation for this function, then it is not hard to transmit it to $N_k(x)$. Since the main term in an asymptotical representation of $N_k(x)$ is deduced from the residues of $F_k(s)$ at the simple poles $s = 1/v$ $(v = k, k+1, \dots, 2k-1)$, the following definition is useful.

Definition 7. 2. *Let $N_k(x)$ and $F_k(s)$ be defined by* (7.1)—(7.4). *Then we write*

$$N_k(x) = \sum_{v=k}^{2k-1} \gamma_{v,k} x^{1/v} + \Delta_k(x)\,, \tag{7.5}$$

where

$$\gamma_{v,k} = \operatorname*{Res}_{s=1/v} F_k(s)/s\,.$$

Moreover, let

$$\lambda_k = \inf\{\varrho_k \colon \Delta_k(x) \ll x^{\varrho_k}\}\,.$$

In Section 7.1.1 we consider the problem of good approximations of $N_k(x)$. A. Ivić showed, what will not be proved here, that we would have under the assumption of the thruth of Lindelöf's hypothesis $\lambda_k \leqq 1/2k$. Consequently, the question arises what is unconditionally possible to prove. We show that the inequality is correct for $k \leqq 4$. Even in the special cases $k = 2, 3$ it is possible to prove much more. Then there arises a new situation, which enables us to consider the number of square-full and cube-full integers in short intervals. Finally, the problem of powerful divisors in investigated.

7.1.1. *The number of powerful integers*

We begin with a useful lemma.

Lemma 7.1. *Let* $\boldsymbol{a} = (a_1, a_2, \dots, a_p)$, $1 \leqq a_1 \leqq a_2 \leqq \dots \leqq a_p$. *Suppose that*

$$D(\boldsymbol{a}; x) = H(\boldsymbol{a}; x) + O(x^{\alpha_p}(\log x)^{\beta_p})\,.$$

where $D(\boldsymbol{a}; x)$ *is defined by Definition* 6.1 *and* $H(\boldsymbol{a}; x)$ *is given by* (6.26). *Let* $c_r(n)$ *denote an arithmetical function with*

$$\sum_{n \leqq x} |c_r(n)| \ll x^{1/r}(\log x)^{\gamma} .$$

If $1/a_p > \alpha_p \geqq 0$, $\beta_p \geqq 0$, $r > a_p$, $\gamma \geqq 0$, *then*

$$\sum_{mn \leqq x} d(\boldsymbol{a}; m)\, c_r(n) = \sum_{n=1}^{\infty} c_r(n)\, H\left(\boldsymbol{a}; \frac{x}{n}\right)$$

$$+ \begin{cases} O(x^{\alpha_p}(\log x)^{\beta_p}) & \text{for} \quad \alpha_p > \dfrac{1}{r}, \\ O(x^{\alpha_p}(\log x)^{\beta_p+\gamma+1}) & \text{for} \quad \alpha_p = \dfrac{1}{r}, \\ O(x^{1/r}(\log x)^{\gamma}) & \text{for} \quad \alpha_p < \dfrac{1}{r}. \end{cases} \tag{7.6}$$

Proof. We have

$$\sum_{mn \leqq x} d(\boldsymbol{a}; m)\, c_r(n) = \sum_{n \leqq x} c_r(n) \left\{ H\left(\boldsymbol{a}; \frac{x}{n}\right) + O\left(\left(\frac{x}{n}\right)^{\alpha_p} \left(\log \frac{x}{n}\right)^{\beta_p}\right)\right\} .$$

For the sake of simplicity assume that $1 \leqq a_1 < a_2 < \ldots < a_p$. Then, by means of (6.27),

$$\sum_{n \leqq x} c_r(n)\, H\left(\boldsymbol{a}; \frac{x}{n}\right) = \sum_{\nu=1}^{p} \prod_{\substack{\mu=1 \\ \mu \neq \nu}}^{p} \zeta\left(\frac{a_\mu}{a_\nu}\right) \sum_{n \leqq x} c_r(n) \left(\frac{x}{n}\right)^{1/a_\nu}$$

$$= \sum_{n=1}^{\infty} c_r(n)\, H\left(\boldsymbol{a}; \frac{x}{n}\right) + O(x^{1/r}(\log x)^{\gamma}) .$$

Clearly, a similar result holds if some of the numbers a_i are equal. In case $\alpha_p > 1/r$ we get for the error term

$$\sum_{n \leqq x} \left(\frac{x}{n}\right)^{\alpha_p} \left(\log \frac{x}{n}\right)^{\beta_p} |c_r(n)| < \sum_{n=1}^{\infty} \left(\frac{x}{n}\right)^{\alpha_p} (\log x)^{\beta_p} |c_r(n)| \ll x^{\alpha_p}(\log x)^{\beta_p} .$$

In case $\alpha_p \leqq 1/r$ result (7.6) follows at once by partial summation.

Theorem 7.1. *Let* $\boldsymbol{a}_{k,m} = (k, k+1, \ldots, k+m_k)$. *Suppose that*

$$D(\boldsymbol{a}_{k,m}; x) = H(\boldsymbol{a}_{k,m}; x) + O(x^{\varrho_k}(\log x)^{\beta_k}) \tag{7.7}$$

holds with $m_k < k$, $\varrho_k < \dfrac{1}{k + m_k}$, $\beta_k \geqq 0$. *Let* $\Delta_k(x)$ *be defined by* (7.5). *If* $m_k = k - 1$, *then*

$$\Delta_k(x) \ll \begin{cases} x^{\varrho_k}(\log x)^{\beta_k} & \text{for } \varrho_k > \dfrac{1}{2k+2}, \\ x^{\varrho_k}(\log x)^{\beta_k + 1} & \text{for } \varrho_k = \dfrac{1}{2k+2}, \\ x^{\frac{1}{2k+2}} & \text{for } \varrho_k < \dfrac{1}{2k+2}. \end{cases} \tag{7.8}$$

If $m_k < k - 1$, *then*

$$\Delta_k(x) \ll \begin{cases} x^{\varrho_k}(\log x)^{\beta_k} & \text{for } \varrho_k > \dfrac{1}{k+m_k+1}, \\ x^{\varrho_k}(\log x)^{\beta_k + 1} & \text{for } \varrho_k = \dfrac{1}{k+m_k+1}, \\ x^{\frac{1}{k+m_k+1}} & \text{for } \varrho_k < \dfrac{1}{k+m_k+1}. \end{cases} \tag{7.9}$$

Proof. From (7.4) it follows that

$$N_k(x) = \sum_{nn' \leqq x} d(\boldsymbol{a}_{k,m}; n)\, c_r(n'),$$

where $r = 2k + 2$ for $m_k = k - 1$ and $r = k + m_k + 1$ for $m_k < k - 1$. Because of

$$\sum_{n' \leqq x} |c_r(n')| \asymp x^{1/r}$$

(7.8) and (7.9) are immediate consequences of Lemma 7.1.

Applying only trivial estimates it is easy to prove by means of Theorem 7.1 that $\lambda_2 \leqq 1/5$, $\lambda_3 \leqq 1/6$, and that

$$\lambda_k \leqq \frac{1}{k + m_k}, \quad m_k = [\sqrt{k}],$$

holds for $k \geqq 4$. Here we prove the following results.

Theorem 7.2.

$$\Delta_2(x) \ll x^{1/6}, \tag{7.10}$$

$$\Delta_3(x) \ll x^{1/8} \log^4 x, \tag{7.11}$$

$$\Delta_4(x) \ll x^{106/913+\varepsilon} \quad \left(\frac{106}{913} = 0{,}1161\ldots\right), \tag{7.12}$$

$$\Delta_5(x) \ll x^{65/622+\varepsilon} \qquad \left(\frac{65}{622} = 0{,}1045\ldots\right), \tag{7.13}$$

$$\Delta_k(x) \ll x^{\varrho_k+\varepsilon}, \quad \varrho_k = \frac{6k+9}{8k^2+26k+36} \quad \text{for} \quad 6 \leqq k \leqq 8 \tag{7.14}$$

$$(\varrho_6 = 0{,}09375\,, \quad \varrho_7 = 0{,}0836\ldots\,, \quad \varrho_8 = 0{,}0753\ldots)\,,$$

$$\lambda_k \leqq \frac{1}{k+m_k}, \quad m_k = [\sqrt{2k}]\,. \tag{7.15}$$

Proof. If $k = 2$, we apply Theorem 5.3. The estimate (5.6) shows that

$$\Delta(2, 3; x) \ll x^{1/7}\,.$$

Now we use Theorem 7.1 with $\varrho_2 = 1/7$, $\beta_2 = 0$. Then (7.10) follows from (7.8).

If $k = 3$, the conditions of Theorem 6.5 are satisfied such that

$$\Delta(3, 4, 5; x) \ll x^{1/8} \log^3 x\,.$$

Applying Theorem 7.1 with $\varrho_3 = 1/8$, $\beta_3 = 3$, (7.11) follows from (7.8).

If $k = 4, 5$, we again apply Theorem 6.5 and we obtain

$$\Delta(k, k+1, k+2; x) \ll x^{\frac{1}{2(k+1)}+\varepsilon_0} \qquad (\varepsilon_0 > 0)\,.$$

Now we use Theorem 6.11, where the conditions (6.41) and (6.42) are satisfied if we take the exponent pair $(k, l) = (2/18, 13/18)$. Condition (6.43) requires the equality

$$\left(\frac{x}{y}\right)^{\frac{2}{3}\frac{1}{k+1}} (x^k y^{2k+3})^{\frac{2}{9}\frac{1}{(2k+3)(k+1)}} = y^{\frac{1}{k+3}} \left(\frac{x}{y}\right)^{\frac{1}{2(k+1)}+\varepsilon_0}$$

In case $k = 4$ this is satisfied by putting

$$y = x^{\frac{343}{913}+\varepsilon_1}\,.$$

It is easily seen that (6.44) is fulfilled, and (7.12) now follows from (6.45) and from Theorem 7.1. In case $k = 5$ the above equality is satisfied by putting

$$y = x^{\frac{118}{325}+\varepsilon_1}\,.$$

(6.44) is fulfilled such that

$$\Delta(5, 6, 7, 8; x) \ll x^{\frac{32}{325}+\varepsilon_2}$$

follows from (6.45). We now apply Theorem 6.8 with $p = 5$, $q = 4$ and

$$a_1 = 5\,, \quad a_5 = 9\,, \quad \alpha_4 = \frac{32}{325} + \varepsilon_2\,, \quad \beta_4 = 0\,.$$

This gives by (6.30)

$$\alpha_5 = \frac{65}{622} + \varepsilon ,$$

and (7.13) follows from Theorem 7.1.

In case $k = 6, 7, 8$ it is easily seen that all conditions of Theorem 6.11 are satisfied, if we take the exponent pair $(k, l) = (2/18, 13/18)$. Then we obtain from (6.46)

$$\Delta(k, k + 1, k + 2, k + 3; x) \ll x^{\frac{4}{3(2k+3)} + \varepsilon_0} .$$

Now we again apply Theorem 6.8 with $p = 5$, $q = 4$. Then there is

$$a_1 = k , \quad a_5 = k + 4 , \quad \alpha_4 = \frac{k}{3(2k + 3)} + \varepsilon_0 , \quad \beta_4 = 0 .$$

This gives by (6.30)

$$\alpha_5 = \frac{6k + 9}{8k^2 + 26k + 36} + \varepsilon$$

and (7.14) follows from Theorem 7.1.

In order to prove (7.15), we apply Theorem 6.10 with $p = m_k + 1$, $q = m_k/2$ for $m_k \equiv 0 \pmod 2$ and $q = (m_k + 1)/2$ for $m_k \equiv 1 \pmod 2$, $a_1 = k$, $a_{q+1} = k + q$. Then (6.34) shows that

$$\Delta(k, k + 1, \ldots , k + m_k; x) \ll x^{\gamma_p + \varepsilon} ,$$

$$\gamma_p = \frac{2m_k + 1}{(2q + 1)\, k + (2(m_k - q) + 3)\,(k + q)}$$

$$= \frac{2m_k + 1}{(2m_k + 4)\, k + (2(m_k - q) + 3)\, q}$$

$$\leq \frac{4m_k + 2}{(4m_k + 8)\, k + m_k^2 + 3m_k} .$$

It easy to prove that this value is less than $1/(k + m_k)$ if we put $m_k = [\sqrt{2k}]$. Then, by means of Theorem 7.1, (7.15) follows at once.

7.1.2. *A historical outline of the development of the problem*

Let λ_k denote the value given by Definition 7.2. Writing

$$\lambda_k \leq \frac{1}{k + m_k} ,$$

there are the following values for m_k:

$m_k = 1$ (P. Erdös — G. Szekeres [1], 1935),

$m_k = [\sqrt{2k}]$ (P. Bateman — E. Grosswald [1], 1958),

$m_k = [\sqrt{8k/3}]$ for $k \geqq 5$ (E. Krätzel [9], 1972),

$$m_k = [\sqrt{2ck + (c - 1/2)^2 - 1} + c - 1/2]$$

with $c = \dfrac{4q}{q+2}$, $q = 2, 3, 4, 5, 6$ for $k \geqq \dfrac{q}{2}(4q - 1)$,

$c \geqq 1$ for $k \geqq (c2^{c+1} + 7)(2^c + 3)$

and

$m_k = [\sqrt{k \log k}]$ *for* $k \geqq e^8$ (E. Krätzel [15], 1983).

Furthermore, we find the following results for small values of k:

$\lambda_2 \leqq \dfrac{1}{6}$, $\lambda_3 \leqq \dfrac{7}{46} = 0{,}1521\ldots$ (P. Bateman — E. Grosswald [1], 1958),

$\lambda_3 \leqq \dfrac{16}{113} = 0{,}1415\ldots$, $\lambda_4 \leqq \dfrac{169}{1360} = 0{,}1242\ldots$,

$\lambda_5 \leqq \dfrac{16188}{151297} = 0{,}1069\ldots$, $\lambda_6 \leqq \dfrac{113}{1173} = 0{,}0963\ldots$,

$\lambda_7 \leqq \dfrac{274}{3213} = 0{,}0852\ldots$ (E. Krätzel [9], 1972),

$\lambda_3 \leqq \dfrac{655}{4643} = 0{,}1410\ldots$, $\lambda_4 \leqq \dfrac{257}{2072} = 0{,}1240\ldots$,

$\lambda_5 \leqq \dfrac{6656613}{62279970} = 0{,}1068\ldots$ (A. Ivić [1], 1978),

$\lambda_3 \leqq \dfrac{577}{4176} = 0{,}1381\ldots$, $\lambda_4 \leqq \dfrac{3187}{25852} = 0{,}1232\ldots$,

$\lambda_5 \leqq \dfrac{124371}{1165874} = 0{,}1066\ldots$ (A. Ivić [5], 1981),

$\lambda_3 \leqq \dfrac{263}{2052} = 0{,}1281\ldots$, $\lambda_4 \leqq \dfrac{3091}{25981} = 0{,}1189\ldots$,

$\lambda_5 \leqq \dfrac{1}{10}$, $\lambda_6 \leqq \dfrac{1}{12}$, $\lambda_7 \leqq \dfrac{1}{14}$ (A. Ivić — P. Shiu [1], 1982),

$$\lambda_4 \leqq \frac{5}{44} = 0{,}1136\ldots, \quad \lambda_5 \leqq \frac{6}{65} = 0{,}0923\ldots,$$

$$\lambda_6 \leqq \frac{13}{162} = 0{,}0802\ldots, \quad \lambda_8 \leqq \frac{1}{16} \quad \text{(E. Krätzel [15], 1983)},$$

$$\lambda_3 \leqq \frac{1}{8} \quad \text{(E. Krätzel [17], 1985)}$$

7.1.3. *Square-full and cube-full numbers*

From representation (7.5) and estimation (7.10) we obtain the following asymptotic representation for the number $N_2(x)$ of square-full integers not exceeding x:

$$N_2(x) = \gamma_{2,2}x^{1/2} + \gamma_{3,2}x^{1/3} + O(x^{1/6}).$$

The coefficients are given by

$$\gamma_{2,2} = \frac{\zeta(3/2)}{\zeta(3)}, \quad \gamma_{3,2} = \frac{\zeta(2/3)}{\gamma(2)}.$$

Here it has to be remarked that to our present knowledge we are not able to improve substantially the estimation of the remainder. That means that we cannot replace the exponent 1/6 by a smaller one. Moreover, P. T. Bateman and E. Grosswald applied deep results of prime number theory in order to give only the improvement of $\Delta_2(x) = o(x^{1/6})$.

A similar situation arises in case of cube-full integers. Here (7.5) and (7.11) show that the following asymptotic representation holds for the number $N_3(x)$ of cube-full integers not exceeding x:

$$N_3(x) = \gamma_{3,3}x^{1/3} + \gamma_{4,3}x^{1/4} + \gamma_{5,3}x^{1/5} + O(x^{1/8}\log^4 x).$$

From (7.3) we get for the coefficients

$$\gamma_{3,3} = \frac{\zeta\left(\frac{4}{3}\right)\zeta\left(\frac{5}{3}\right)}{\zeta\left(\frac{8}{3}\right)} \sum_{n=1}^{\infty} \frac{c_9(n)}{n^{1/3}}, \qquad \gamma_{4,3} = \frac{\zeta\left(\frac{3}{4}\right)\zeta\left(\frac{5}{4}\right)}{\zeta(2)} \sum_{n=1}^{\infty} \frac{c_9(n)}{n^{1/4}},$$

$$\gamma_{5,3} = \frac{\zeta\left(\frac{3}{5}\right)\zeta\left(\frac{4}{5}\right)}{\zeta\left(\frac{8}{5}\right)} \sum_{n=1}^{\infty} \frac{c_9(n)}{n^{1/5}}.$$

Whereas the preceding problem is connected with the divisor problem $D(2, 3; x)$, here we have to consider the divisor problem $D(3, 4, 5; x)$ in order to prove

$\Delta_3(x) = o(x^{1/8})$. This is not easy, since the best known estimate of the remainder of $D(3, 4, 5; x)$ is

$$\Delta(3, 4, 5; x) \ll x^{1/8} \log^3 x\,,$$

coming from Theorem 6.5. But, because of Theorem 6.6 we can be in the hope of proving that $\Delta(3, 4, 5) \ll x^{1/8-\varepsilon}$.

We need the following deep result of prime number theory, which we cannot prove here. Let $M(x)$ denote the sum

$$M(x) = \sum_{n \leqq x} \mu(n)\,,$$

$\mu(n)$ is the Möbius function, then

$$M(x) \ll x\, e^{-A\delta(x)}\,,$$

where A is a positive constant and $\delta(x)$ denotes a certain function which tends to infinity with x. Due to A. Walfisz [4] the best known estimate is given by the function

$$\delta(x) = (\log x)^{3/5} (\log \log x)^{-1/5}\,. \tag{7.16}$$

The following lemma gives the connection with our problem.

Lemma 7.2. *Let $f(n)$ be an arithmetical function such that*

$$\sum_{n \leqq x} f(n) = \sum_{\nu=1}^{r} c_\nu x^{\alpha_\nu} + O(x^\alpha)\,, \qquad \sum_{n \leqq x} |f(n)| \ll x^{\alpha_1}\,,$$

where $\alpha_1 > \alpha_2 > \ldots > \alpha_r > \frac{1}{k}\alpha \geqq 0$, and k is a fixed positive integer. If

$$h(n) = \sum_{t^k / n} \mu(t)\, f(n/t^k)\,,$$

where $\mu(n)$ denotes the Möbius function, then

$$\sum_{n \leqq x} h(n) = \sum_{\nu=1}^{r} \frac{c_\nu}{\zeta(k\alpha_\nu)} x^{\alpha_\nu} + O(x^{1/k}\, e^{-A_k \delta(x)})\,, \tag{7.17}$$

where A_k is a positive constant depending only on k, and $\delta(x)$ is given by (7.16).

Proof. Let y be a suitably chosen value with $1 < y^k \leqq x$. Then

$$\sum_{n \leqq x} h(n) = \sum_{m^k n \leqq x} \mu(m)\, f(n)$$

$$= \Bigg\{ \sum_{\substack{m^k n \leqq x \\ m \leqq y}} + \sum_{\substack{m^k n \leqq x \\ y^k n \leqq x}} - \sum_{\substack{y^k n \leqq x \\ m \leqq y}} \Bigg\} \mu(m)\, f(n)$$

$$= \sum_{m \leqq y} \mu(m) \sum_{m^k n \leqq x} f(n) + \sum_{y^k n \leqq x} f(n)\, M\left(\left(\frac{x}{n}\right)^{1/k}\right) + M(y) \sum_{y^k n \leqq x} f(n)$$

$$= \sum_{m \leqq y} \mu(m) \left\{ \sum_{\nu=1}^{r} c_\nu \left(\frac{x}{m^k}\right)^{\alpha_\nu} + O\left(\left(\frac{x}{m^k}\right)^{\alpha}\right)\right\}$$

$$+ O\left(\sum_{y^k n \leqq x} |f(n)| \left(\frac{x}{n}\right)^{1/k} \mathrm{e}^{-A\delta(y)}\right) + O\left(y \left(\frac{x}{y^k}\right)^{\alpha_1} \mathrm{e}^{-A\delta(y)}\right)$$

$$= \sum_{\nu=1}^{r} \frac{c_\nu}{\zeta(k\alpha_\nu)}\, x^{\alpha_\nu} + O\left(y\left(\frac{x}{y^k}\right)^{\alpha}\right) + O\left(y\left(\frac{x}{y^k}\right)^{\alpha_1} \mathrm{e}^{-A\delta(y)}\right)$$

Putting

$$y^k = x\, \mathrm{e}^{-\frac{A}{\alpha_1 - \alpha}\delta(x^{1/2k})},$$

(7.17) follows for sufficiently large x.

Now we are in a position to prove the following theorem.

Theorem 7.3. *Let A_2 and A_3 be some positive constants, and let $\delta(x)$ be defined by* (7.16). *Then*

$$\Delta_2(x) \ll x^{1/6}\, \mathrm{e}^{-A_2\delta(x)}, \tag{7.18}$$

$$\Delta_3(x) \ll x^{1/8}\, \mathrm{e}^{-A_3\delta(x)}. \tag{7.19}$$

Proof. In proving (7.18) we use Theorems 5.1 and 5.2. Then

$$D(2, 3; x) = \zeta(3/2)\, x^{1/2} + \zeta(2/3)\, x^{1/3} + O(x^{1/7}).$$

Now we apply Lemma 7.2 with

$$f(n) = d(2, 3; n), \quad \gamma = 2, \quad \alpha_1 = \frac{1}{2}, \quad \alpha_2 = \frac{1}{3}, \quad \alpha = \frac{1}{7}, \quad k = 6.$$

(7.18) now follows immediately from (7.17).

In order to obtain the estimate (7.19), we first prove that

$$\Delta(3, 4, 5; x) \ll x^{22/177} \log^3 x. \tag{7.20}$$

Let the functions $S(u, v, w; x)$ and $S(u, v, w; M, N; x)$ be defined as before by (6.7) and (6.10); respectively. Since the triplet (3, 4, 5) does not satisfy the conditions of Theorem 6.6, which would give a sufficiently good estimation, the six permutations of the numbers 3, 4, 5 are considered separately. First of all it is seen from the proof of Theorem 6.6 that

$$S(u, v, w; x) \ll x^{25/204} \log^3 x \tag{7.21}$$

if

$$18w \geqq 7(u + v), \quad 2(v + w) \geqq 3u.$$

These conditions are satisfied by the three triplets

$$(u, v, w) = (3, 5, 4)\,,\ (4, 3, 5)\,,\ (3, 4, 5)\,.$$

We now consider the other three triplets

$$(u, v, w) = (5, 4, 3)\,,\ (4, 5, 3)\,,\ (5, 3, 4)\,.$$

Each sum

$$S(u, v, w; x) = \sum_{\substack{n^u m^{v+w} \leq x \\ n \leq m}} \psi\left(\left(\frac{x}{n^u m^v}\right)^{1/w}\right)$$

will be divided into two subsums corresponding to $n \geqq z$ and $n < z$, where z is a suitable value. In case $n \geqq z$ the estimate (6.18) may be applied. Because of $z \ll N \ll M \ll (xN^{-u})^{1/(v+w)}$ we obtain

$$\begin{aligned}
S(5, 4, 3; M, N; x) &\ll (x^7 M^{-4} N^{-5})^{1/51} \log x \ll (x^7 z^{-9})^{1/51} \log x\,,\\
S(4, 5, 3; M, N; x) &\ll (x^7 M^{-11} N^2)^{1/51} \log x \ll (x^7 z^{-9})^{1/51} \log x\,,\\
S(5, 3, 4; M, N; x) &\ll (x^7 M^{11} N^5)^{1/68} \log x\\
&\ll (x^7 M^{-4} N^{-5})^{1/51} \log x \ll (x^7 z^{-9})^{1/51} \log x\,.
\end{aligned}$$

In case $n < z$ the estimate (6.17) is used. Because of $N \ll z$, $N \ll M \ll (xN^{-u})^{1/(v+w)}$ we obtain

$$\begin{aligned}
S(5, 4, 3; M, N; x) &\ll (x^2 M N^2)^{1/18} \log x \ll (x^5 z^3)^{1/42} \log x\,,\\
S(4, 5, 3; M, N; x) &\ll (x^2 M^{-1} N^4)^{1/18} \log x \ll (x^2 z^3)^{1/18} \log x\\
&\ll (x^5 z^3)^{1/42} \log x \quad \text{if} \quad z \leqq x^{1/12}\,,\\
S(5, 3, 4; M, N; x) &\ll (x M^3 N^3)^{1/12} \log x\\
&\ll (x^2 M N^2)^{1/18} \log x \ll (x^5 z^3)^{1/42} \log x\,.
\end{aligned}$$

If we put $z = x^{13/177}$, we get for all permutations of u, v, w

$$S(u, v, w; M, N; x) \ll x^{22/177} \log x\,.$$

Hence

$$S(u, v, w; x) \ll x^{22/177} \log^3 x\,.$$

This result and (7.21) now prove (7.20). Thus we have

$$D(3, 4, 5; x) = b_1 x^{1/3} + b_2 x^{1/4} + b_3 x^{1/5} + O(x^{22/177} \log^3 x)$$

with well-known coefficients b_1, b_2, b_3. Applying Lemma 7.1 with

$$a_1 = 3\,,\quad a_2 = 4\,,\quad a_3 = 5\,,\quad \alpha_3 = \frac{22}{177}\,,\quad \beta_3 = 3\,,\quad r = 9\,,\quad \gamma = 0\,,$$

we obtain

$$\sum_{nn' \leqq x} d(3, 4, 5; n)\, c_9(n') = b_1' x^{1/3} + b_2' x^{1/4} + b_3' x^{1/5} + O(x^{22/177} \log^3 x)$$

with some well-defined coefficients b_1', b_2', b_3'. Now we may use (7.4) and Lemma 7.2 with

$$f(n) = \sum_{n'n''=n} d(3, 4, 5; n'')\, c_9(n'), \qquad r = 3, \qquad k = 8,$$

$$\alpha_1 = \frac{1}{3}, \quad \alpha_2 = \frac{1}{4}, \quad \alpha_3 = \frac{1}{5}, \quad \alpha = \frac{22}{177} + \varepsilon .$$

(7.19) now follows from (7.17). This completes the proof of the theorem.

The results of Theorem 7.3 suggest that in cases of square-full and cube-full numbers we should ask for their distribution in short intervals $(x, x + h]$, $h = o(x)$. Neglecting the exponential factors in (7.18) and (7.19), which are meaningless hereafter, we get from (7.5)

$$N_2(x; h) = N_2(x + h) - N_2(x) = \frac{1}{2} \gamma_{2,2} \frac{h}{\sqrt{x}} (1 + o(1)) + O(x^{1/6}),$$

$$N_3(x; h) = N_3(x + h) - N_3(x) = \frac{1}{3} \gamma_{3,3} \frac{h}{x^{2/3}} (1 + o(1)) + O(x^{1/8}).$$

Indeed, we can substantially improve these estimations.

Theorem 7.4. *Let $\Delta(2, 3; x)$ and $\Delta(2, 3, 6; x)$ be the error terms in the divisor problems related to the divisor functions $d(2, 3; n)$ and $d(2, 3, 6; n)$. Suppose that*

$$\Delta(2, 3; t) \ll \Delta^*(2, 3; x) = x^{\alpha_2} (\log x)^{\beta_2},$$

$$\Delta(2, 3, 6; t) \ll \Delta^*(2, 3, 6; x) = x^{\alpha_3} (\log x)^{\beta_3}$$

for $1 \leqq t \leqq x + h$. If $\alpha_2 < \alpha_3 < 1/6$ *and* $0 < h = o(x)$, *then*

$$N_2(x; h) = \frac{1}{2} \gamma_{2,2} \frac{h}{\sqrt{x}} (1 + o(1)) + O(\Delta^*(2, 3, 6; x)). \tag{7.22}$$

In particular, this representation holds with

$$\Delta^*(2, 3, 6; x) = x^{\frac{577}{3828}} \log^2 x \left(\frac{577}{3828} = 0{,}1507\ldots \right). \tag{7.23}$$

Similarly, let $\Delta(3, 4, 5; x)$ and $\Delta(3, 4, 5, 8; x)$ be the error terms in the divisor problems related to the divisor functions $d(3, 4, 5; n)$ and $d(3, 4, 5, 8; n)$. Suppose that

$$\Delta(3, 4, 5; t) \ll \Delta^*(3, 4, 5; x) = x^{\gamma_3} (\log x)^{\delta_3},$$

$$\Delta(3, 4, 5, 8; t) \ll \Delta^*(3, 4, 5, 8; x) = x^{\gamma_4} (\log x)^{\delta_4}$$

for $1 \leqq t \leqq x + h$. *If* $\gamma_3 < \gamma_4 < 1/8$, $\gamma_4 < 1/9$ *and* $0 < h = o(x)$, *then*

$$N_3(x; h) = \frac{1}{3}\gamma_{3,3}\frac{h}{x^{2/3}}(1 + o(1)) + O(\Delta^*(3, 4, 5, 8; x))\,. \tag{7.24}$$

In particular, this representation holds with

$$\Delta^*(3, 4, 5, 8; x) = x^{\frac{6451}{51756}+\varepsilon}\left(\frac{6451}{51756} = 0{,}1246\ldots\right), \qquad \varepsilon > 0\,. \tag{7.25}$$

Proof. Let $y = y(x)$ be a suitably chosen positive function of x, which tends to infinity with x. To prove (7.22) we use the representation (7.2). Then

$$\begin{aligned}
N_2(x; h) &= \sum_{x<nm^6\leqq x+h} d(2, 3; n)\,\mu(m) \\
&= \sum_{m\leqq y}\mu(m)\left\{D\left(2, 3; \frac{x+h}{m^6}\right) - D\left(2, 3; \frac{x}{m^6}\right)\right\} \\
&\quad + \sum_{\substack{x<nm^6\leqq x+h\\ m>y}} d(2, 3; n)\,\mu(m) \\
&= \sum_{m\leqq y}\mu(m)\left\{\zeta\left(\frac{3}{2}\right)\left(\left(\frac{x+h}{m^6}\right)^{1/2} - \left(\frac{x}{m^6}\right)^{1/2}\right)\right. \\
&\quad + \zeta\left(\frac{2}{3}\right)\left(\left(\frac{x+h}{m^6}\right)^{1/3} - \left(\frac{x}{m^6}\right)^{1/3}\right) + \Delta\left(2, 3; \frac{x+h}{m^6}\right) \\
&\quad \left. - \Delta\left(2, 3; \frac{x}{m^6}\right)\right\} + O\left(\sum_{\substack{x<nm^6\leqq x+h\\ m>y}} d(2, 3; n)\right).
\end{aligned}$$

It is known from Theorems 5.3 and 5.8 that $x^{1/10} \ll |\Delta(2, 3; x)| \ll x^{1/7}$. Hence

$$\begin{aligned}
N_2(x; h) &= \frac{1}{2}\gamma_{2,2}\frac{h}{\sqrt{x}}(1 + o(1)) + O\left(y\Delta^*\left(2, 3; \frac{x}{y^6}\right)\right) \\
&\quad + O\left(\sum_{\substack{x<nm^6\leqq x+h\\ m\leqq y}} d(2, 3; n)\right).
\end{aligned}$$

We can now choose y in the first error term such that

$$y\Delta^*\left(2, 3; \frac{x}{y^6}\right) \ll \Delta^*(2, 3, 6; x)\,.$$

The sum in the second error term is directly connected with the divisor problem related to the arithmetical function $d(2, 3, 6; n)$:

$$\sum_{\substack{x < nm^6 \leqq x+h \\ m > y}} d(2, 3; n) = D(2, 3, 6; x+h) - D(2, 3, 6; x)$$

$$- \sum_{m \leqq y} \left\{ D\left(2, 3; \frac{x+h}{m^6}\right) - D\left(2, 3; \frac{x}{m^6}\right) \right\}$$

$$= \zeta\left(\frac{3}{2}\right) \zeta(3) \left((x+h)^{1/2} - x^{1/2}\right)$$

$$+ \zeta\left(\frac{2}{3}\right) \zeta(2) \left((x+h)^{1/3} - x^{1/3}\right)$$

$$+ \zeta\left(\frac{1}{3}\right) \zeta\left(\frac{1}{2}\right) \left((x+h)^{1/6} - x^{1/6}\right)$$

$$+ \Delta(2, 3, 6; x+h) - \Delta(2, 3, 6; x)$$

$$- \sum_{m \leqq y} \left\{ \zeta\left(\frac{3}{2}\right) \left(\left(\frac{x+h}{m^6}\right)^{1/2} - \left(\frac{x}{m^6}\right)^{1/2} \right) \right.$$

$$+ \zeta\left(\frac{2}{3}\right) \left(\left(\frac{x+h}{m^6}\right)^{1/3} - \left(\frac{x}{m^6}\right)^{1/3} \right)$$

$$\left. + \Delta\left(2, 3; \frac{x+h}{m^6}\right) - \Delta\left(2, 3; \frac{x}{m^6}\right) \right\}$$

$$= \Delta(2, 3, 6; x+h) - \Delta(2, 3, 6; x) + o\left(\frac{h}{\sqrt{x}}\right)$$

$$+ O\left(y \Delta^*\left(2, 3; \frac{x}{y^6}\right)\right)$$

$$= O(\Delta^*(2, 3, 6; x)) + o\left(\frac{h}{\sqrt{x}}\right).$$

This completes the proof of (7.22).

We can easily deduce estimation (7.23) from Theorems 5.3 and 6.2. Using the exponent pair $\left(\frac{97}{696}, \frac{480}{696}\right)$, the conditions of Theorem 6.2 are satisfied by the triplet $(a, b, c) = (2, 3, 6)$. Thus, (7.23) follows from (6.12).

The proof of (7.24) stands in analogy to the proof of (7.22) and is left to the reader. To prove (7.25) we use the estimate (7.20) and Theorem 6.11 with the exponent $\left(\frac{1}{14}, \frac{11}{14}\right)$. Then the conditions (6.41) and (6.42) are satisfied, and equation (6.43) reads as follows:

$$\left(\frac{x}{y}\right)^{\frac{1}{6}} (xy^3)^{\frac{5}{308}} = y^{\frac{1}{8}} \left(\frac{x}{y}\right)^{\frac{22}{177} + \varepsilon_1}.$$

This equation holds for $y = x^{\frac{2130}{4313}+\varepsilon_2}$, where ε_2 depends on ε_1. Clearly, condition (6.44) is satisfied and (7.25) follows from (6.45).

7.1.4. *The number of powerful divisors*

Let t_k denote a k-full divisor of integer n. Then we define the divisor function

$$\alpha_k(n) = \#\{t_k : t_k/n\}$$

such that $\alpha_k(n)$ *represents the number of k-full divisors of n.* Obviously, we have

$$\sum_{n \leq x} \alpha_k(n) = \sum_{t_k d \leq x} 1 \quad \text{and} \quad \sum_{n=1}^{\infty} \frac{\alpha_k(n)}{n^s} = \zeta(s)\, F_k(s)\,,$$

where $F_k(s)$ is given by (7.1) and (7.3). The reader will have no difficulty in proving the trivial estimate

$$\sum_{n \leq x} \alpha_k(n) = F_k(1)\, x + \zeta\left(\frac{1}{k}\right) \gamma_{k,k} x^{1/k} + O(x^{1/(k+1)}),$$

where $\gamma_{k,k}$ is defined as before. Furthermore it is seen that the representation

$$\sum_{n=1}^{\infty} \frac{\alpha_k(n)}{n^s} = \zeta(s)\, \zeta(ks)\, \zeta((k+1)\, s)\, B_k(s) = \sum_{n=1}^{\infty} \frac{d(1, k, k+1; n)}{n^s} B_k(s)$$

$$B_k(s) = \begin{cases} \dfrac{1}{\zeta(6s)} & \text{for} \quad k = 2, \\ \displaystyle\sum_{n=1}^{\infty} \frac{c_{k+2}(n)}{n^s} & \text{for} \quad k > 2. \end{cases}$$

Hence, the problem of estimating $\sum_{n \leq x} \alpha_k(n)$ is closely related to the corresponding one of $D(1, k, k+1; x)$. A. Ivić [1] was the first who investigated this problem, and he proved

$$\Delta(1, k, k+1; x) \ll x^{z_k},$$

where $z_3 = 8/35$, $z_4 = 28/149$, which come from the authors paper [9], and

$$z_k = \left(1 - \frac{6}{2k^2 + 9k + 13}\right) \frac{1}{k+1}$$

holds with

for $k \geqq 5$. The case $k = 2$ connected with the problem of enumerating the finite non-isomorphic Abelian groups, and we refer to the next section. For $k > 2$ we can use the simple estimation

$$\Delta(a, b, c; x) \ll x^{\frac{2}{2a+b+c}} \log^2 x$$

such that we obtain the better result

$$\Delta(1, k, k+1) \ll x^{\frac{2}{2k+3}} \log^2 x .$$

In the following theorem we give an improvement of this result for $k \geqq 3$ based on Theorem 6.4. The case $k = 2$ may be allowed, but later a better estimation will be proved.

Theorem 7.5. *The estimate*

$$\sum_{n \leqq x} \alpha_k(n) = F_k(1)\, x + \zeta\left(\frac{1}{k}\right) \gamma_{k,k} x^{1/k} + \zeta\left(\frac{1}{k+1}\right) \gamma_{k+1,k} x^{1/(k+1)} + O(x^{z_k} \log^4 x) \tag{7.26}$$

holds with $z_k = \left(1 - \dfrac{2(\varkappa - 1)}{3K - 4}\right) \dfrac{1}{k+1}$, $K = 2^\varkappa, \varkappa \geqq 3$ *under the condition* $3K - 2\varkappa - 6 \geqq 4k$.

Proof. Theorem 6.4 shows that

$$\Delta(1, k, k+1; x) \ll x^{z_k} \log^3 x$$

holds under the condition of the theorem. It is seen from the Dirichlet series for $\alpha_k(n)$ that Lemma 7.1 can now be used with $p = 3$ and

$$a_1 = 1\,, \quad a_2 = k\,, \quad a_3 = k+1\,, \quad \alpha_3 = z_k\,, \quad \beta_3 = 3\,, \quad \gamma = 0 \text{ or } 1\,,$$

$$r = \begin{cases} 6 & \text{for} \quad k = 2\,, \\ k+2 & \text{for} \quad k > 2\,. \end{cases}$$

Now (7.26) follows immediately from (7.6).

We notice the values for $\varkappa = 3, 4, 5$:

$\varkappa$	$1 - \dfrac{2(\varkappa - 1)}{3K - 4}$	condition
3	$\frac{4}{5}$	$2 \leqq k \leqq 3$
4	$\frac{19}{22}$	$3 < k \leqq 8$
5	$\frac{21}{23}$	$9 < k \leqq 20$

For large k we put

$$\varkappa = \left[\frac{\log\left(\frac{4k}{3}\left(1 + \frac{4 \log k}{k}\right)\right)}{\log 2} \right] + 1\,.$$

Then, if $k \geqq 3$, we have

$$3K - 2\varkappa - 6 > 4k\left(1 + \frac{4\log k}{k}\right) - \frac{2}{\log 2}\log\left(\frac{4k}{3}\left(1 + \frac{4\log k}{k}\right)\right) - 6$$

$$> 4k + 13\log k - 10 > 4k$$

such that the condition of the theorem is satisfied. Now

$$1 - \frac{2(\varkappa - 1)}{3K - 4} < 1 - \frac{2\left(\log\frac{4k}{3}\left(1 + \frac{4\log k}{k}\right)\right)}{\log 2\left(8k\left(1 + \frac{4\log k}{k}\right) - 4\right)}$$

$$< 1 - \frac{\log(k + 4\log k)}{(k + 4\log k)\log 16}.$$

Therefore, the estimate (7.26) *holds for* $k \geqq 3$ *with*

$$z_k = \left(1 - \frac{\log(k + 4\log k)}{(k + 4\log k)\log 16}\right)\frac{1}{k+1}.$$

7.2. FINITE ABELIAN GROUPS

It is a well-known fact that every finite Abelian group can be represented as a direct product of cyclic groups of prime power order. Moreover, the representation of an Abelian group of prime power order as a direct product of cyclic groups is unique except for possible rearrangements of the factors. Let the arithmetical function $a(n)$ denote the number of non-isomorphic Abelian groups of order n. Then it follows that $a(n)$ is equal to the number of ways in which n can be expressed as the product of powers of primes, where the order of factors is irrelevant. Obviously, the function $a(n)$ is multiplicative, that is

$$a(mn) = a(m)\,a(n) \quad \text{if} \quad (m, n) = 1.$$

Therefore, the function $a(n)$ is uniquely determined, when $a(n)$ is known for prime powers $n = p^\nu$. It turns out that

$$a(p^\nu) = P(\nu),$$

the number of unrestricted partitions of ν. Hence, if $\mathrm{Re}(s) > 1$,

$$\sum_{n=1}^{\infty}\frac{a(n)}{n^s} = \prod_p\left(\sum_{\nu=0}^{\infty}\frac{a(p^\nu)}{p^{\nu s}}\right) = \prod_p\left(\sum_{\nu=0}^{\infty}\frac{P(\nu)}{p^{\nu s}}\right)$$

$$= \prod_p\prod_{\nu=1}^{\infty}(1 - p^{-\nu s})^{-1} = \prod_{\nu=1}^{\infty}\prod_p(1 - p^{-\nu s})^{-1}$$

$$= \prod_{\nu=1}^{\infty}\zeta(\nu s), \tag{7.27}$$

where $\zeta(s)$ denotes Riemann's zeta-function. Furthermore, it is seen that $a(n)$ is equal to the number of ways in which n can be expressed by $n = n_1 n_2^2 n_3^3 \ldots$, where $n_1, n_2, \ldots$ are positive integers. This shows that $a(n)$ is connected with some divisor functions.

The problem of estimating the average order of $a(n)$ was raised by P. Erdös and G. Szekeres in 1935. Since that time a great number of papers have been written to this topic. In Section 7.2.1 questions of this type are considered.

D. G. Kendall and R. A. Rankin investigated the asymptotic frequency distribution of values of $a(n)$ in 1947. Section 7.2.2 is devoted to this problem.

7.2.1. *The average order of* $a(n)$

Throughout this section let c_n ($n = 1, 2, \ldots$) denote the values

$$c_n = \prod_{\substack{\nu=1 \\ \nu \neq n}}^{\infty} \zeta\left(\frac{\nu}{n}\right).$$

As has already been mentioned, P. Erdös and G. Szekeres were the first who investigated the sum

$$A(x) = \sum_{n \leq x} a(n).$$

They proved that $a(n)$ has the average order c_1 and more precisely that

$$A(x) = c_1 x + O(\sqrt{x}).$$

The first improvement of this result was given by D. G. Kendall and R. A. Rankin who proved by applying a famous theorem of E. Landau that

$$A(x) = c_1 x + c_2 x^{1/2} + O(x^{1/3} \log x),$$

which can now be also proved by elementary methods. After the result of D. G. Kendall and R. A. Rankin many improvements were found by others. The first essential progress was made by H.-E. Richert who presented a third main term of order $x^{1/3}$ and an error term of order less than $x^{1/3}$. All the other following improvements lead to error terms of type $O(x^\alpha (\log x)^\beta)$ with $1/4 < \alpha < 1/3$. Moreover it will be seen from the next theorem that the problem of estimating $A(x)$ is connected with the divisor problem of $D(1, 2, 3; x)$. Then P. G. Schmidt achieved the most important progress by a more symmetric treatment of the function $D(1, 2, 3; x)$, which we could see in Theorem 6.1. Let $\Delta(1, 2, 3; x)$ be defined by (6.1). Then he first proved by an elementary method of I. M. Vinogradov that

$$\Delta(1, 2, 3; x) \ll x^{2/7} \log^2 x.$$

This result also follows from the general estimate

$$\Delta(a, b, c; x) \ll x^{\frac{2}{2a+b+c}} \log^2 x.$$

Then he proved by van der Corput's method of exponent pairs that

$$\Delta(1, 2, 3; x) \ll x^{34/123} \log^2 x ,$$

which we here obtain from Theorem 6.2 using the exponent pair (11/82, 57/82). The estimate

$$\Delta(1, 2, 3; x) \ll x^{4/15} \log^4 x$$

which is a special case of (7.26) demonstrates that the simplest application of the two-dimensional method of exponential sums certainly leads to a sharper result than all the foregoing ones applying van der Corput's one-dimensional method. Here we prove the following result (7.29). All the further improvements by P. G. Schmidt, B. R. Srinivasan and G. Kolesnik are based on refinements of the two-dimensional method.

Theorem 7.6. *Let $\Delta(1, 2, 3; x)$ be the error term in the divisor problem related to the divisor function $d(1, 2, 3; n)$. If*

$$\Delta(1, 2, 3; x) \ll \Delta^*(1, 2, 3; x) = x^{\alpha}(\log x)^{\beta}$$

with $1/4 < \alpha < 1/3$, $\beta \geqq 0$, then

$$A(x) = c_1 x + c_2 x^{1/2} + c_3 x^{1/3} + O(\Delta^*(1, 2, 3; x)) . \qquad (7.28)$$

In particular, the estimate holds with

$$\Delta^*(1, 2, 3; x) = x^{11/42} \log^3 x \left(\frac{11}{42} = 0{,}2619\ldots\right). \qquad (7.29)$$

Proof. From representation (7.27) it follows that

$$A(x) = \sum_{mn \leqq x} d(1, 2, 3; m)\, c_4(n) .$$

We now apply Lemma 7.1 with $p = 3$ and $a_1 = 1$, $a_2 = 2$, $a_3 = 3$, $\alpha_3 = \alpha$, $\beta_3 = \beta$, $r = 4$. Then (7.28) follows at once from (7.6).

In order to prove estimate (7.29), we consider the functions $S(u, v, w; x)$ and $S(u, v, w; M, N; x)$, as defined by (6.7) and (6.10), separately for the six permutations of the numbers 1, 2, 3. First of all it is seen from the proofs of Lemma 6.2 and Theorem 6.3 that

$$S(u, v, w; x) \ll x^{22/87} \log^2 x \qquad (7.30)$$

if $3w \geqq u + v$, where the exponent pair (2/18, 13/18) is used. This condition is satisfied by the triplets

$$(u, v, w) = (1, 2, 3) , \quad (1, 3, 2) , \quad (2, 1, 3) , \quad (3, 1, 2) .$$

In the remaining cases of

$$(u, v, w) = (3, 2, 1) , \quad (2, 3, 1)$$

we divide each sum

$$S(u, v, w; x) = \sum_{\substack{n^u m^{v+w} \leqq x \\ n \leqq m}} \psi\left(\left(\frac{x}{n^u m^v}\right)^{1/w}\right)$$

into two subsums corresponding to $mn \geqq y$ and $mn < y$, where y is a suitable value. In case $mn \geqq y$ estimate (6.13) again may be applied with the exponent pair (2/18, 13/18). Because of $N \ll M$, $MN \gg y$ we obtain

$$S(3, 2, 1; M, N; x) \ll (xM^{-1}N^{-1})^{11/29} \ll (xy^{-1})^{11/29},$$

$$S(2, 3, 1; M, N; x) \ll (xM^{-2})^{11/29} \ll (xM^{-1}N^{-1})^{11/29} \ll (xy^{-1})^{11/29}.$$

In case $mn < y$ estimate (6.15) is used with $k = 3$. Because of $N \ll M$, $MN \ll y$ we obtain

$$S(3, 2, 1; M, N; x) \ll (xMN)^{1/5} \log x \ll (xy)^{1/5} \log x,$$

$$S(2, 3, 1; M, N; x) \ll (xN^2)^{1/5} \log x \ll (xMN)^{1/5} \log x \ll (xy)^{1/5} \log x.$$

If we put $y = x^{13/42}$, we get for the two triplets

$$S(u, v, w; M, N; x) \ll x^{11/42} \log x$$

not only for $MN \ll y$ but also for $MN \gg y$. Hence

$$S(u, v, w; x) \ll x^{11/42} \log^3 x.$$

This result and (7.30) now prove (7.29).

A historical outline of the development of the problem. Let

$$\vartheta = \inf\{\alpha: \Delta(1, 2, 3; x) \ll x^\alpha\},$$

then

$\vartheta \leqq \frac{1}{2} = 0{,}5$ (P. Erdös — G. Szekeres [1], 1935),

$\vartheta \leqq \frac{1}{3} = 0{,}\overline{3}$ (D. G. Kendall — R. A. Rankin [1], 1947),

$\vartheta \leqq \frac{3}{10} = 0{,}3$ (H.-E. Richert [1], 1952),

$\vartheta \leqq \frac{6105117}{20354782} = 0{,}2999\ldots$ (R. A. Rankin [1], 1955),

$\vartheta \leqq \frac{20}{69} = 0{,}2898\ldots$ (W. Schwarz [1], 1966),

$\vartheta \leqq \frac{2}{7} = 0{,}2857\ldots$ (P. G. Schmidt [2], 1968),

$$\vartheta \leqq \frac{34}{123} = 0{,}2764 \ldots \qquad \text{(P. G. Schmidt [2], 1968)},$$

$$\vartheta \leqq \frac{7}{27} = 0{,}\overline{259} \qquad \text{(P. G. Schmidt [3], 1968)},$$

$$\vartheta \leqq \frac{105}{407} = 0{,}2579 \ldots \qquad \text{(B. R. Srinivasan [3], 1973)},$$

$$\vartheta \leqq \frac{97}{381} = 0{,}2545 \ldots \qquad \text{(G. Kolesnik [5], 1981)}.$$

In the other direction W. Schwarz [1] proved in 1967 that

$$A(x) = \sum_{\nu=1}^{6} c_\nu x^{1/\nu} + \Omega(x^{1/6-\varepsilon})$$

for every $\varepsilon > 0$, provided that the Riemann hypothesis is true. R. Balasubramanian and K. Ramachandra [1] even showed in 1981 that

$$A(x) = \sum_{\nu=1}^{6} c_\nu x^{1/\nu} + \Omega(x^{1/6}(\log x)^{1/2})$$

without condition.

7.2.2. *The distribution of values of $a(n)$*

In order to investigate the distribution of values of $a(n)$, we use the following definition.

Definition 7.3. *Let $f(n)$ be any non-negative, integer-valued, arithmetical function. The asymptotic frequency distribution of the values of $f(n)$ is defined as the sequence of numbers*

$$P_k = \lim_{x\to\infty} \frac{1}{x} \sum_{\substack{n \leqq x \\ f(n)=k}} 1 \qquad (k = 0, 1, 2, \ldots)$$

if these limits exist.

Thus P_k denotes the limit of the frequency with which $f(n)$ assumes the value k $(1 \leqq n \leqq x \to \infty)$. If we take $f(n) = a(n)$ to be the number of Abelian groups of order n, D. G. Kendall and R. A. Rankin proved the remarkable fact that the values of $a(n)$ possess an asymptotic frequency distribution. In order to calculate these limits, we need some preliminary arrangements.

We define for $k = 1, 2, \ldots$ the Dirichlet series

$$\Phi_k(s) = \sum_{\substack{n=1 \\ a(n)=k}}^{\infty} \frac{1}{n^s} \tag{7.31}$$

which is absolutely convergent for $\mathrm{Re}(s) > 1$. Furthermore, we define an arithmetical function $g_k(n)$ by

$$\Phi_k(s) = \zeta(s) \sum_{n=1}^{\infty} \frac{g_k(n)}{n^s} \tag{7.32}$$

such that $g_k(n)$ itself is represented by

$$g_k(n) = \sum_{\substack{t \mid n \\ a(t)=k}} \mu\left(\frac{n}{t}\right), \tag{7.33}$$

where $\mu(n)$ denotes the Möbius function.

Suppose now that $n = mn_2$ $(m, n_2) = 1$, where m is a square-free number and n_2 a square-full number. If $m > 1$, put $t = dd'$ such that d runs over the divisors of m and d' over those of n_2. Since $a(n)$ is multiplicative and since $a(m) = 1$ if m is a square-free number, we obtain

$$g_k(mn_2) = \sum_{d \mid m} \mu\left(\frac{m}{d}\right) \sum_{\substack{d' \mid n_2 \\ a(d')=k}} \mu\left(\frac{n_2}{d'}\right) = 0 .$$

Hence $g_k(n) = 0$ if n is not a square-full number. For square-full numbers n_2 we get the inequality

$$|g_k(n_2)| \leqq \sum_{t \mid n_2} \left|\mu\left(\frac{n_2}{t}\right)\right| = 2^{\omega(n_2)}, \tag{7.34}$$

where $\omega(n)$ denotes the number of distinct prime factors of n. Since

$$\sum_{n_2=1}^{\infty} 2^{\omega(n_2)} n_2^{-s} = \prod_p \left(1 + 2\sum_{\nu=2}^{\infty} p^{-\nu s}\right) = \zeta^2(2s) \sum_{n=1}^{\infty} \frac{c_3(n)}{n^s},$$

where $c_3(n)$ again denotes an arithmetical function such that the associated Dirichlet series is absolutely convergent for $\mathrm{Re}(s) > 1/3$, we obtain the trivial estimate

$$\sum_{n_2 \leqq x} |g_k(n_2)| \leqq \sum_{n_2 \leqq x} 2^{\omega(n_2)} \ll \sqrt{x} \log x . \tag{7.35}$$

This shows that the Dirichlet series for $g_k(n)$ in (7.32) is absolutely convergent for $\mathrm{Re}(s) > 1/2$.

Another relation for $g_k(n)$ will be of interest later on. Let p be a prime not dividing the square-full number n_2. Because of $a(p^2) = 2$ and $a(p^3) = 3$ representation (7.33) shows that

$$g_k(p^3 n_2) = g_{k/3}(n_2) - g_{k/2}(n_2) .$$

Hence, in case $k \equiv \pm 1 \pmod 6$, we have $g_k(p^3 n_2) = 0$. If m, n are square-free numbers and n_4 is a powerful number of type 4 such that m, n, n_4 are pairwise coprime, then

$$g_k(n^2 m^3 n_4) = \begin{cases} 0 & \text{for } m > 1, \\ g_k(n^2 n_4) & \text{for } m = 1. \end{cases} \tag{7.36}$$

We are now in a position to prove the following theorems.

Theorem 7.7. *For the values k assumed by a(n) the asymptotic frequency distribution is given by*

$$P_0 = 0, \qquad P_k = \sum_{n=1}^{\infty} \frac{g_k(n)}{n} \qquad (k \geqq 1).$$

More precisely, if $k \geqq 1$, *then*

$$A_k(x) = \sum_{\substack{n \leqq x \\ a(n) = k}} 1 = P_k x + O(\sqrt{x} \log x). \tag{7.37}$$

Proof. $P_0 = 0$ is clear. From (7.31) and (7.32) it follows that

$$A_k(x) = \sum_{mn \leqq x} g_k(n)$$

for $k \geqq 1$. Summing over m and applying (7.35), then

$$A_k(x) = x \sum_{n \leqq x} \frac{g_k(n)}{n} + O\Big(\sum_{n \leqq x} |g_k(n)|\Big)$$

$$= P_k x + O\Big(\sum_{n_2 > x} \frac{x}{n_2} g_k(n_2)\Big) + O\Big(\sum_{n_2 \leqq x} |g_k(n_2)|\Big)$$

$$= P_k x + O(\sqrt{x} \log x).$$

This proves (7.37).

For the distribution of the values of $a(n)$ in short intervals $(x, x + h]$, $h = o(x)$, it is possible to prove much more. From (7.37) we have

$$A_k(x; h) = A_k(x + h) - A_k(x) = P_k h + O(\sqrt{x} \log x).$$

It will be seen in the next theorem that we can obtain a considerably smaller error term.

Theorem 7.8. *Let* $\Delta(1, 2; x)$ *and* $\Delta(1, 2, 3; x)$ *be the error terms in the divisor problems related to the divisor functions* $d(1, 2; n)$ *and* $d(1, 2, 3; n)$. *Let*

$$\Delta(1, 2; t) \ll \Delta^*(1, 2; x) = x^{\alpha} \qquad \left(\frac{1}{5} < \alpha < \frac{1}{2}\right),$$

$$\Delta(1, 2, 3; t) \ll \Delta^*(1, 2, 3; x)$$

for $1 \leqq t \leqq x + h$. *Suppose that* $x^\gamma < h = o(x)$, $\gamma > 0$. *Then, for every* $\varepsilon > 0$,

$$A_k(x; h) = (P_k + o(1))\, h + O(\Delta^*(1, 2, 3; x)\, x^\varepsilon)\,. \tag{7.38}$$

In particular, the asymptotic representation

$$A_k(x; h) = (P_k + o(1))\, h + O(x^{11/42 + \varepsilon}) \tag{7.39}$$

holds.

In case $k \equiv \pm 1 \pmod 6$ *even the asymptotic representation*

$$A_k(x; h) = (P_k + o(1))\, h + O(\Delta^*(1, 2; x)\, x^\varepsilon) \tag{7.40}$$

holds. In particular, we have

$$A_k(x; h) = (P_k + o(1))\, h + O(x^{2/9 + \varepsilon})\,. \tag{7.41}$$

Proof. Let $x^\delta < y = o(h)$, $\delta > 0$. Then

$$A_k(x; h) = \sum_{\substack{x < n \leqq x+h \\ a(n) = k}} 1 = \sum_{x < mn \leqq x+h} g_k(n)$$

$$= \sum_{n \leqq y} g_k(n) \left(\left[\frac{x+h}{n} \right] - \left[\frac{x}{n} \right] \right) + \sum_{\substack{x < mn \leqq x+h \\ n > y}} g_k(n)$$

$$= \sum_{n \leqq y} g_k(n) \left(\frac{h}{n} + O(1) \right) + O\Bigg(\sum_{\substack{x < mn \leqq x+h \\ n > y}} |g_k(n)| \Bigg)$$

We know that $g_k(n) = 0$ if $n \neq n_2$ and that

$$\sum_{n \leqq x} |g_k(n)| \ll \sqrt{y} \log y = o(h)\,.$$

Hence

$$A_k(x; h) = (P_k + o(1))\, h + O(R)\,, \tag{7.42}$$

where R is given by $R = \sum_{\substack{x < mn_2 \leqq x+h \\ n_2 > y}} |g_k(n_2)|\,.$

To prove (7.38), we first use the inequality (7.34) and the well-known fact that $2^{\omega(n)} \ll n^\varepsilon$ for every $\varepsilon > 0$. Thus

$$R \ll x^\varepsilon \sum_{\substack{x < mn_2 \leqq x+h \\ n_2 > y}} 1\,.$$

A square-full number n_2 can be represented by $n_2 = n^2 n'^3$, where n' is square-free. Therefore

$$R \ll x^{\varepsilon} \sum_{\substack{x < mn^2 n'^3 \leqq x+h \\ n^2 n'^3 > y}} 1 \ll x^{\varepsilon} \sum_{\substack{x < mn \leqq x+h \\ n > y}} d(2, 3; n)$$

$$= x^{\varepsilon} \left\{ \sum_{x < n \leqq x+h} d(1, 2, 3; n) - \sum_{\substack{x < mn \leqq x+h \\ n \leqq y}} d(2, 3; n) \right\}$$

$$= x^{\varepsilon} \left\{ D(1, 2, 3; x + h) - D(1, 2, 3; x) - \sum_{n \leqq y} d(2, 3; n) \left(\frac{h}{n} + O(1) \right) \right\}$$

$$= x^{\varepsilon} \left\{ \zeta(2)\, \zeta(3)\, h + O\left(\frac{h}{\sqrt{x}} \right) + O(\Delta^*(1, 2, 3; x)) \right.$$

$$\left. - \sum_{n=1}^{\infty} d(2, 3; n) \frac{h}{n} + O\left(\frac{h}{\sqrt{y}} \right) + O(\sqrt{y}) \right\}.$$

We may choose y such that $x^{\varepsilon}/\sqrt{y} = o(1)$ and $x^{\varepsilon} \sqrt{y} = o(h)$. Then

$$R \ll x^{\varepsilon} \Delta^*(1, 2, 3; x) + o(h)\,.$$

Using this estimate in (7.42) we obtain (7.38). The representation (7.39) now follows from (7.29).

In order to prove (7.40), we use (7.36). If n is square-free and $(n, n_4) = 1$, we obtain

$$R = \sum_{\substack{x < mn^2 n_4 \leqq x+h \\ n^2 n_4 > y}} |g_k(n^2 n_4)|$$

and, without restriction for n, n_4,

$$R \leqq \sum_{\substack{x < mn^2 n_4 \leqq x+h \\ n^2 n_4 > y}} |\mu(n)|\, |g_k(n^2 n_4)| \ll x^{\varepsilon} \sum_{\substack{x < mn^2 n_4 \leqq x+h \\ n^2 n_4 > y}} |\mu(n)|\,.$$

A powerful number n_4 can be represented by $n_4 = a_0^4 a_1^5 a_2^6 a_3^7$ with square-free numbers a_1, a_2, a_3 and $(a_1, a_2) = (a_1, a_3) = (a_2, a_3) = 1$. Hence

$$R \ll x^{\varepsilon} \Sigma_1 |\mu(n)| \ll x^{\varepsilon} \Sigma_2 |\mu(n)|\,,$$

$$SC(\Sigma_1)\colon x < mn^2 a_0^4 a_1^5 a_2^6 a_3^7 \leqq x + h;\ n^2 a_0^4 a_1^5 a_2^6 a_3^7 > y;$$

$$a_1, a_2, a_3 \text{ square-free}, (a_1, a_2) = (a_1, a_3) = (a_2, a_3) = 1\,,$$

$$SC(\Sigma_2)\colon x < mn^2 a_0^4 a_1^5 a_2^6 a_3^7 \leqq x + h;\ n^2 a_0^4 a_1^5 a_2^6 a_3^7 > y\,.$$

Because of $\sum_{na_0^2=b} |\mu(n)| = 1$ we obtain

$$R \ll x^{\varepsilon} \sum_{\substack{x<mn\leq x+h \\ n>y}} d(2, 5, 6, 7; n)$$

$$= x^{\varepsilon} \left\{ \sum_{x<n\leq x+h} d(1, 2, 5, 6, 7; n) - \sum_{\substack{x<mn\leq x+h \\ n\leq y}} d(2, 5, 6, 7; n) \right\}$$

$$= x^{\varepsilon} \left\{ D(1, 2, 5, 6, 7; x + h) - D(1, 2, 5, 6, 7; x) \right.$$

$$\left. - \sum_{n\leq y} d(2, 5, 6, 7; n) \left(\frac{h}{n} + O(1) \right) \right\}.$$

If we suppose that

$$D(1, 2; x) = \zeta(2)\, x + \zeta\left(\frac{1}{2}\right) x^{1/2} + O(x^{\alpha})$$

with $1/5 < \alpha < 1/2$, we may apply Lemma 7.1 with $c_r(n) = d(5, 6, 7; n)$, $r = 5$, $\gamma = 0$, $p = 3$, $\alpha_3 = \alpha$, $\beta_3 = 0$. Then

$$D(1, 2, 5, 6, 7; x) = \sum_{n=1}^{\infty} d(5, 6, 7; n) \left\{ \zeta(2) \frac{x}{n} + \zeta\left(\frac{1}{2}\right) \left(\frac{x}{n}\right)^{1/2} \right\} + O(x^{\alpha}).$$

This gives

$$R \ll x^{\varepsilon} \left\{ \zeta(2)\, \zeta(5)\, \zeta(6)\, \zeta(7)\, h + O\left(\frac{h}{\sqrt{x}}\right) + O(\Delta^*(1, 2; x)) \right.$$

$$\left. - \sum_{n=1}^{\infty} d(2, 5, 6, 7; n) \frac{h}{n} + O\left(\frac{h}{\sqrt{y}}\right) + O(\sqrt{y}) \right\}.$$

We again choose y such that $x^{\varepsilon}/\sqrt{y} = o(1)$ and $x^{\varepsilon}\sqrt{y} = o(h)$. Then

$$R \ll x^{\varepsilon} \Delta^*(1, 2; x) + o(h).$$

Using this estimate in (7.42), we obtain (7.40). The representation (7.41) now follows from (5.40).

NOTES ON CHAPTER 7

Section 7.1

Note that the results of Theorem 7.2 for $\Delta_k(x)$ are only good for $k = 2, 3$ and in a qualified sense for $k = 4, 5$. For larger k it is much better to apply complex methods in connection with the theory of Riemann's zeta-function than the method of exponential sums.

Lemma 7.2 is a special case of a general result due to A. Ivić [2]. Estimate (7.18) in Theorem 7.3 was proved by P. T. Bateman and E. Grosswald [1] and (7.19) by E. Krätzel [17]. Theorems 7.4 and 7.5 are unpublished results of the author.

P. Shiu [2] proved in 1984 that

$$\Delta^*(2, 3, 6; x) = x^{0{,}1526}$$

is a possible value. Recently P. G. Schmidt [4], [5] found the improvements

$$\Delta^*(2, 3, 6; x) = x^{\frac{68}{451}}$$

and

$$\Delta^*(2, 3, 6; x) = x^{\frac{7561}{53025}}.$$

Section 7.2

It may be noted that D. G. Kendall and R. A. Rankin [1] proved not only that the values of $a(n)$ possess an asymptotic frequency distribution, but also that

$$\sum_{k=0}^{\infty} P_k = 1\,, \qquad \sum_{k=0}^{\infty} kP_k = \lim_{x\to\infty} \frac{1}{x} \sum_{n\leq x} a(n)\,.$$

The asymptotic representation (7.37) was first proved by A. Ivić [3]. E. Krätzel [13] found the improvement

$$A_k(x) = P_k x + O\left(\sqrt{x}\,\frac{(\log\log x)^{d-1}}{\log x}\right)$$

if $k = 2^d k'$ with $k' \equiv 1 \pmod 2$ and

$$A_k(x) = P_k x + O(\sqrt{x}\,\mathrm{e}^{-A\delta(x)})$$

if $k \equiv 1 \pmod 2$. Note that a more explicit expression for P_k than the given one is possible.

A. Ivić [5] was the first who investigated the distribution of the values of finite Abelian groups in short intervals and E. Krätzel [10] found some improvements. However, Theorem 7.8 which represents an unpublished result of the author yields much better estimates.

References

Abljalimov, S. B.

[1] Integral points in perturbed circles (Russian), Dokl. Akad. Nauk SSSR **180** (1968), 263—265.

[2] The number of integral points in ovals (Russian), Izv. Akad. Nauk Kaz SSR, Ser. Fiz-Mat., 1970, no. 3, 30—37.

[3] A short proof of a theorem on lattics points (Russian), Izv. Vysš. Učebn. Zaved. Matematika 1977, no. 11, 3—6.

Atkinson, F. V.

[1] A divisor problem, Quart. J. Math. Oxford Ser. **12** (1941), 193—200.

Balasubramanian, R. — Ramachandra, K.

[1] Some problems of analytic number theory III, Hardy-Ramanujan-J. **4** (1981), 13—40.

Bateman, P. T.

[1] The Erdös-Fuchs-Theorem on the square of a power series, J. Number Theory **9** (1977), 330—337.

Bateman, P. T. — Grosswald, E.

[1] On a theorem of Erdös and Szekeres, Ill. J. Math. **2** (1958), 88—98.

Bateman, P. T. — Kohlbecker, E. E. — Tull, J. P.

[1] On a theorem of Erdös and Fuchs in additive number theory, Proc. Amer. Math. Soc. **14** (1963), 278—284.

Bellman, R. E.

[1] The Dirichlet divisor problem, Duke Math. J. **14** (1947), 411—417.

Berndt, B. C.

[1] On the average order of some arithmetical functions, Bull. Amer. Math. Soc. **76** (1970), 856—859.

[2] On the average order of a class of arithmetical functions I, II, J. Number Theory **3** (1971), 184—203, 288—305.

[3] The Voronoi summation formula, Lecture Notes in Math **251** (1972) 21—36.

Bleicher, M. N. — Knopp, M. I.

[1] Lattice points in a sphere, Acta Arithm. **10** (1964/65), 369—376.

[2] Lattice points in a sphere, J. Res. Nat. Bur. Standards, Sect. B, **69B** (1965), 265—270.

Cauer, D.

[1] Neue Anwendungen der Pfeifferschen Methode zur Abschätzung zahlentheoretischer Funktionen, Diss. Göttingen 1914.

Chaix, H.

[1] Démonstration d'un théorème de van der Corput, C. R. Acad. Sci. Paris, Sér. A, **275** (1972) 883—885.

Chandrasekharan, K.
[1] Introduction to analytic number theory, Die Grundlehren der mathematischen Wissenschaften in Einzeldarstellungen, Bd. 148, Berlin—Heidelberg—New York 1968.
[2] Arithmetical functions, Die Grundlehren der mathematischen Wissenschaften in Einzeldarstellungen, Bd. 167, Berlin—Heidelberg—New York 1970.
[3] Exponential sums in the development of number theory, Proc. Steklov-Inst. Math. **132** (1973), 3—24.

Chandrasekharan, K. — Narasimhan, R.
[1] Hecke's functional equation and the average order of arithmetical functions, Acta Arithm. **6** (1961), 487—503.
[2] Functional equations with multiple gamma factors and average order of arithmetical functions, Ann. Math. **76** (1962), 93—136.
[3] On the mean value of the error term for a class of arithmetical functions, Acta Math. **112** (1964), 41—67.

Chen Jing-run
[1] The number of lattice points in a given region, Acta Math. Sinica **12** (1962), 408—420 (Chinese); translation. Chinese Math. **3** (1963), 439—452.
[2] Improvement of asymptotic formulas for the number of lattice points in a region of three dimensions, Sci. Sinica **12** (1963), 151—161.
[3] The lattice-points in a circle, Sci. Sinica **12** (1963), 633—649.
[4] The lattice points in a circle, Acta Math. Sinica **13** (1963), 299—313 (Chinese); translation: Chinese Math. **4** (1963), 322—339.
[5] Improvement on the asymptotic formulas for the number of lattice points in a region of three dimensions II, Sci. Sinica **12** (1963), 751—764.
[6] An improvement of asymptotic formula for $\Sigma_{n \leq x} d_3(n)$, where $d_3(n)$ denotes the number of solutions of $n = pqr$, Sci. Sinica **13** (1964), 1185—1188.
[7] On the divisor problem for $d_3(n)$, Acta Math. Sinica **14** (1964), 549—558 (Chinese); translation: Chinese Math. **5** (1964), 591—601.
[8] On the divisor problem for $d_3(n)$, Sci. Sinica **14** (1965), 19—29.
[9] On the order of $\zeta(1/2 + it)$, Acta Math. Sinica **15** (1965), 159—173 (Chinese); translation: Chinese Math. **6** (1965), 463—478.

Chih Tsung-tao
[1] A divisor problem, Acad. Sinica Science Record **3** (1950), 177—182.
[2] The Dirichlet's divisor problem, Sci. Rep. National Tsing-Hua Univ., Ser. A, **5** (1950), 402—427.

Colin de Verdière, Y.
[1] Nombres de points entiers dans une famille homothétique des domaines de R^n, Ann. Sci. Ecole Norm. Sup. Sér. 4, **10** (1977), 559—576.

Copson, E. T.
[1] Asymptotic expansions, Cambridge 1965.

van der Corput, J. G.
[1] Over roosterpunkten in het plate vak, Diss. Gronigen 1919.
[2] Über Gitterpunkte in der Ebene, Math. Ann. **81** (1920), 1—20.
[3] Zahlentheoretische Abschätzungen, Math. Ann. **84** (1921), 53—79.
[4] Zahlentheoretische Abschätzungen nach der Piltzschen Methode, Math. Z. **10** (1921), 105—120.
[5] Verschärfung der Abschätzung beim Teilerproblem, Math. Ann. **87** (1922), 39—65.
[6] Neue zahlentheoretische Abschätzungen, Math. Ann. **89** (1923), 215—254.

[7] Zahlentheoretische Abschätzungen mit Anwendung auf Gitterpunktprobleme, Math. Z. **17** (1923), 250—259.

[8] Zahlentheoretische Abschätzungen mit Anwendung auf Gitterpunktprobleme (II), Math. Z. **28** (1928), 301—310.

[9] Zum Teilerproblem, Math. Ann. **98** (1928), 697—716; correction ibid. **100** (1928), 480.

[10] Neue zahlentheoretische Abschätzungen (II), Math. Z. **29** (1929), 397—426.

van der Corput, J. G. — Landau, E.

[1] Über Gitterpunkte in ebenen Bereichen, Nachr. K. Gesellschaft Wiss. Göttingen, Math.-Phys. Klasse 1920, 135—171.

Corrádi, K. — Kátai, I.

[1] A comment on K. S. Gangadharan's paper entitled "Two classical lattice point problems" (Hungarian, English summary), Magyar Tud. Akad. Mat. Fiz. Oszt. Közl. **17** (1967), 89—97.

Cramér, H.

[1] Über zwei Sätze des Herrn G. H. Hardy, Math. Z. **15** (1922), 201—210.

Dirichlet, P. G. L.

[1] Über die Bestimmung der mittleren Werte der Zahlentheorie, Abh. Königl. Preuss. Akad. Wiss. 1849, 69—83.

Duttlinger, J.

[1] Über die Anzahl Abelscher Gruppen gegebener Ordnung, J. Reine Angew. Math. **273** (1972), 61—75.

Duttlinger, J. — Schwarz, W.

[1] Über die Verteilung der Pythagoräischen Dreiecke, Coll. Math. **43** (1980), 365—372.

Edgorov, Ž. — Lavrik, A. F.

[1] On the divisor problem in arithmetical progressions (Russian). Izv. Nauk USSR, Ser. fiz.-mat. Nauk, **17** (1973), 14—18.

Erdös, P.

[1] On the integers of the form $x^k + y^k$, J. London Math. Soc. **14** (1939), 194—198.

Erdös, P. — Fuchs, W. H. J.

[1] On a problem of additive number theory, J. London Math. Soc. **31** (1956), 67—73.

Erdös, P. — Szekeres, G.

[1] Über die Anzahl der Abelschen Gruppen gegebener Ordnung und über ein verwandtes zahlentheoretisches Problem, Acta sci. Math. Szeged, tom. VII, fasc. 11 (1934), 95—102.

Erdös, P. — Turán, P.

[1] On a problem of Sidon in additive number theory, and on some related problems, J. London Math. Soc. **16** (1941), 212—215.

Esterman, T.

[1] On the divisor-problem in a class of residues, J. London Math. Soc. **3** (1928), 247—250.

Fine, B.

[1] A note on the two-square theorem, Canad. Math. Bull. **20** (1977), 93—94.

Fischer, K.-H.

[1] Eine Bemerkung zur Verteilung der pythagoräischen Zahlentripel, Monatsh. Math. **87** (1979), 269—271.

[2] Über die Gitterpunktanzahl auf Kreisen mit quadratfreien Radien, Archiv Math. **33** (1979), 150—154.

Fogels, E.

[1] On average values of arithmetic functions, Proc. Cambridge Philos. Soc. **37** (1941), 358—372.

Fomenko, O. M.
[1] On a problem of Gauss, Acta Arith. **6** (1961), 277—284.
Fraser, W. C. G. — Gotlieb, C. C.
[1] A calculation of the number of lattice poits in the circle and sphere, Math. Comp. **16** (1962), 282—290.
Freeden, W.
[1] Über gewichtete Gitterpunktsummen in kreisförmigen Bereichen, Mitteilungen Math. Seminar Giessen **132** (1978), 1—22.
[2] Eine Verallgemeinerung der Hardy-Landauschen Identität, Manuscr. Math. **24** (1978), 205—216.
Fricker, F.
[1] Über die Verteilung der pythagoreischen Zahlentripel, Archiv Math. **28** (1977), 491—494.
[2] Einführung in die Gitterpunktlehre, Basel—Boston—Stuttgart 1982.
[3] Gelöste und ungelöste Gitterpunktprobleme, Jahrbuch Überblicke Math. 1983, 117—135.
Fuji, A.
[1] On the problem of divisors, Acta Arith. **31** (1976), 355—360.
Gangadharan, K. S.
[1] Two classical lattice point problems, Proc. Cambridge Philos. Soc. **57** (1961), 699—721.
Gelfond, A. O. — Linnik, Y. V.
[1] Elementary methods in analytic number theory, Chicago 1965.
Golomb, S. W.
[1] Powerful numbers, Amer. Math. Monthly **77** (1970), 848—855.
Haberland, K.
[1] Bemerkungen zu Gitterpunktsproblemen in R^2, Forschungsergebnisse FSU Jena N/85/19 (1985).
Hafner, J. L.
[1] New omega theorems for two classical lattice point problems, Invent. Math. **63** (1981), 181—186.
[2] On the average order of a class of arithmetical functions, J. Number Theory **15** (1982), 36—76.
[3] New omega results in a weighted divisor problem, Preprint 1986.
Halberstam, H. — Roth, K. F.
[1] On the gaps between consecutive k-free integers, J. London Math. Soc. **26** (1951), 268—273.
[2] Sequences I, Oxford 1966.
Hamm, G.
[1] Ein $(m + 1)$-dimensionales Gitterpunktproblem, Math. Nachr. **114** (1983), 87—96.
Hammer, J.
[1] Unsolved problems concerning lattice points, London 1977.
Haneke, W.
[1] Verschärfung der Abschätzung von $\zeta(1/2 + it)$, Acta Arith. **8** (1962/63), 357—430.
Hardy, G. H.
[1] On the expression of a number as the sum of two squares, Quart. J. Math. **46** (1915), 263—283.
[2] Sur le probleme des diviseurs de Dirichlet, Comptes Rendus **160** (1915), 617—619.
[3] On Dirichlet's divisor problem, Proc. London Math. Soc. (2) **15** (1916), 1—25.
[4] The average order of the arithmetical functions $P(x)$ and $\Delta(x)$, Proc. London Math. Soc. (2) **15** (1916), 192—213.
[5] Additional note on two problems in the analytic theory of numbers, Proc. London Math. Soc. (2) **18** (1920), 201—204.
[6] The lattice points of a circle, Proc. Royal Soc., A. **107** (1925), 623—635.
[7] Collected Papers, vol. I/II, Oxford 1966/67.

Hardy, G. H. — Landau, E.

[1] The lattice points of a circle, Proc. Royal Soc., A. **105** (1924), 244—258.

Hardy, G. H. — Littlewood, J. E.

[1] The trigonometrical series associated with the elliptic ϑ-function, Acta Math. **37** (1914), 193—239.

[2] The approximate functional equation in the theory of the zeta-function, with applications to the divisor problems of Dirichlet and Piltz, Proc. London Math. Soc. (2) **21** (1. 22), 39—74.

Hayashi, E. K.

[1] Omega theorems for the iterated additive convolution of a nonnegative arithmetic function, dissertation, Urbana, Illinois 1973.

[2] An elementary method for estimating error terms in additive number theory. Proc. Amer. Math. Soc. **52** (1975), 55—59.

[3] Omega theorems for the iterated additive convolution of a nonnegative arithmetic function, J. Number Theory **13** (1981), 176—191.

Heath-Brown, D. R.

[1] The twelfth power moment of the Riemann-function, Quart. J. Math. Oxford Ser. (2) **29** (1978), 443—462.

[2] Mean values of the zeta function and divisor problems, Recent progress in analytic number theory, Vol. 1 (Durham 1979), 115—119, Academic Press, London 1981.

Hua Loo-Keng

[1] The lattice points in a circle, Quart. J. Math., Oxford Ser., **13** (1942), 18—29.

[2] On some problems of the geometrical theory of numbers, Acad. Sinica Science Record **1** (1942), 19—21.

[3] Die Abschätzung von Exponentialsummen und ihre Anwendung in der Zahlentheorie, Enzyklopädie der mathematischen Wissenschaften mit Einschluß ihrer Anwendungen, Bd. 1, 2. Teil, D Analytische Zahlentheorie, Leipzig 1959.

Huxley, M. N. — Watt, N.

(1) Exponential sums and the Riemann zeta function, Coll. on Number Theory, 20.—25. July 1987, Budapest.

Ingham, A. E.

[1] On two classical lattice point problems, Proc. Cambridge Phil. Soc. **36** (1940), 131—138.

Ivić, A.

[1] On the asymptotic formulas for powerful numbers, Publ. Inst. Math. Belgrade, vol. **23** (37) (1978), 85—94.

[2] A convolution theorem with applications to some divisor functions, Publ. Inst. Math. Belgrade **24** (38) (1978), 67—78.

[3] The distribution of values of the enumerating function of non-isomorphic abelian groups of finite order, Archiv Math. **30** (1978), 374—379.

[4] Exponent pairs and the zeta-function of Riemann, Studia Sci. Math. Hung. **15** (1980), 157—181.

[5] On the number of finite non-isomorphic abelian groups in short intervals, Math. Nachr. **101** (1981), 257—271.

[6] On the number of abelian groups of a given order and on certain related multiplicative functions, J. Number Theory **16** (1983), 119—137.

[7] Large values of the error term in the divisor problem, Invent. Math. **71** (1983), 513—520.

[8] On the number of abelian groups of a given order and the number of prime factors of an integer, Research Faculty of Science-University Novi Sad **13** (1983), 41—50.

[9] Topics in recent zeta-function theory, Publications Mathematics d'Orsay, Université de Paris-Sud, 1983.

[10] The Riemann zeta-function, John Wiley & Sons, New York 1985.

Ivić, A. — Shiu, P.

[1] The distribution of powerful numbers, Ill. J. Math. **26** (1982), 576—590.

Iwaniec, H. — Mozzochi, J.

(1) On the divisor and circle problems, to appear.

Kalecki, M.

[1] On the sum $\sum_{n \leq x} \left\{ f\left(\frac{x}{n}\right) \right\}$, Prace Mat. **11** (1968), 189—191.

Kanold, H.-J.

[1] Über die Anzahl der natürlichen Zahlen, welche jeden Primteiler in mindestens h-ter Potenz enthalten, Abhandlungen der Braunschweigischen Wiss. Ges. 1979, 1—4.

Karacuba, A. A.

[1] The Dirichlet divisor problem in number fields, Soviet Math. **13** (1972), 697—698.

[2] A uniform estimate for the remainder term in Dirichlet's problem of divisors (Russian), Izv. Akad. Nauk SSSR, Ser. Mat., **36** (1972), 475—483.

Kátai, I.

[1] The number of lattice points in a circle (Russian), Ann. Univ. Sci. Budapest. Eötvös Sect. Math. **8** (1965), 39—60.

Keller, H. B. — Swenson, J. R.

[1] Experiments on the lattice problem of Gauss, Math. Comp. **17** (1963), 223—230.

Kendall, D. G.

[1] On the number of lattice points inside a random oval, Quart. J. Math., Oxford Ser., **19** (1948), 1—26.

Kendall, D. G. — Rankin, R. A.

[1] On the number of Abelian groups of a given order, Quart. J. Math., Oxford Ser., **18** (1947), 197—208.

Kolesnik, G.

[1] An improvement of the remainder term in the divisor problem (Russian), Mat. Zametki **6** (1969), 545—554.

[2] An estimate for certain trigonometric sums, Acta Arith. **25** (1973/74), 7—30.

[3] On the order of Dirichlet L-functions, Pac. J. Math. **82** (1972), 479—484.

[4] On the estimation of multiple exponential sums, Recent progress in analytic number theory, Vol. 1 (Durham 1979), 231—246, Academic Press, London 1981.

[5] On the number of abelian groups of a given order, J. Reine Angew. Math. **329** (1981), 164—175.

[6] On the order of $\zeta\left(\frac{1}{2} + it\right)$ and $\Delta(R)$, Pac. J. Math. **98** (1982) 1, 107—122.

[7] On the method of exponent pairs, Acta Arith. **45** (1985), 115—143.

Krätzel, E.

[1] Ein Gitterpunktsproblem, Acta Arith. **10** (1964), 215—223.

[2] Eine Verallgemeinerung des Kreisproblems, Archiv Math. **18** (1967), 181—187.

[3] Identitäten für die Anzahl der Gitterpunkte in bestimmten Bereichen, Math. Nachr. **36** (1968), 179—191.

[4] Bemerkungen zu einem Gitterpunktproblem, Math. Ann. **179** (1969), 90—96.

[5] Ein Teilerproblem, J. Reine Angew. Math. **235** (1969), 150—174.

[6] Teilerprobleme in drei Dimensionen, Math. Nachr. **42** (1969), 275—288.

[7] Mittlere Darstellung natürlicher Zahlen als Differenz zweier k-ter Potenzen, Acta Arith. **16** (1969), 111—121.

[8] Mittlere Darstellungen natürlicher Zahlen als Summe von n k-ten Potenzen, Czechoslovak. Math. J. **23** (98) (1973), 57—73.

[9] Zahlen k-ter Art, Amer. J. Math. **44** (1972) 1, 309—328.

[10] Die Werteverteilung der Anzahl der nicht-isomorphen Abelschen Gruppen endlicher Ordnung in kurzen Intervallen, Math. Nachr. **98** (1980), 135—144.

[11] Zahlentheorie, Berlin 1981.

[12] The number of non-isomorphic abelian groups of a given order, Forschungsergebnisse FSU Jena N/82/55 (1982).

[13] Die Werteverteilung der Anzahl der nicht-isomorphen abelschen Gruppen endlicher Ordnung und ein verwandtes zahlentheoretisches Problem, Publ. Inst. Math. **31** (45) (1982), 93—101.

[14] Squarefree numbers as sums of two squares, Archiv Math. **39** (1982), 28—31.

[15] Divisor problems and powerful numbers, Math. Nachr. **114** (1983), 97—104.

[16] Ω-Estimates for the number of lattice-points in n-dimensional domains, Colloquia Math. Soc. János Bolyai, 34. Topics in classical Number Theory, Budapest 1981, 979—993.

[17] Zweifache Exponentialsummen und dreidimensionale Gitterpunktprobleme, Elementary and Analytic Theory of Numbers, Banach Center Publications **17**, PWN — Polish Scientific Publishers, Warsaw 1985, 337—369.

[18] Zur Anwendung der Methode von Titchmarsh auf dreidimensionale Gitterpunktprobleme, Forschungsergebnisse FSU Jena N/83/28 (1983), Math. Nachr. **123** (1985), 197—204.

Krätzel, E. — Menzer, H.

[1] Verallgemeinerte Hankel-Funktionen, Publ. Math. Debrecen **18** (1971), 139—147.

Landau, E.

[1] Über die Anzahl der Gitterpunkte im gewissen Bereichen, Nachr. K. Gesellschaft Wiss. Göttingen, Math.-Phys. Klasse 1912, 687—771.

[2] Die Bedeutung der Pfeifferschen Methode für die analytische Zahlentheorie, Sitzungsber. kais. Akad. Wiss. Wien, Math.-Naturwiss. Klasse 1912, 2298—2328.

[3] Über die Gitterpunkte in einem Kreise (Erste, zweite Mitteilung), Nachr. K. Gesellschaft Wiss. Göttingen, Math.-Phys. Klasse 1915, 148—160, 161—171.

[4] Über Dirichlets Teilerproblem, Sitzungsber. Math.-Phys. Klasse Königl. Bayer. Akad. Wiss. 1915, 317—328.

[5] Zur analytischen Zahlentheorie der definiten quadratischen Formen (Über die Gitterpunkte in einem mehrdimensionalen Ellipsoid), Sitzungsber. Königl. Preuß. Akad. Wiss. **31** (1915), 458 bis 476.

[6] Über die Anzahl der Gitterpunkte in gewissen Bereichen. (Zweite, dritte Abhandlung), Nachr. K. Gesellschaft Wiss. Göttingen, Math.-Phys. Klasse 1915, 209—243, 1917, 96—101.

[7] Über die Gitterpunkte in einem Kreise, Math. Z. **5** (1919), 319—320.

[8] Über Dirichlets Teilerproblem, Nachr. K. Gesellschaft Wiss. Göttingen, Math.-Phys. Klasse 1920, 13—32.

[9] Über die Gitterpunkte in einem Kreise (Dritte Mitteilung), Nachr. K. Gesellschaft Wiss. Göttingen, Math.-Phys. Klasse 1920, 109—134.

[10] Über Dirichlets Teilerproblem (Zweite Mitteilung), Nachr. K. Gesellschaft Wiss. Göttingen, Math.-Phys. Klasse 1922, 8—16.

[11] Über die Gitterpunkte in einem Kreise (Vierte, fünfte Mitteilung), Nachr. K. Gesellschaft Wiss. Göttingen, Math.-Phys. Klasse 1923, 58—65, 1924, 135—136.

[12] Über die Anzahl der Gitterpunkte in gewissen Bereichen (Vierte Abhandlung), Nachr. Gesellschaft Wiss. Göttingen, Math.-Phys. Klasse 1924, 137—150.

[13] Über die Gitterpunkte in mehrdimensionalen Ellipsoiden, Math. Z. **21** (1924), 126—132.
[14] Note on the preceding paper, Proc. Royal Soc., A, **106** (1924), 487—488.
[15] Die Bedeutungslosigkeit der Pfeifferschen Methode für die analytische Zahlentheorie, Monatsh. Math. Phys. **34** (1925), 1—36.
[16] Über Gitterpunkte in mehrdimensionalen Ellipsoiden. Zweite Abhandlung, Math. Z. **24** (1925), 299—310.
[17] Über das Konvergenzgebiet einer mit der Riemannschen Zetafunktion zusammenhängenden Reihe, Math. Ann. **97** (1926), 251—290.
[18] Vorlesungen über Zahlentheorie. Zweiter Band. Aus der analytischen und geometrischen Zahlentheorie, Leipzig 1927.
[19] Ausgewählte Abhandlungen zur Gitterpunktlehre, Berlin 1962 (presented by A. Walfisz).

Langmann, K.
[1] Eine endliche Formel für die Anzahl der Teiler von n, J. Number Theory **11** (1979), 116—127.

Lavrik, A. F.
[1] On the problem of divisors in segments of arithmetical progressions (Russian), Dokl. Akad. Nauk SSSR **164** (1965), 1232—1234.

Lavrik, A. F. — Edgorov, Ž.
[1] On the problem of divisors in arithmetical progressions (Russian), Izv. Akad. Nauk SSSR, Ser. fiz.-mat. Nauk, **17** (1973), 14—18.

Lavrik, A. F. — Israilov, M. I. — Edgorov, Ž.
[1] Integrals containing the remainder term of the divisor problem (Russian), Acta Arith. **37** (1980), 381—389.

Li Zhong-Fu — Yin Wen-Lin
[1] An improvement on the estimate for the error term in the divisor problem for $d_3(n)$ (Chinese), Acta Math. Sinica **24** (1981), 865—878.

Littlewood, J. E. — Walfisz, A.
[1] The lattice points of a cricle, Proc. Roy. Soc. London, Ser. (A), **106** (1925), 478—488.

Loxton, J.-H.
[1] The graphs of exponential sums, Mathematika **30** (1983) 2, 125—135.

Menzer, H.
[1] Ein dreidimensionales Gitterpunktproblem, Math. Nachr. **126** (1986), 35—44.
[2] On the number of primitive Pythagorean triangles, Math. Nachr. **128** (1986), 129—133.

Min, S.-H.
[1] On the order of $\zeta(1/2 + it)$, Trans. Amer. Math. Soc. **65** (1949), 448—472.

Mitchell, W. C.
[1] The number of lattice points in a k-dimensional hypersphere, Math. Comp. **20** (1966), 300—310.

Montgomery, H. L. — Vaughan, R. C.
[1] The distribution of squarefree numbers, Recent progress in analytic number theory, Vol. 1 (Durham 1979), 247—256, Academic Press, London 1981.

Moroz, B. Z.
[1] On the number of primitive lattice points in plane domains, Monatsh. Math. **99** (1985), 37—42.

Narkiewicz, W. — Śliva, J.
[1] Finite abelian groups and factorization problems II, Colloq. Math. **46** (1982) 1, 115—122.

Newman, D. J.
[1] A simplified proof of the Erdös-Fuchs Theorem, Proc. Amer. Math. Soc. **75** (1979) 2, 209—210.

Nieland, L. W.

[1] Zum Kreisproblem, Math. Ann. **98** (1928), 717—736.

Nowak, W.-G.

[1] Einige Verallgemeinerungen des Gaußschen Kreisproblems, Anz. Österreich. Akad. Wiss., Math.-Naturwiss. Kl. 1978, no. 2, 45—51.

[2] Über die Anzahl der Gitterpunkte in verallgemeinerten Kreissektoren, Monatsh. Math. **87** (1979), 297—307.

[3] A non-convex generalization of the circle problem, J. Reine Angew. Math. **314** (1980), 136—145.

[4] Gitterpunkte in Bereichen $x^n + y^m \leqq R$, $x \geqq 0$, $y \geqq 0$, Monatsh. Math. **89** (1980) 3, 223—233.

[5] Ein dreidimensionales Gitterpunktproblem, Manuscr. Math. **33** (1980/81) 1, 63—80.

[6] Ein Satz zur Behandlung dreidimensionaler Gitterpunktprobleme, J. Reine Angew. Math. **329** (1981), 125—142.

[7] Eine Ω-Abschätzung zu einem zweidimensionalen Gitterpunktproblem, Monatsh. Math. **94** (1982), 143—147.

[8] Gitterpunkte in Lemniskatenscheiben, Manuscr. Math. **39** (1982), 277—285.

[9] Eine Bemerkung zum Kreisproblem in der p-Norm, Österreich. Akad. Wiss. Math.-Naturwiss. Kl., Sitzungsberichte II, **191** (1982) 1—3, 125—132.

[10] Ein nicht-konvexes Analogon zum mehrdimensionalen Kugelproblem, J. Reine Angew. Math. **338** (1983), 149—165.

[11] Zur Gitterpunktlehre der euklidischen Ebene, Nederl. Akad. Wetensch. Indag. Math. **46** (1984), 209—223.

[12] Lattice points in a circle and divisors in arithmetic progressions, Manuscr. Math. **49** (1984), 195—205.

Odoni, R. W. K.

[1] A problem of Erdös on sums of two squarefull numbers, Acta Arithm. **39** (1981), 145—162.

Pfeiffer, E.

[1] Über die Periodicität in der Teilbarkeit der Zahlen und über die Verteilung der Klassen positiver quadratischer Formen auf ihre Determinanten, Jahresbericht der Pfeiffer'schen Lehr- und Erziehungsanstalt zu Jena 1886, 1—21.

Phillips, E.

[1] The zeta function of Riemann; Further developments of van der Corput's method, Quart. J. Math. (Oxford) **4** (1933), 209—225.

Piltz, A.

[1] Über das Gesetz, nach welchem die mittlere Darstellbarkeit der natürlichen Zahlen als Produkte einer gegebenen Anzahl Faktoren mit der Größe der Zahlen wächst, Diss., Berlin 1881.

Pjateckii-Šapiro, I. I.

[1] On an asymptotic formula for the number of Abelian groups whose order does not exceed n (Russian), Mat. Sb. **26** (68) (1950), 479—486.

Pomerance, C. — Suryanarayana, D.

[1] On a problem of Evelyn-Linfoot and Page in additive number theory, Publ. Math. Debrecen **26** (1979), 3—4.

Popov, V. N.

[1] The number of integer points under a parabola, Math. Notes **18** (1975), 1007—1010.

Randol, B.

[1] A lattice-point problem (I), Trans. Amer. Math. Soc. **121** (1966), 257—268.

[2] A lattice-point problem (II), Trans. Amer. Math. Soc. **125** (1966), 101—113.

Rankin, R. A.

[1] Van der Corput's method and the theory of exponent pairs, Quart. J. Math. (2) **6** (1955), 147—153.

Rao, V. Venugopal

[1] Lattice point problems and quadratic forms, Math. Student **27** (1959), 137—152.

Richert, H.-E.

[1] Über die Anzahl Abelscher Gruppen gegebener Ordnung I, Math. Z. **56** (1952), 21—32; II, ibid. **58** (1953), 71—84.

[2] Ein Gitterpunktproblem, Math. Ann. **125** (1953), 467—471.

[3] Verschärfung der Abschätzung beim Dirichletschen Teilerproblem, Math.-Z. **58** (1953), 204 bis 218.

[4] On the difference between consecutive squarefree numbers. J. London Math. Soc. **29** (1954), 16—20.

[5] Einführung in die Theorie der starken Rieszschen Summierbarkeit von Dirichletreihen, Nachr. Akad. Wiss. Göttingen, Math.-Phys. Kl. II, 1960, 17—75.

[6] Zur multiplikativen Zahlentheorie, J. Reine Angew. Math. **206** (1961), 31—38.

Rogosinski, W.

[1] Neue Anwendung der Pfeifferschen Methode bei Dirichlets Teilerproblem, Diss. Göttingen 1922.

Roth, K. F.

[1] On the gaps between squarefree numbers, J. London Math. Soc. **26** (1951), 263—268.

Sárközy, A.

[1] On a theorem of Erdös and Fuchs, Acta Arith. **37** (1980), 333—338.

Schaal, W.

[1] Übertragung des Kreisproblems auf reell-quadratische Zahlkörper, Math. Ann. **145** (1961/62), 273—284.

[2] Der Satz von Erdös-Fuchs in reell-quadratischen Zahlkörpern, Acta Arith. **32** (1977), 147—156.

Schierwagen, A.

[1] Über ein Teilerproblem, Math. Nachr. **72** (1974), 151—168.

[2] Bemerkungen zu einem Teilerproblem, Publ. Math. Debrecen **25** (1978), 41—46.

Schmidt, P. G.

[1] Abschätzungen bei unsymmetrischen Gitterpunktproblemen, Diss. Göttingen 1964.

[2] Zur Anzahl Abelscher Gruppen gegebener Ordnung I, J. Reine Angew. Math. **229** (1968), 34—42.

[3] Zur Anzahl Abelscher Gruppen gegebener Ordnung II, Acta Arith. **13** (1968), 405—417.

[4] Zur Anzahl quadratvoller Zahlen in kurzen Intervallen, Acta Arith. **46** (1986), 159—164.

[5] Zur Anzahl quadratvoller Zahlen in kurzen Intervallen und ein verwandtes Gitterpunktproblem, Acta Arith. (to appear).

Schnabel, L.

[1] Über eine Verallgemeinerung des Kreisproblems, Diss. Jena 1981; Wiss. Z. FSU Jena, Math.-Nat. R., **31** (1982), 667—681.

Schwarz, W.

[1] Über die Anzahl Abelscher Gruppen gegebener Ordnung I, Math. Z. **92** (1966), 314—320; II, J. Reine Angew. Math. **228** (1967), 133—138.

Shiu, P.

[1] On the number of square-full integers between successive squares, Mathematika **27** (1980), 171—178.

[2] On square-full integers in a short interval, Glasgow Math. J. **25** (1984) 1, 127—134.

Sierpiński, W.

[1] O pewnem zagadnieniu z rachunku funckcyj asymptotycznych, Prace mat.-fiz. **17** (1906), 77—118.

[2] O pewnej summie potrónjaj, C. Rendus Soc. Sc. Lettres Varsovie (3) **2** (1909), 117—120.

Sitaramaiah, V. — Suryanarayana, D.

[1] On certain divisor sums over square-full integers, Proc. Conf. on Number Theory (Mysore, 1979), 98—109, Matscience Rep. 101, Inst. Math. Sci., Madras 1980.

Smith, R. A.

[1] The circle problem in an arithmetic progression, Canad. Math. Bull. **11** (1968), 175—184.

Srinivasan, B. R.

[1] Lattice points in a circle, Proc. Nat. Inst. Sci. India, Part A, **29** (1963), 332—346.

[2] The lattice point problem of many-dimensional hyperboloids (I), Acta Arith. **8** (1962/63), 153—172; (II), ibid. **8** (1962/63), 173—204; (III), Math. Ann. **160** (1965), 280—311.

[3] On the number of Abelian groups of a given order, Acta Arith. **23** (1973), 195—205.

Steinhaus, H.

[1] Sur un théorème de M. V. Jarník, Colloq. Math. **1** (1947), 1—5.

Suryanarayana, D.

[1] On a paper of S. Chowla and H. Walum concerning the divisor problem, J. Indian Math. Soc. **41** (1977), 293—299.

[2] On the order of the error function of the square-full integers, Period. Math. Hungar. **10** (1979), 261—271.

[3] Some more remarks on uniform O-estimate for k-free integers, Ind. J. Pure Appl. Math. **12** (1981), 1420—1424.

[4] On the order of the error function of the square-full integers, Period. Math. Hungar. **14** (1983) 1, 69—75.

Szegö, G.

[1] Beiträge zur Theorie der Laguerreschen Polynome II: Zahlentheoretische Anwendungen, Math. Z. **25** (1926), 388—404.

Szegö, G. — Walfisz, A.

[1] Über das Piltzsche Teilerproblem in algebraischen Zahlkörpern (Zweite Abhandlung), Math. Z. **26** (1927), 467—486.

Tarnopolska-Weiss, M.

[1] On the number of lattice points in planar domains, Proc. Amer. Math. Soc. **69** (1978), 308—311.

Titchmarsh, E. C.

[1] On van der Corput's method and the zeta-function of Riemann (I), Quart. J. Math. (Oxford) **2** (1931), 161—173; (II), ibid. **2** (1931), 313—320; (III), ibid. **3** (1932), 133—141; (IV), ibid. **5** (1934), 98—105; (V), ibid. **5** (1934), 195—210; (VI), ibid. **6** (1935), 106—112.

[2] On Epstein's zeta-function, Proc. London Math. Soc. (2) **36** (1934), 485—500.

[3] The lattice-points in a circle, Proc. London Math. Soc. (2) **38** (1934), 96—115; Corrigendum 555.

[4] On the order of $\zeta\left(\frac{1}{2}+it\right)$ Quart. J. Math. (Oxford) **13** (1942), 11—17.

[5] The theory of Riemann zeta-function, Oxford 1951.

Tong Kwang-chang

[1] On divisor problems (Chinese, English summary), J. Chinese Math. Soc. **2** (1953), 258—266.

[2] On division problems I (Chinese, English summary), Acta Math. Sinica **5** (1955), 313—324.

[3] On divisor problems II, III (Chinese, English summary), Acta Math. Sinica **6** (1956), 139—152, 515—541.

Vaughan, R. C.

[1] On the addition of sequences of integers, J. Number Theory **4** (1972), 1—16.

Vinogradov, I. M.

[1] On the number of lattice points in a sphere (Russian), Trudy Mat. Steklov Akad. Nauk. SSSR **9** (1935), 17—38.

[2] Izbrannye trudy, Izdat. Akad. Nauk SSSR, Moscow 1952.

[3] Improvement of asymptotic formulas for the number of lattice points in a region of three-dimensions (Russian), Izv. Akad. Nauk SSSR Ser. Mat. **19** (1955), 3—10.

[4] On the number of integral points in a given domain (Russian), Izv. Akad. Nauk SSSR Ser. Mat. **24** (1960), 777—786.

[5] On the number of integral points in a three-dimensional domain (Russian), Izv. Akad. Nauk SSSR, Ser. Mat., **27** (1963), 3—8.

[6] On the number of integral points in a sphere (Russian), Izv. Akad. Nauk SSSR, Ser. Mat., **27** (1963), 957—968.

[7] Special variants of the method of trigonometric sums (Russian), Moscow 1976.

Vogts, M.

[1] Teilerprobleme in drei Dimensionen, Math. Nachr. **101** (1981), 243—256.

[2] Many-dimensional generalized divisor problems, Math. Nachr. **124** (1985), 103—121.

Voronoi, G.

[1] Sur un problème du calcul des fonctions asymptotiques, J. Math. **126** (1903), 241—282.

[2] Sur le développement, à l'aide des fonctions cylindriques, des sommes doubles $\Sigma f(pm^2 + 2qmn + rn^2)$, où $pm^2 + 2qmn + rn^2$ est une forme positive à coefficients entiers, Verhandlungen des dritten Int. Math.-Kongresses in Heidelberg 1905, 241—245.

Walfisz, A.

[1] Über zwei Gitterpunktprobleme, Math. Ann. **95** (1926), 69—83.

[2] Teilerprobleme, Math. Z. **26** (1927), 66—88.

[3] Gitterpunkte in mehrdimensionalen Kugeln, Warsaw 1957.

[4] Weylsche Exponentialsummen in der neueren Zahlentheorie, Berlin 1963.

Weyl, H.

[1] Über die Gleichverteilung der Zahlen mod. Eins, Math. Ann. **77** (1916), 313—352.

Wild, R. E.

[1] On the number of lattice points in $x^t + y^t = n^{1/2}$, Pac. J. Math. **8** (1958), 929—940.

Wilton, J. R.

[1] The lattice points of a circle: an historical account of the problem, Mess. Math. **48** (1928), 67—80.

Wu Fang

[1] The lattice-points in an ellipse, Acta Math. Sinica **13** (1963), 238—253 (Chinese); translation Chinese Math. **4** (1963), 260—274.

Wu Fang — Yuh Ming-i

[1] On the divisor problem for $d_3(n)$, Sci. Sinica **11** (1962), 1055—1060.

[2] On the divisor problem for $d_3(n)$, Acta Math. Sinica **12** (1962), 170—174 (Chinese); translation: Chinese Math. **3** (1963), 184—189.

Yin Wen-lin

[1] On Dirichlet's divisor problem, Sci. Record (Peking) **3** (1959), 6—8.

[2] Piltz's divisor problem for $k = 3$, Sci. Record (Peking) **3** (1959), 169—173.

[3] On divisor problem for $d_3(n)$, Acta Sci. Nat. Univ. Pekinensis **5** (1959), 193—196.

[4] The lattice-points in a circle, Sci. Sinica **11** (1962), 10—15.
Yuh Ming-i
[1] A divisor problem, Sci. Record (Peking) **2** (1958), 326—328.
[2] A divisor problem (Chinese, English summary), Acta Math. Sinica **8** (1958), 496—506.

Index of Names

Subject Index